LONDON MATHEMATICAL SOCIETY LECTURE NOTE SERIES

Managing Editor: Professor J.W.S. Cassels, Department of Pure Mathematics and Mathematical Statistics, University of Cambridge, 16 Mill Lane, Cambridge CB2 1SB, England

The titles below are available from booksellers, or, in case of difficulty, from Cambridge University Press.

London Mathematical Society Lecture Note Series. 224

Computability, Enumerability, Unsolvability

Directions in recursion theory

Edited by

S.B. Cooper
University of Leeds

T.A. Slaman
University of Chicago

S.S. Wainer
University of Leeds

CAMBRIDGE
UNIVERSITY PRESS

CAMBRIDGE UNIVERSITY PRESS
Cambridge, New York, Melbourne, Madrid, Cape Town, Singapore, São Paulo, Delhi

Cambridge University Press
The Edinburgh Building, Cambridge CB2 8RU, UK

Published in the United States of America by Cambridge University Press, New York

www.cambridge.org
Information on this title: www.cambridge.org/9780521557368

First published 1996

A catalogue record for this publication is available from the British Library

ISBN 978-0-521-55736-8 paperback

Transferred to digital printing 2009

Contents

Preface

This volume is a collection of refereed research articles commemorating the Leeds Recursion Theory Year 1993-94. The year was funded principally by the (then) UK Science and Engineering Research Council, with additional support from the London Mathematical Society, European Twinning/Human Capital and Mobility Networks on 'Complexity, Logic and Recursion Theory', and on 'Proof Theory and Computation', a MURST-British Council travel grant, an EC PECO visiting fellowship, and with the backing of the Leeds University Department of Pure Mathematics. We thank them all for enabling an invigorating year.

It is fifteen years since the publication of the last Leeds Recursion Theory volume in this same series (LMS Lecture Notes 45). In that time the subject has made great strides. New methods have been developed and out of the immense technical machinery have finally emerged solutions to long-standing problems which originally motivated the pioneers some forty years ago, notably on definability, decidability and automorphisms for recursion theoretic structures. In addition the fundamental ideas concerning computation and recursion have naturally found their place at the interface between logic and theoretical computer science, and the feedback continues to motivate mathematical research in a variety of new directions. Thus the following contributions provide a picture of current ideas and methods in the ongoing investigations of the structure of the computable and non-computable universe. A number of the articles contain introductory and background material, which it is hoped will make the volume an invaluable source of information for specialist and non-specialist alike. All but four of the authors visited Leeds during the year and Slaman in particular was present throughout, as a SERC Visiting Fellow.

We dedicate this volume to the memory of Stephen Cole Kleene, the father-figure of modern recursion theory, who died on 25th January 1994.

S. Barry Cooper (Leeds)
Theodore A. Slaman (Chicago)
Stanley S. Wainer (Leeds)

Resource-Bounded Genericity*

Klaus Ambos-Spies
Mathematisches Institut
Universität Heidelberg
D-69120 Heidelberg, Germany
ambos@math.uni-heidelberg.de

Abstract

Resource-bounded genericity concepts have been introduced by Ambos-Spies, Fleischhack and Huwig [AFH84], [AFH88], Lutz [Lu90], and Fenner [Fe91]. Though it was known that some of these concepts are incompatible, the relations among these notions were not fully understood. Here we survey these notions and clarify the relations among them by specifying the types of diagonalizations captured by the individual concepts. Moreover, we introduce two new, stronger resource-bounded genericity concepts corresponding to fundamental diagonalization concepts in complexity theory. First we define general genericity, which generalizes all of the previous concepts and captures both, standard finite extension arguments and slow diagonalizations. The second new concept, extended genericity, actually is a hierarchy of genericity concepts for a given complexity class which extends general genericity and in addition captures delayed diagonalizations. Moreover, this hierarchy will show that in general there is no strongest genericity concept for a complexity class. A similar hierarchy of genericity concepts was independently introduced by Fenner [Fe95].

Finally we study some properties of the Baire category notions on **E** induced by the genericity concepts and we point out some relations between resource-bounded genericity and resource-bounded randomness.

*A preliminary, short version of this paper appeared in the proceedings of the Tenth Annual IEEE Conference on Structure in Complexity Theory. This research was supported in part by the Human Capital and Mobility Program of the European Community under grant CHRX-CT93-0415 (COLORET).

1   Introduction

The finite extension method is a central diagonalization technique in computability theory (see e.g. [Ro67], [Od89], [Le83], [So87]). In a standard finite extension argument a set A of strings (or equivalently of numbers) is inductively defined by specifying longer and longer initial segments of it. The global property to be ensured for A by the construction is split into countably many subgoals, given as a list $\{R_e : e \geq 0\}$ of so called requirements. Furthermore, for each requirement R_e, there is a finite extension strategy f_e which extends every initial segment $X|x$ to a longer initial segment $X|y$ such that any set X extending $X|y$ will meet the requirement R_e.

As observed already in the early days of recursion theory, finite extension arguments are closely related to the topological concept of Baire category. The class of all sets which meet a single requirement is open and dense, whence the property ensured by a finite extension argument is shared by a comeager class. The advantage of the category approach is its modularity and combinability. Since the intersection of any countable family of comeager classes is comeager again, hence nonempty, any countable number of finite extension arguments are compatible with each other, so that there is a set having all of the corresponding properties. This observation has been used to give elegant proofs of some of the basic facts on the degree structures in recursion theory (see e.g. [Ro67],[Od89]).

The disadvantage of the category approach is its nonconstructivity. In computation theory, in general we are not only interested in the mere existence of a set with a certain property, but we also want to know from what complexity level on we may find a set with this property. Since most of the interesting classes in recursion theory (and all complexity classes) are countable, hence meager, the category approach cannot provide this information.

This shortcoming of category can be overcome in part, however, by introducing a bounded category concept corresponding to a given complexity class C. Since the complexity of finite extension strategies used in diagonalizations related to C is determined by this class, we may define a Baire category concept for C by considering only those open dense classes which are determined by one of these C-extension strategies. Moreover, by countability of C there are only countably many C-strategies too, whence there will be a "generic" finite extension argument for C which is using all of them. So the "generic" set for C constructed by this argument, will have all properties provable by finite extension arguments corresponding to this class. In other words, any class which is provably comeager by C-extensions, contains all generic sets for C. So the complexity of any C-generic set will give an upper bound on the complexity of the easiest members of such a comeager class.

These ideas have been implemented in recursion theory in the 60s and

70s: By refining the arithmetical forcing concept of Feferman [Fe65], Hinman [Hi69] introduced the n-generic sets, which are generic for the n-th level Σ_n of the arithmetical hierarchy $(n \geq 1)$. Interesting applications of n-genericity to degree theory have been given by Jockusch [Jo80] and his students. For a recent survey of this topic, see Kumabe [Ku95].

In the 80s these ideas have been adopted in computational complexity theory too, and genericity notions for complexity classes, mainly for deterministic time or space classes were introduced. The first of these genericity concepts were introduced by Ambos-Spies, Fleischhack and Huwig in 1984, first defined only for tally sets ([AFH84],[AFH87]) but later also extended to the standard binary sets ([AFH88],[Fl85]). It followed a resource-bounded Baire category concept by Lutz in [Lu90], which, shortly later, was generalized by Fenner [Fe91].

Though it was known that the concepts of Ambos-Spies et al. are incompatible with the ones of Lutz and Fenner, the relations among these concepts had not been fully analyzed. One of the goals of this paper is to clarify these relations. Our answer to this question will be embedded in the more general attempt to develop resource-bounded genericity notions capturing the most fundamental diagonalization techniques in structural complexity theory.

Similar attempts have been made in recursion theory: E.g. Maass [Ma82] and Jockusch [Jo85] defined genericity concepts formalizing the finite injury method, a fundamental diagonalization technique for the construction of recursively enumerable sets.

Before we will give an outline of the contents of this paper, we have to look at a problem arising for the implementation of bounded Baire category and genericity in complexity theory, which is not present in recursion theory. Since finite extension functions are defined on finite initial segments, the length of the input of an extension strategy and the length of the corresponding input strings of the constructed set differ by an exponential factor. Hence a $t(n)$-time bounded extension strategy will correspond to the deterministic time class $DTIME(t(2^n))$ and not to $DTIME(t(n))$. In particular, n^k-time bounded extension strategies will correspond to $DTIME(2^{kn})$ $(k \geq 1)$, whence polynomial time bounded extension strategies capture **E**, the class of the linear exponential time sets. Since for the standard Turing machine concept sublinear time bounds do not make sense, this leads to serious obstacles for defining genericity concepts for subexponential time classes. The first genericity concept introduced by Ambos-Spies, Fleischhack and Huwig in 1984 [AFH84] solved this problem in a rather crude way. There only such diagonalizations are considered which yield tally diagonals. Since, for tally sets, the above described length problem does not arise, in this restricted setting genericity concepts can be defined for arbitrary time classes. Though, as shown in [AFH84] and [AFH87], this genericity captures many interesting diagonalization results, the limitation to tally sets is certainly not desirable. It seems, however, that no attempts have been made to get more satisfactory solutions to this problem, though some recent attacks on the corresponding problem for resource-bounded measure by Mayordomo ([Ma94], [Ma94b]) and by Allender and Strauss [AS94] may help to solve this problem in the category setting too.

Here we will not further discuss this problem, since our interest goes in a different direction. As mentioned before, our main goal is the classification of different types of genericity notions for a given complexity class according to their intrinsic power. This can be demonstrated by taking any sufficiently closed complexity class as an example. Though we will define the genericity concepts which we will discuss for all time bounds, we will develop the corresponding Baire category concepts only for the class **E** and all of our applications of genericity will be limited to this class. The extension of the category concepts to other (sufficiently closed) deterministic time and space classes extending **E** is routine, when following the more systematic approach developed by Lutz [Lu90] and Fenner [Fe91] for their concepts. So we will mainly deal with polynomial time computable extension functions. (For simplicity, in the following an n^k-extension function will be an extension function computable in time n^k and a p-extension function will be an extension function computable in polynomial time.) We will say that the corresponding genericity concepts will pertain to **E**, if, first, n^k-genericity captures (some class of) diagonalizations over $DTIME(2^{kn})$ so that, in particular, there are no n^k-generic sets in $DTIME(2^{kn})$ (whence p-generic sets capture diagonalizations over **E**); and, second, these genericity concepts do not capture diagonalizations over larger classes. The latter can be rephrased by requiring that, for all $k \geq 1$, n^k-generic sets can be uniformly constructed in $DTIME\left(2^{(k+c)n}\right)$ (for some constant c) whence p-generic sets exist in all sufficiently closed classes containing a universal set for **E**.

In Section 2, where we shortly review classical Baire category and its relation to finite extension arguments, we will introduce a frame work for bounded genericity and category notions to be used in the following sections.

Then we start our discussion of resource-bounded genericity by reviewing Lutz's category concept ([Lu90]), here called L-category, in Section 3. L-category is defined in terms of standard n^k-extension functions mapping initial segments to initial segments. This concept covers diagonalizations of the standard finite-extension type over exponential time *sets*. It fails, however, in the case of diagonalizations over exponential time *functions*, or more generally, over exponential time bounded *reductions*. As observed by Fenner [Fe91], this failure is related to the length problem discussed above: An n^k-extension function f can map an initial segment $X|x$ only to some extension $X|y$ where $|y| \leq k|x|$, whence p-extension functions capture only such diagonalizations which require only a linear look ahead. Such diagonalizations, however, do not even suffice for diagonalizing against polynomial time reductions. We demonstrate this limitation by showing that there are L-n^k-generic sets which are complete for **E**, whence the class of the **P**-m-complete sets for **E** is not L-meager.

In [Fe91], Fenner proposed a solution to this problem: he replaces the classical extension functions by functions which do not completely specify the extension, but only determine the values of some strings relevant for the diagonalization step. This F-extension function concept, which we present in Section 4, successfully eliminates the above problem, and Fenner's F-category concept for **E** covers the standard finite extension arguments over **E**.

Still there are diagonalizations over **E** or even **P** not covered by Fenner's concept. Mayordomo [Ma94] pointed out that there are F-p-generic sets which are not **P**-bi-immune, though **P**-bi-immune sets can be obtained by diagonalization in **E** (see [BS85]). This shortcoming of F-category can be explained by the observation, that, though a **P**-bi-immune set can be constructed by a standard finite extension argument, the resources required by the construction will not be recursively bounded. The construction of a **P**-bi-immune set of low complexity requires a different type of finite-extension argument, namely a *slow diagonalization* or *wait-and-see* argument. A slow diagonalization is a kind of effective priority argument, in which the requirements are not anymore met in the given order. Slow diagonalizations are a fundamental proof technique in structural complexity theory. E.g., bi-immunity results proved by slow diagonalizations are used for the proofs of hierarchy theorems for almost-everywhere complexity (see e.g. [GHS87]) and for the study of complexity cores (see e.g. [BDG90]).

Slow diagonalizations are based on the observation, that sometimes simple extensions suffice if we wait for the right moment. So slow diagonalizations can be described by partially defined extension functions. Using this observation, in Section 5 we introduce two new genericity concepts, called general genericity, or shortly G-genericity, and a slightly weaker variant of it, G_w-genericity, which generalize Fenner's concept in such a way, that slow diagonalizations over **E** are covered too. Both concepts are based on partially defined F-extension functions, but the requirements on the density of the domains slightly differ for the two concepts. As we will show, these new concepts coincide on the recursive sets, whence they yield the same Baire category theory on **E**, but differ in general.

Section 6 is devoted to the genericity concept of Ambos-Spies, Fleischhack and Huwig. More generally, we study restrictions of the genericity concepts introduced in the preceding sections, by considering only those F-extension functions which, on each input, specify only a constant number of strings. Restrictions of this sort correspond to diagonalizations over sets, many-one reductions, and bounded truth-table (bounded query) reductions. While for F-genericity, the restrictions of varying strength will yield a proper hierarchy, for general genericity all these restrictions lead to the same concept, of which we show that it coincides with the genericity concept of [AFH88]. Though these concepts are all strictly weaker than G-genericity and, in general, do not allow diagonalizations over truth-table or Turing reductions, they deserve some interest, since, as shown in [ANT94], these category concepts are compatible with the resource-bounded measure of Lutz [Lu92], what is not the case for the other category concepts discussed in this paper.

In Section 7 we address the question of still stronger category concepts for **E**. As, independently observed by Fenner [Fe95], the F-extension-function concept has a principal limitation. It requires the computation of the complete diagonalization step in advance, whence the total length of the diagonalization step will be bounded by the resource-bound on the extension function. For a resource-bounded diagonalization, however, it suffices that all the necessary action along the diagonalization can be *locally* done within the resource-bound (i.e., the admissible resources are measured in the length of the string currently processed), so that, with growing length of the diagonalization step, also the available resources are growing. This observation is not superficial, since one of the fundamental diagonalization techniques in structural complexity is based on this phenomenon, namely the *delayed diagonalization* or *looking back* technique (see e.g. [La75] or [BDG95]). There a diagonalization step is performed by simulating a given recursive diagonal until, by looking back, a witness for the desired diagonalization action is found.

In Section 7, we introduce a hierarchy of genericity concepts for **E**, called *extended genericity* (shortly E-genericity), which extend G-genericity and which in addition capture delayed diagonalizations over **E**. (Fenner [Fe95] independently introduced a corresponding hierarchy for F-genericity, which also captures delayed diagonalizations but still fails to capture slow diagonalizations.) These new genericity concepts depend on an additional parameter f, namely a recursive bound f on the length of the diagonalization steps. Though, for fixed recursive f, the (E, f)-p-genericity concept, i.e. the polynomial-time bounded version of E-genericity with length bound f, will pertain to **E**, there will be no recursive set which is (E, f)-p-generic for all recursive f. In other words, the fact, that there is no universal recursive function, will imply that, in contrast to the standard finite-extension arguments and the slow diagonalizations, there is no uniform way for doing all delayed diagonalizations over **E** effectively. While these observations show, that there is no strongest genericity concept pertaining to **E**, still - as observed by Fenner [Fe95] - the hierarchy of the (E, f)-n^k-generic sets gives rise to a natural, very strong concept of Baire category on **E**.

In Section 8 we further illustrate the power of the individual genericity

concepts discussed in this paper, by studying the sizes of the lower and upper
P-m-spans of sets in **E** for the corresponding Baire category concepts on **E**.
In classical Baire category, all nontrivial lower and upper spans are meager.
For the **E**-bounded case the full analogue only holds for the category concept
of E-genericity for variable recursive length bound f. For the other category
concepts, the analogy fails in quite different ways.

We conclude this section by introducing some basic notation.

Let $2^{<\omega}$ or $\Sigma^* = \{0,1\}^*$ denote the set of binary strings. For a string x,
$x(m)$ denotes the $(m+1)$th bit in x, i.e., $x = x(0)...x(n-1)$, where $n = |x|$
is the length of x. λ is the empty string. $x \subseteq y$ denotes that the string x
is extended by y and $x \subset y$ denotes that this extension is proper. We often
identify strings with numbers, by letting n be the $(n+1)$th string under the
canonical length-lexicographical ordering. Note that $|n| \approx \log(n)$. Lower
case letters $...,x,y,z$ from the end of the alphabet will be used to denote
strings, while the other letters denote numbers, with the exception of f, g, h
and s, t which also denote functions. N denotes the set of natural numbers.

A set of strings is called a *problem* or shortly a *set*, while sets of sets are
called *classes*, and sets of classes are called *families*. Capital letters denote
sets, boldface capital letters classes. We identify a set with its infinite characteristic string, i.e., $x \in A$ iff $A(x) = 1$ and $x \notin A$ iff $A(x) = 0$, so that
2^ω, the set of infinite binary sequences, is identified with the power class
of $2^{<\omega}$. We let $A|n$ denote the initial segment $A(0)...A(n-1) \in 2^{<\omega}$ of
A of length n. Similarly $A\lceil n$ will denote the set determined by $A|n$, i.e.,
$A\lceil n = \{x : x < n \ \& \ x \in A\}$. We write $x \subseteq A$ and say that A *extends* x if
x is a finite initial segment of A. The class of all sets extending a string x is
denoted by $\mathbf{B}_x = \{A : x \subseteq A\}$.

Note that we use strings in two different meanings: as elements of sets and
as finite initial segments of sets. In an attempt to avoid confusion, usually
we will use the notation $X|x$ for strings intended to denote initial segments.
Then $X|x$ denotes a string of length x and, for $y < x$, $X(y)$ or $(X|x)(y)$ will
denote the $(y+1)$th bit of $X|x$.

For the following it will be important to note the difference in the length
of an initial segment $A|x$ and the length of its bound x, namely

(1.1) $2^{|x|} - 1 \le |A|x| \le 2^{|x|+1} - 1,$

whence, for any number $k \ge 1$, $|A|x|^k \approx 2^{k \cdot |x|}$. These inequalities will be
crucial for our investigations. Our complexity theoretic notation is mainly
standard. For the fundamental complexity theoretic concepts used but not
explained in this paper we refer the reader to [HU79] and [BDG95]. In particular we assume the reader to be familiar with the basic results on time
constructible functions and recursively presentable (r.p.) classes, i.e. recursive classes of recursive sets. We will consider the deterministic time classes

$$\mathbf{P} = DTIME(poly) = \bigcup_{k \geq 1} DTIME\left(n^k\right),$$
$$\mathbf{E} = DTIME\left(2^{lin}\right) = \bigcup_{k \geq 1} DTIME\left(2^{kn}\right), and$$
$$\mathbf{E_2} = DTIME\left(2^{poly}\right) = \bigcup_{k \geq 1} DTIME\left(2^{n^k}\right).$$

Moreover, in our notation we will not distinguish between complexity classes for sets and functions. For any reducibility r (like 1 (= one-one), m (= many-one), $k - tt$ (= k-query truth-table), btt (= bounded truth-table), tt (= truth-table) and T (= Turing)), we let $\mathbf{P}_r^{\leq}(A)$ or shortly $\mathbf{P}_r(A)$ be the class $\{B : B \leq_r^P A\}$ of sets which can be reduced to A by an r-reduction in polynomial time. (For the definition of these reducibilities, see [LLS75].) Note that $\mathbf{P}_T(A) = \mathbf{P}(A)$. Also note that $\mathbf{E_2}$ is the downward closure of $\mathbf{E}$ under any of the polynomial reducibilities, whence a set A is $\mathbf{P}$-r-complete for $\mathbf{E}$ iff $A \in \mathbf{E}$ and A is $\mathbf{P}$-r-complete for $\mathbf{E_2}$.

2   Baire Category, Finite Extension Arguments, and Genericity

We first shortly review the concept of Baire category for the Cantor space. Based on the standard topology on this space, this concept gives a classification of the classes in small (=meager) classes, large (=comeager) classes, and classes of intermediate or not measurable size. For more details on this concept and the topological background see e.g. [Ox80].

Definition 2.1 *(a) For any string x, the class $\mathbf{B}_x = \{A : x \subseteq A\}$ is* basic open. *A class is* open *if it is the union of basic open classes.*

(b) A class $\mathbf{A}$ is dense *if $\mathbf{A}$ intersects all open classes; and $\mathbf{A}$ is* open-dense *if $\mathbf{A}$ contains an open and dense class. $\mathbf{A}$ is* nowhere dense *if $\mathbf{A}$ is contained in the complement of an open and dense class. $\mathbf{A}$ is* meager, *if $\mathbf{A}$ is the countable union of nowhere dense classes; and $\mathbf{A}$ is* comeager, *if the complement of $\mathbf{A}$ is meager or, equivalently, if $\mathbf{A}$ contains the countable intersection of open-dense classes.*

Note that for any class $\mathbf{A}$,

(2.1) $\mathbf{A}$ dense $\Leftrightarrow \forall x \; \exists A \in \mathbf{A} \; (x \subseteq A)$

(2.2) $\mathbf{A}$ open-dense $\Leftrightarrow \forall x \; \exists y \supseteq x \; (\mathbf{B_y} \subseteq \mathbf{A})$

(2.3) $\mathbf{A}$ nowhere dense $\Leftrightarrow \forall x \; \exists y \supseteq x \; (\mathbf{A} \cap \mathbf{B_y} = \emptyset)$

We will use these characterizations in the following tacitly. Baire's Theorem, which asserts that

(2.4) 2^ω is not meager,

immediately implies that no class can be both, meager and comeager. Together with the following basic properties of Baire category, which easily follow from the definitions, this shows that the interpretation of meager (comeager) classes as small (large) classes gives a sound measure system on the Cantor space.

(2.5)   For any class **A**, **A** is comeager iff $\overline{\textbf{A}}$ is meager.

(2.6)   For any set $A, \{A\}$ is meager.

(2.7)   For any classes **A** and **A'** , if **A** is meager and $\textbf{A'} \subseteq \textbf{A}$ then **A'** is meager too; and if **A** is comeager and $\textbf{A} \subseteq \textbf{A'}$ then **A'** is comeager too.

(2.8)   Let $\{\textbf{A}_e : e \geq 0\}$ be a countable family of classes. If the classes $\textbf{A}_e$ are meager then $\bigcup_{e \geq 0} \textbf{A}_e$ is meager too; and if the classes $\textbf{A}_e$ are comeager then $\bigcap_{e \geq 0} \textbf{A}_e$ is comeager too.

Note that, by (2.6) and (2.8), every countable class is meager. In the following lemma we give an example of an uncountable meager class which we will need later.

Lemma 2.2 *The class* $\textbf{C}_* = \{A : \exists m \, \forall n \geq m \, (A_{=n} \neq \emptyset)\}$, *where* $A_{=n} = \{x : |x| = n \,\&\, x \in A\}$, *is meager.*

Proof. Since $\textbf{C}_* = \bigcup_{m \geq 0} \textbf{C}_m$, where $\textbf{C}_m = \{A : \forall n \geq m \, (A_{=n} \neq \emptyset)\}$, it suffices to show that the classes $\textbf{C}_m$ are nowhere dense. So fix m and take any string $X|x$. We have to find a string $X|y$ extending $X|x$ which is not extended by any set in the class $\textbf{C}_m$. Obviously, $X|0^{n+1}$, where $n = max(|x| + 1, m)$ and $X \upharpoonright x = X \upharpoonright 0^{n+1}$, has this property.

Baire category is closely related to the finite extension method in computability theory. A *finite extension argument* is a diagonalization argument by which a set A is constructed to have a certain (global) property P. This global property P is split into countably many local properties $R_s \, (s \geq 0)$, called *requirements*, which together imply P. The set A is inductively defined in stages $s \geq 0$, i.e., there are strings $x_{-1} = \lambda < x_0 < x_1 < x_2 < \ldots$ such that x_{s-1} and $A|x_{s-1}$ are given at the beginning of stage s, x_s is defined during stage s and $A|x_{s-1}$ is extended to $A|x_s$ at this stage. In a *standard* finite extension argument, $A|x_s$ is chosen so that (no matter how A will be defined on the input strings $x \geq x_s$) A will meet requirement R_s (i.e. $A|x_s$ "forces" R_s). Moreover, for every requirement R_s, there is an extension strategy which tells us how any given initial segment $X|x$ can be extended to a finite initial segment $X|y$ such that all sets extending $X|y$ will meet the requirement R_s. The latter can be expressed in terms of Baire category by saying that there is an open and dense class such that all sets in this class meet requirement R_s. It follows that the class of the sets which meet R_s is comeager, so that (by (2.8)) the class of sets with property P is comeager too.

The proof of Baire's Theorem (2.4) can be viewed as a standard finite extension argument too. Given a meager class $\mathbf{C}$, say $\mathbf{C} = \bigcup_{e \geq 0} \mathbf{C}_e$ for the nowhere dense classes $\mathbf{C}_e$, we have to define a set $A \notin \mathbf{C}$. Since $\overline{\mathbf{C}}_s$ contains an open dense class, for given $A|x_{s-1}$, the extension $A|x_s$ can be chosen so that it forces that $A \notin \mathbf{C}_s$.

The close relation between Baire category and finite extension arguments shows that any countable number of standard finite extension arguments are compatible. So Baire category allows a modular approach to this method, which can greatly simplify the presentation of proofs. As an example we prove a theorem on the sizes of the lower and upper spans of the polynomial reducibilities, which gives a very simple proof for the existence of minimal pairs. (This result holds for all reducibilities defined by recursive operators and it was established first for the unbounded Turing reducibility in recursion theory (see [Od89]).)

Theorem 2.3 *For any polynomial time reducibility $\leq_r^P$ and any set $A \notin \mathbf{P}$, the following classes are comeager:*

1. $\mathbf{P}_r^{\not\leq}(A) = \{B : B \not\leq_r^P A\}$

2. $\mathbf{P}_r^{\not\geq}(A) = \{B : A \not\leq_r^P B\}$

3. $\mathbf{P}_r^{|}(A) = \{B : A \mid_r^P B\}$

4. $\mathbf{P}_r^{MP}(A) = \{B : A, B \text{ is a } \mathbf{P}\text{-}r\text{-minimal pair, i.e., } \forall C \leq_r^P A, B \ (C \in \mathbf{P})\}$

Proof. It suffices to give the proof for $r = T$. Let $\{M_e : e \geq 0\}$ be an enumeration of the polynomial-time bounded oracle Turing machines and let p_e be a polynomial bound for M_e. For a proof of part 1 note that

$$\mathbf{P}_T^{\leq}(A) = \{B : B \leq_T^P A\} = \{M_e^A : e \geq 0\}$$

is countable, hence meager. For a proof of part 2, fix e. It suffices to show that $\{B : A = M_e^B\}$ is nowhere dense. An extension function f witnessing this fact can be defined as follows. Given any string $X|x$, the corresponding set $X{\restriction}x$ is finite, whence $M_e^{X{\restriction}x} \in \mathbf{P}$. So, by assumption that $A \notin \mathbf{P}$, $A \neq M_e^{X{\restriction}x}$ whence we may fix y minimal with $A(y) \neq M_e^{X{\restriction}x}(y)$. Then, for any set Y with $Y{\restriction}0^{p_e(|y|)} = X{\restriction}x$, $A(y) \neq M_e^Y(y)$. So, for $z = max(x, 0^{p_e(|y|)})$,

$f(X|x) = \hat{X}|z$ will do, where $\hat{X} \restriction z = X \restriction x$. Part 3 follows from parts 1 and 2 since $\mathbf{P}_T^|(A) = \mathbf{P}_T^{\not\leq}(A) \cap \mathbf{P}_T^{\not\geq}(A)$. Finally, for a proof of 4, note that

$$\mathbf{P}_T^{MP}(A) = \bigcap \{\mathbf{P}_T^|(B) : B \in \mathbf{P}_T^{\leq}(A) - \mathbf{P}\}.$$

So, by the above, $\mathbf{P}_T^{MP}(A)$ is the countable intersection of comeager classes, hence comeager itself.

The disadvantage of the category approach is that it does not give any information on the complexity of the members of a comeager class. So Theorem 2.3 tells us that, for any set $A \notin \mathbf{P}$ there are sets B and B' such that B is $\mathbf{P}$-T-incomparable with A and A and B' form a $\mathbf{P}$-T-minimal pair, but it does not give us any information how the complexity of these sets depends on that of A, e.g. whether for recursive A we may choose B and B' to be recursive too. In fact it has been shown by other methods that the complexity of B and B' may be very different: For a $\mathbf{P}$-T-complete set A for $\mathbf{E}$, a corresponding B can be found in $DTIME(2^{2^n})$ but any corresponding B' will not be elementary recursive (see [Am87a]).

This shortcoming of classical Baire category can be overcome in part, by considering resource-bounded category concepts. The goal of this paper is to discuss such concepts for the deterministic time classes where we will focus on the class $\mathbf{E}$ of the linear exponential time sets. All of the concepts which we will discuss, will be introduced along the following lines. First, for any recursive time bound $t(n)$, we will specify a class $\mathbf{F}_t = \{f_{t,e} : e \geq 0\}$ of $t(n)$−time bounded extension strategies, so that the class $\mathbf{D}_{t,e}$ of the sets on which strategy $f_{t,e}$ succeeds will be open-dense. These strategies will operate on initial segments, whence, by (1.1), the $t(n)$-time bounded strategies will correspond to the satisfaction of the individual requirements in diagonalizations over $DTIME(t(2^n))$. Then, $t(n)$-time bounded category for these concepts will be defined by saying that a class is $t(n)$-open-dense if it contains one of the open-dense classes $\mathbf{D}_{t,e}$ ($e \geq 0$). The other concepts like $t(n)$-nowhere density, $t(n)$-meagerness and $t(n)$-comeagerness are derived from this concept as in the classical theory, where in case of meagerness and comeagerness, however, only $DTIME(t(n))$-uniform unions and intersections, respectively, are admitted. In fact, here we will develop the corresponding Baire category concepts only for the class $\mathbf{E}$, i.e., for polynomial time bounded extension strategies. For a more complete treatment we refer to Lutz [Lu90]. The framework developed there can be easily adapted to the new concepts introduced here.

The phenomenon we will focus on is genericity. By countability of $\mathbf{F}_t$ there will be sets on which all of the extension strategies $f_{t,e}, e \geq 0$, succeed, i.e., sets which have all properties which can be proved by diagonalizations over $DTIME(t(2^n))$ formalized by the strategies in $\mathbf{F}_t$. So the n^k-generic sets will reflect the diagonalizations over $DTIME(2^{kn})$ possible in the given frame work.

Next we introduce the notion of an abstract category and genericity concept for deterministic time, which will provide notations and basic facts for the concrete concepts introduced later.

Definition 2.4 *(a) An* abstract category concept for deterministic time *is a sequence* $\mathcal{D} = <\mathcal{D}(t) : t \ recursive>$ *of countable families* $\mathcal{D}(t)$ *of open-dense classes such that, for all recursive functions t and t':*

(2.9)   *For any class* $\mathbf{D} \in \mathcal{D}(t)$, *co-$\mathbf{D} = \{\overline{A} : A \in \mathbf{D}\} \in \mathcal{D}(t)$ too.*

(2.10)   *If* $t \leq t'$ *a.e. then* $\mathcal{D}(t)$ *is contained in* $\mathcal{D}(t')$.

(b) Let $\mathcal{D}$ be an abstract category concept for deterministic time. A class $\mathbf{A}$ is $\mathcal{D}$-$t(n)$-open-dense *if it contains one of the classes in* $\mathcal{D}(t)$, *and a set A is* $\mathcal{D}$-$t(n)$-generic *if A is a member of all classes in* $\mathcal{D}(t)$.

Note that (2.9) and (2.10) reflect two basic properties of the deterministic time classes, namely that $DTIME(t(n))$ is closed under complements and that, for $t \leq t'$, $DTIME(t(n))$ is contained in $DTIME(t'(n))$. The following observations are immediate by definition.

Proposition 2.5 *Let $\mathcal{D} = <\mathcal{D}(t) : t \ recursive>$ be an abstract category concept for deterministic time and let t, t' be any recursive functions.*

1. *There is a $\mathcal{D}$-$t(n)$-generic set; in fact, the class of $\mathcal{D}$-$t(n)$-generic sets is comeager.*

2. *For any set A, A is $\mathcal{D}$-$t(n)$-generic iff $\overline{A}$ is $\mathcal{D}$-$t(n)$-generic.*

3. *If $t \leq t'$ a.e. then any $\mathcal{D}$-t'-generic set is $\mathcal{D}$-t-generic too.*

Proof. Since A is $\mathcal{D}$-$t(n)$-generic iff $A \in \cap \mathcal{D}(t)$ and since $\mathcal{D}(t)$ is a countable family of open-dense classes, part 1 is immediate by Definition 2.1, while parts 2 and 3 follow from (2.9) and (2.10), respectively.

Next we will focus on the polynomial time bounded strategies corresponding to $\mathbf{E}$, for which we also introduce the corresponding Baire category concepts.

Definition 2.6 *Let $\mathcal{D} = <\mathcal{D}(t) : t \ recursive>$ be an abstract category concept for deterministic time.*

1. *A set A is $\mathcal{D}$-p-generic if A is $\mathcal{D}$-n^k-generic for all $k \geq 1$.*

2. *$\mathcal{D}(p)$ pertains to $\mathbf{E}$ if there is a constant c such that for all $k \geq 1$*

(2.11)   *there is no $\mathcal{D}$-n^k-generic set in $DTIME(2^{kn})$, whence there is no $\mathcal{D}$-p-generic set in $\mathbf{E}$,* *but*

(2.12)   there is a $\mathcal{D}$-n^k-generic set in $DTIME(2^{(k+c)n})$ and

(2.13)   there is an $\mathcal{D}$-p-generic set in $DTIME(2^{n^2})$.

3. *A class $\mathbf{A}$ is $\mathcal{D}$-p-open-dense if $\mathbf{A}$ is $\mathcal{D}$-n^k-open-dense for some k. $\mathbf{A}$ is $\mathcal{D}$-p-nowhere dense if $\mathbf{A}$ is contained in the complement of an $\mathcal{D}$-p-open-dense class. $\mathbf{A}$ is $\mathcal{D}$-p-comeager if $\mathbf{A}$ contains the countable intersection of some $\mathcal{D}$-n^k-open-dense classes for some (fixed) $k \geq 1$. $\mathbf{A}$ is $\mathcal{D}$-p-meager if the complement of $\mathbf{A}$ is $\mathcal{D}$-p-comeager.*

4. *A class $\mathbf{A}$ is $\mathcal{D}$-meager in $\mathbf{E}$, if $\mathbf{A} \cap \mathbf{E}$ is $\mathcal{D}$-p-meager and $\mathbf{A}$ is $\mathcal{D}$-comeager in $\mathbf{E}$ if $\overline{\mathbf{A}}$ is $\mathcal{D}$-meager in $\mathbf{E}$.*

The intuition behind Definition 2.6.2 is as follows. We want that the $\mathcal{D}$-$t(n)$-generic sets are uniformly (in $t(n)$) defined and that they capture (a certain class of) diagonalizations over $DTIME(t(2^n))$ (cf. (1.1)), so that $\mathcal{D}$-n^k-genericity pertains to $DTIME(2^{kn})$ and $\mathcal{D}$-p-genericity pertains to $\mathbf{E}$. Now (2.11) expresses that at least the most simple diagonalizations over these classes are covered. The fact that in general diagonalizations over larger classes are not captured, can be enforced by requiring that there are $\mathcal{D}$-n^k-generic and $\mathcal{D}$-p-generic sets which are not more complex (modulo some small overhead) than universal sets for the classes $DTIME(2^{kn})$ and $\mathbf{E}$, respectively. To avoid technicalities we have rephrased this by (2.12) and (2.13) in a somewhat more liberal form.

Similarly, in part 3 of the definition, the fact that $\mathcal{D}$-p-comeager classes are defined by *p-uniform* intersections of $\mathcal{D}$-p-open-dense classes is stated indirectly in a technically simpler form. Since a p-uniform sequence can contain only $\mathcal{D}$-n^k-open-dense sets for a fixed level k of $\mathbf{P}$, and since there is a uniform enumeration of $DTIME(n^k)$ in $DTIME(n^{k+1})$, hence in $\mathbf{P}$, any p-uniform sequence of $\mathcal{D}$-p-open-dense classes will be contained in a sequence of $\mathcal{D}$-n^k-open-dense classes (for some k), while the sequence of all of the basic $\mathcal{D}$-n^k-open-dense classes will be p-uniform by uniformity of the $\mathcal{D}$-category concept.

The following characterization of p-(co)meagerness in terms of n^k-genericity easily follows from the definition. It shows that the investigation of the $\mathcal{D}$-category concept on $\mathbf{E}$ can be reduced to the investigation of the $\mathcal{D}$-n^k-generic sets in $\mathbf{E}$.

Proposition 2.7 *For any class $\mathbf{A}$ the following hold.*

1. *$\mathbf{A}$ is $\mathcal{D}$-p-meager iff, for some $k \geq 1$, $\mathbf{A}$ does not contain any $\mathcal{D}$-n^k-generic set.*

2. **A** *is not* $\mathcal{D}$-p-meager iff, for all $k \geq 1$, **A** contains some $\mathcal{D}$-n^k-generic set.

3. **A** *is* $\mathcal{D}$-p-comeager iff, for some $k \geq 1$, **A** contains all $\mathcal{D}$-n^k-generic sets.

Proposition 2.8 *Let* $\mathcal{D} = < \mathcal{D}(t) : t$ *recursive$>$ be an abstract category concept for deterministic time such that* $\mathcal{D}(p)$ *pertains to* **E**.

1. *For any* $k \geq 1$, $DTIME(2^{kn})$ *is* $\mathcal{D}$-p-meager, hence $\mathcal{D}$-meager in **E**. So, in particular, for any $A \in$ **E**, $\{A\}$ is $\mathcal{D}$-p-meager, hence $\mathcal{D}$-meager in **E**.

2. **E** *is not* $\mathcal{D}$-p-meager, hence not $\mathcal{D}$-meager in **E**.

Proof. By Proposition 2.7, this follows from (2.11) and (2.12).

Note that, for a category concept $\mathcal{D}$ where $\mathcal{D}(p)$ pertains to **E**, by Proposition 2.8 and by definition, the fundamental properties (2.4) - (2.7) of Baire category hold for the $\mathcal{D}$-category on **E** too, if the universe 2^ω is replaced by **E**. Moreover, (2.8) carries over too, if countability is replaced by p-uniformity. So $\mathcal{D}$-category on **E** resembles the classical category concept.

We conclude this section with a short discussion of Lebesgue measure, which is an alternative approach to define a measure system on the Cantor space. Here the large classes are the measure 1-classes and the small classes are the measure 0-classes. The Lebesgue measure is the product measure on 2^ω induced by $\mu(0) = \mu(1) = 1/2$. Alternatively, the measure 0-classes can be characterized by martingales (cf. Lutz [Lu92]). A *martingale* is a function d from $2^{<\omega}$ into the nonnegative reals satisfying $d(x0) + d(x1) \leq 2d(x)$ for all x. A martingale d *succeeds* on a set A if $\limsup_x d(A|x) = \infty$. Then a class $\mathbf{C}$ has Lebesgue measure 0 iff there is a martingale d which succeeds on all sets in $\mathbf{C}$. Lutz used this characterization to introduce a resource-bounded measure.

Definition 2.9 *(Lutz [Lu92]).* A class $\mathbf{C}$ has $t(n)$-measure 0 if there is a martingale $d : 2^{<\omega} \rightarrow Q^+$, where Q^+ is the set of the nonnegative rationals, such that $d \in DTIME(t(n))$ and d succeeds on all sets in $\mathbf{C}$. The class $\mathbf{C}$ has $t(n)$-measure 1 if $\overline{\mathbf{C}}$ has $t(n)$-measure 0. A set A is $t(n)$-random if A is an element of all $t(n)$-measure 1-classes. The class $\mathbf{C}$ has p-measure 0 (1) if $\mathbf{C}$ has n^k-measure 0 (1) for some k.

This definition is a slight simplification of Lutz's original definition. The equivalence to the original definition for the p-measure has been shown in [ATZ94]. It is well known that classical Baire category and Lebesgue measure are incompatible, i.e., there are classes which are large in one concept but are small in the other concept. By Lemma 2.2 and the following lemma, the class $\mathbf{C}_*$ is an example for this phenomenon. In the following we will use this example to prove some incompatibility results for the resource-bounded case too.

Lemma 2.10 *The class* $\mathbf{C}_* = \{A : \exists m\, \forall n \geq m\, (A_{=n} \neq \emptyset)\}$ *has measure 1, in fact* n^2*-measure 1.*

Proof. A martingale $d \in DTIME(n^2)$ which succeeds on all sets in $\overline{\mathbf{C}}_*$ is defined by $d(\lambda) = 1$ and by letting $d((X|x)i) = a_i \cdot d(X|x)$ where $a_i = (1-i) + 1/2$ if, for all $y < x$ with $|y| = |x|$, $(X|x)(y) = 0$, and where $a_i = 1$ otherwise $(i = 0,1)$. Note that $d(A|0^{n+1}) = (3/2)^{2^n} \cdot d(A|0^n)$ if $A_{=n} = \emptyset$, and $d(A|0^{n+1}) \geq \frac{1}{2} \cdot d(A|0^n)$ otherwise. Hence, for n with $A_{=n} = \emptyset$, $d(A|0^{n+1}) \geq g(n)$ where $g(n) = (1/2)^n \cdot (3/2)^{2^n}$, and, as one can easily check, $\lim_n g(n) = \infty$.

3 Lutz's Genericity Concept

The resource-bounded category and genericity concepts of Lutz [Lu90], which we want to discuss first, are based on the standard extension function concept. This is the, probably, most straightforward way to introduce resource-bounded category notions. By imposing resource bounds on the admissible extension functions, Lutz defined category notions for deterministic time and space classes.

Definition 3.1 *(Lutz [Lu90]).* *(a) An* extension function *is a total function* $f : 2^{<\omega} \to 2^{<\omega}$ *such that* $x \subseteq f(x)$ *for all* x. *If, moreover,* $f \in DTIME(t(n))$ *then* f *is a* $t(n)$-*extension function, and* f *is a* p-*extension function if* f *is an* n^k-*extension function for some* $k \geq 1$.
(b) A set A meets *the extension function* f *if* $f(A|x) \subseteq A$ *for some* x. *Otherwise* A avoids f. *Let* $\mathbf{D}_f$ *be the class of all sets which meet* f.

In the following we call Lutz's resource-bounded category concept based on $t(n)$-extension functions *L-category*.

Definition 3.2 *L-category is the category concept* $\mathcal{D} =\; <\mathcal{D}(t) : t\ recursive>$, *where* $\mathcal{D}(t) = \{\mathbf{D}_f :\; f\ t(n)$-*extension function*$\}$. *In particular,* A *is* L-$t(n)$-*generic if* A *meets all* $t(n)$-*extension functions.*

It easily follows from (2.2) that, in classical Baire category, an open class A is dense iff there is an extension function f such that $\mathbf{D}_f \subseteq \mathbf{A}$, whence the Baire category concepts can be defined in terms of Definition 3.1. Moreover, for any extension function f, the class $\mathbf{D}_f$ is open, whence Definition 3.2 yields a category concept for deterministic time in the sense of Definition 2.4:

Proposition 3.3 *L-category is an abstract category concept.*

Proof. Straightforward.

The following two theorems of Lutz show that the polynomial section of *L*-category pertains to **E**, whence, by Proposition 2.8, Lutz's concept gives a consistent Baire category notion on **E**.

Theorem 3.4 *(Lutz). There is an L-n^k-generic (L-p-generic) set A in* $DTIME(2^{(k+2)n})(DTIME(2^{n^2}))$.

Proof. The proof is a standard finite extension argument. We consider the case of *L*-n^k-genericity and leave the similar case of *L*-*p*-genericity to the reader. Since there is a universal function for $DTIME(n^k)$ in $DTIME(n^{k+1})$ we may fix a recursive enumeration $\{f_e : e \geq 0\}$ of the n^k-extension functions such that, for x with $|x| \geq e$, $f_e(x)$ can be computed in $|x|^{k+1}$ steps (uniformly in e and x). Then A is constructed in stages, where at stage s we ensure that A meets the s-th extension function $f_s \in DTIME(n^k)$: Given $A|x_{s-1}$ (where $x_{-1} = \lambda$) let $A|x_s = f_s(A|x_{s-1})$ if $f_s(A|x_{s-1})$ properly extends $A|x_{s-1}$, and let $A|x_s = (A|x_{s-1})0 = f_s(A|x_{s-1})0$ otherwise. Note that $|A|x_{s-1}| \geq s$, whence, by our assumptions on the functions f_e and by (1.1), $A \in DTIME(2^{(k+2)n})$.

Theorem 3.5 *(Lutz). Let A be L-n^k-generic ($k \geq 1$). Then $A \notin DTIME(2^{kn})$. Hence, for any L-p-generic set A, $A \notin$ **E**.*

Proof. Given any set $B \in DTIME(2^{kn})$, to show that $A \neq B$ define the n^k-extension function f by $f(X|x) = (X|x)(1 - B(x))$. Then, by *L*-n^k-genericity, A meets f. Obviously this implies $A \neq B$.

Corollary 3.6 *(Lutz). L-p-category pertains to **E** in the sense of Definition 2.6. So **E** is not L-p-meager, hence not L-meager in **E**.*

Proof. Theorems 3.4 and 3.5 imply the first part of the corollary. The second part follows with Proposition 2.8.

Lutz has also shown that *L*-*n*-generic sets cannot be sparse. Here we will prove a similar result which will show that *L*-genericity and randomness are incompatible.

Lemma 3.7 *Let* $\mathbf{C}_* = \{A : \exists m \, \forall n \geq m \, (A_{=n} \neq \emptyset)\}$. *For any L-$n^2$-generic set A, $A \notin \mathbf{C}_*$, whence $\mathbf{C}_*$ is L-p-meager.*

Proof. The proof of Lemma 2.2 actually gives n^2-extension functions f_m, such that the sets in the classes $\mathbf{C}_m$ defined there avoid f_m.

Theorem 3.8 *For $t(n), t'(n) \geq n^2$, there is no set which is both, L-t(n)-generic and t'(n)-random. Hence, for $t(n) \geq n^2$, the class of L-t(n)-generic sets has n^2-measure 0 (hence p-measure 0 and classical measure 0) and the class of t(n)-random sets is L-p-meager (hence meager in the classical sense).*

Proof. By Lemmas 2.10 and 3.7.

Though L-p-genericity is a useful tool for diagonalizing over exponential time sets, it does not capture diagonalizations over exponential time reductions, in fact not even over polynomial time reductions. As observed by Fenner [Fe91], the reason for this limitation is the shortness of the p-extensions.

Proposition 3.9 *(Fenner). For any $t(n)$-extension function f and for any initial segment $X|x$, $|f(X|x)| \leq t(2^{|x|+1})$ whence $f(X|x) = \widehat{X}|\widehat{x}$ for some initial segment $\widehat{X}|\widehat{x}$ where $|\widehat{x}| \leq \log t(2^{|x|+1})$. In particular, for $f \in DTIME(n^k)$, $|f(X|x)| \leq 2^{k(|x|+1)}$ whence $f(X|x) = \widehat{X}|\widehat{x}$ for some initial segment $\widehat{X}|\widehat{x}$ where $|\widehat{x}| \leq k(|x|+1)$.*

Proof. This easily follows from (1.1).

So, given an initial segment $X|x$, a p-extension $f(X|x)$ of $X|x$ can determine only strings of length linear in the length of x, whereas a polynomial time reduction can relate x to strings of length polynomial in $|x|$. Fenner [Fe91] exploited this observation to show that there is an L-p-generic set A such that $\mathbf{P}(A) = \mathbf{NP}(A)$. In the following we prove this result and the general failure of L-p-genericity to diagonalize over polynomial reductions by analyzing the freedom which we have in the construction of an L-n^k-generic (or, more generally, an L-$t(n)$-generic) set.

In the construction of an L-n^k-generic set A (cf. the proof of Theorem 3.4) it suffices to ensure that, for all n^k-extension functions $f_e(e \geq 0)$, $f_e(A|x_e) \subseteq A$ for some string x_e. There are no (nontrivial) restrictions on the choice of the strings x_e used for satisfying these genericity requirements. So we can make these strings easily recognizable and we may introduce arbitrarily large (recursively bounded) gaps between the "generic" parts of A. On the other hand, by Proposition 3.9, the generic parts are very short, namely contained in intervals $[0^n, 0^{(k+1)n}]$. So the major portions of the generic set can be freely determined. Only small, easily recognizable parts are not cotrollable. This allows the efficient coding of any given set in an L-n^k-generic set, so that, for instance, there are $\mathbf{P}$-m-complete sets for $\mathbf{E}$ which are L-n^k-generic.

To state these observations more formally, we need the following notation. For any set B let $DTIME^{(B,\leq)}(t(n))$ denote the class of sets which can be recognized by a deterministic $t(n)$ time bounded oracle Turing machine M with oracle B such that M queries only strings $y \leq x$ on input x. Then the general existence theorem for L-$t(n)$-generic sets is as follows.

Theorem 3.10 *Let B be any set, let $C \subseteq N$ be infinite such that $\{0^n : n \in C\} \in \mathbf{P}$, and let t, t' be nondecreasing recursive time bounds such that $n \leq t(n) \leq t'(n)$ for all n and there is a universal function for $DTIME(t(n))$ in $DTIME(t'(n))$. There is an L-$t(n)$-generic set $A \in DTIME^{(B,\leq)}(2^n t'(2^{n+1}))$ such that*

(3.1) $\forall x (A(x) \neq B(x) \;\Rightarrow\; \exists n \in C(x \in [0^n, 0^{\log t(2^{n+1})+1}]))$.

In particular, for $k \geq 2$, there is an L-n^k-generic set $A \in DTIME^{(B,\leq)}(2^{(k+2)n})$ such that

(3.2) $\forall x (A(x) \neq B(x) \Rightarrow \exists n \in C(x \in [0^n, 0^{kn+k+1}]))$

and, there is an L-p-generic set $A \in DTIME^{(B,\leq)}(2^{n^2})$ such that

(3.3) $\forall x (A(x) \neq B(x) \Rightarrow \exists n \in C(x \in [0^n, 0^{n^2}]))$.

Proof (sketch). Let $\{f_e : e \geq 0\}$ be a recursive enumeration of the $t(n)$-extension functions such that, for x with $|x| \geq e$, $f_e(x)$ can be computed in $t'(|x|)$ steps and fix a polynomial time computable subset $\{0^{n_e} : e \geq 0\}$ of $\{0^n : n \in C\}$ such that $\log t(2^{n_e+1}) + 1 < n_{e+1}$ for all e. Then define $A|0^{n_e}$ by induction on e as follows. For $e = 0$ let $A|0^{n_0} = B|0^{n_0}$. For the inductive step assume that $A|0^{n_e}$ is given. Then, by Proposition 3.9, $f_e(A|0^{n_e}) = X|x$ for some string $X|x$ with $|x| \leq \log t(2^{n_e+1})$, whence, by choice of the numbers n_e, $x < 0^{n_{e+1}}$. So, if we define $A|0^{n_{e+1}}$ by letting $A(y) = f_e(A|0^{n_e})(y)$ for $y < x$ and $A(y) = B(y)$ for $x \leq y < 0^{n_{e+1}}$, then this is compatible with (3.1) and at the same time ensures that A meets f_e. Finally, the desired time bound for A (relative to B) easily follows from the choice of the enumeration $\{f_e : e \geq 0\}$ and (1.1).

Corollary 3.11 *For every set B there is an L-n^k-generic set ($k \geq 2$) $A \in DTIME^{(B,\leq)}(2^{(k+2)n})$ (L-p-generic set $A \in DTIME^{(B,\leq)}(2^{n^2})$) such that B is reducible to A by a polynomial time bounded, length-increasing one-one reduction.*

Proof. We sketch the proof for n^k-genericity. Let ϵ be the iterated exponential function defined by $\epsilon(0) = 2^0$ and $\epsilon(n+1) = 2^{\epsilon(n)}$. Then, for $C = \{\epsilon(n) : n \geq k\}$ and $G = \{x : \exists n \in C(x \in [0^n, 0^{kn+k+1}])\}$, $\{0^n : n \in C\} \in \mathbf{P}$ and $G \in \mathbf{P}$. Moreover, for every string x, $0x \notin G$ or $1^{k(|x|+2)}0x \notin G$. Now let

$$B' = \{0x : x \in B\} \cup \{1^{k(|x|+2)}0x : x \in B\}$$

and apply the second part of Theorem 3.10 to B' and C. We claim that the resulting L-n^k-generic set A has the required properties. Since, by definition of B',

$$DTIME^{(B',\leq)}(2^{(k+2)n}) \subseteq DTIME^{(B,\leq)}(2^{(k+2)n}),$$

it suffices to show that $B \leq^P_{1-li} A$. By choice of G and by (3.2), this holds via the reduction

$$g(x) = \text{ if } 0x \notin G \text{ then } 0x \text{ else } 1^{k(|x|+2)}0x.$$

Corollary 3.12 *There is an L-p-generic set which is* **P**-1-*li-complete for* $\mathbf{E}_2$; *and, for* $k \geq 1$, *there is an L-n^k-generic set which is* **P**-1-*li-complete for* **E**.

Proof. Apply Corollary 3.11 to a **P**-1-*li*-complete set B for $\mathbf{E}_2$ and **E**, respectively.

Corollary 3.13 *There is an L-p-generic set* $A \in \mathbf{E}_2$ *(L-n^c-generic set* $A \in \mathbf{E}$ $(c \geq 1))$ *such that*

$$(3.4) \qquad \begin{aligned} \mathbf{P}_{1-li}(A) &= \mathbf{P}_1(A) = \mathbf{P}_m(A) = \mathbf{P}_{k-tt}(A)(k \geq 1) = \mathbf{P}_{btt}(A) = \\ \mathbf{P}_{tt}(A) &= \mathbf{P}(A) = \mathbf{NP}(A). \end{aligned}$$

Proof. Since $\mathbf{E}_2$ is closed under the **NP**-operator, (3.4) holds for any **P**-1-*li*-complete set A for $\mathbf{E}_2$. So the claim follows from Corollary 3.12.

Corollary 3.14 *The class of* **P**-*m-complete sets for* **E** *is not L-p-meager, hence not L-meager in* **E**.

Proof. By the second part of Corollary 3.12 and by Proposition 2.7.

4 Fenner's Genericity Concept

In [Fe91] Fenner presented a generalization of Lutz's resource-bounded category concept which overcomes the limitations of Lutz's concept due to the size restraints of the extension functions. Fenner introduced a new type of extension functions which do not necessarily completely specify the extension but determine it only on certain strings. This parallels the situation in a resource-bounded reduction, where also the oracle is queried only on certain strings, not necessarily on a full initial segment. Though Fenner introduced only total extension functions of this new type, for later use we will extend the definition to partial functions.

Definition 4.1 *(Fenner [Fe91])(a) A (partial) F-extension function is a (partial) function* f *which maps strings to finite sequences of pairs of strings and bits such that, for any input* $X|x$ *(on which* f *is defined),* $f(X|x) = (x_0, i_0), \ldots,$
(x_n, i_n) *where* $n \geq 0, x_0, \ldots, x_n$ *are strings satisfying* $x \leq x_0 < \ldots < x_n$, *and* $i_0, \ldots, i_n \in \{0, 1\}$.
(b) A (partial) F-t(n)-extension function is a (partially defined) F-extension function f *such that* $(dom(f) \in DTIME(t(n))$ *and for* $X|x \in dom(f))$ $f(X|x)$ *can be computed by a deterministic* $t(n)$-*time bounded Turing machine.* f *is a (partial) F-p-extension function if* f *is a (partial) F-n^k-extension function for some* $k \geq 1$.
(c) Let f *be a (partial) F-extension function and let* $f(X|x) = (x_0, i_0), \ldots,$

(x_n, i_n). *A string* $Y|y$ *is* consistent with f *on* $X|x$ *if* $Y|y$ *extends* $X|x$ *and* $Y(x_m) = i_m$ *for all* m *with* $0 \leq m \leq n$ *and* $x_m < y$; $Y|y$ *extends* f *on* $X|x$ *if* $Y|y$ *is consistent with* f *on* $X|x$ *and* $x_n < y$. *A set* A *extends* f *on* $X|x$ *if* $A|y$ *extends* f *on* $X|x$ *for some* y. *A* meets f, *if* A *extends* f *on some string* $X|x$; *and* A avoids f *otherwise.* $\mathbf{D}_f$ *is the class of all sets which meet* f.

In the remainder of this section all F-extension functions are total. Here we call Fenner's category concept based on total F-extension functions F-category.

Definition 4.2 F-category *is the category concept* $\mathcal{D} =< \mathcal{D}(t) : t \ recursive>$ *where*

$$\mathcal{D}(t) = \{\mathbf{D}_f : f \ total \ F\text{-}t(n)\text{-}extension \ function\}.$$

In particular, A *is* F-$t(n)$-generic *if* A *meets all total* F-$t(n)$-extension func-tions.

We should remark, that Fenner defined several genericity concepts in [Fe91]. The concept here is equivalent to the most powerful concept introduced there, namely St, Lp-genericity. (Fenner's original definition was based on F-extension functions defined on all finite subsequences of the characteristic sequence of a set, not just on its initial segments. It can be easily shown, however, that the induced genericity concepts are the same.)

Note that for any F-extension function f, the class $\mathbf{D}_f$ is open and dense. So it is easy to show that F-category is a category concept for deterministic time in the sense of Definition 2.4.

Proposition 4.3 F-category *is a category concept in the sense of Definition 2.4.*

Also note that any standard extension function can be rephrased as an F-extension function in a straightforward way, whence F-category generalizes L-category (modulo a small overhead due to the differences in the represen-tations):

Proposition 4.4 *(Fenner). Every* F-n^{k+1}-generic *(*F-p-generic*) set is* L-n^k-generic *(*L-p-generic*).*

Proof. Let f be any n^k-extension function. Then f', defined by

$$f'(X|x) = (x, Y(x)), \ldots, (y - 1, Y(y - 1))$$

where $f(X|x) = Y|y$, is a total F-n^{k+1}-extension function and, for any set X, X meets f if and only if X meets f'.

Proposition 4.4 and Theorem 3.5 imply that there are no F-p-generic sets in $\mathbf{E}$. Due to the linear overhead in Propopsition 4.4, we have to show the corresponding result for the individual levels of $\mathbf{E}$ directly.

Lemma 4.5 *(Fenner) Let A be F-n^k-generic ($k \geq 1$). Then $A \notin DTIME(2^{kn})$.*

Proof. In the proof of Theorem 3.5 replace f by the equivalent F-n^k-extension function $f(X|x) = (x, 1 - B(x))$.

So, to show that the polynomial part of F-category pertains to **E**, it suffices to establish the existence of F-n^k-generic sets in **E**.

Theorem 4.6 *(Fenner). There is an F-n^k-generic (F-p-generic) set in $DTIME(2^{(k+2)n})$ $(DTIME(2^{n^2}))$.*

Proof. Note that there is a uniform enumeration of the F-n^k-extension functions in $DTIME(n^{k+1})$. So the desired generic sets can be constructed by a standard finite extension argument like in the proof of Theorem 3.4: Given $A|x_{s-1}$, the extension $A|x_s$ which ensures that A meets the s-th F-n^k-extension function f_s is as follows. For $f_s(A|x_{s-1}) = (y_0, i_0), \ldots, (y_n, i_n)$ let $x_s = y_n + 1$, let $A(y_m) = i_m$ for $0 \leq m \leq n$ and, for all other y with $x_{s-1} \leq y < x_s$ let $A(y) = 0$.

Corollary 4.7 *(Fenner) F-p-category pertains to **E** in the sense of Definition 2.6. So **E** is not F-p-meager, hence not F-meager in **E**.*

Proof. By Lemma 4.5, Theorem 4.6, and Proposition 2.8.

The next theorem shows that, in contrast to L-category, F-category can separate the standard exponential, hence polynomial, reducibilities.

Theorem 4.8 *(Fenner, Ambos-Spies, Wang). Let A be F-n^2-generic. Then A separates all the standard polynomial time reducibilities. In particular,*

$$\mathbf{P}_1(A) \subset \mathbf{P}_m(A) \subset \mathbf{P}_{k-tt}(A) \subset \mathbf{P}_{(k+1)-tt}(A) \subset \mathbf{P}_{btt}(A) \subset \mathbf{P}_{tt}(A) \subset \mathbf{P}_T(A) \subset \mathbf{NP}(A).$$

Similarly, if A is F-p-generic, then A separates the corresponding deterministic exponential time reducibilities.

The separation of $\mathbf{P}(A)$ and $\mathbf{NP}(A)$ is due to Fenner [Fe91], while $\mathbf{P}_{tt}(A) \neq \mathbf{P}_T(A)$ was established first by Yongge Wang (private communication).

Proof. The proofs rephrase, in terms of F-extension functions, the oracle separation of **P** and **NP** by Baker, Gill and Solovay [BGS75] and the separation results of Ladner, Lynch and Selman [LLS75] for the polynomial time reducibilities, respectively. Here we give the proofs for two of the separations, namely (a) $\mathbf{P}_m(A) \neq \mathbf{P}_{1-tt}(A)$ and (b) $\mathbf{P}(A) \neq \mathbf{NP}(A)$. More separations are proved in Section 6 where we analyze restricted forms of F-genericity. The extension from the polynomial to the exponential case for F-p-genericity is straightforward.

(a) To show that $\mathbf{P}_m(A) \neq \mathbf{P}_{1-tt}(A)$ it suffices to show that A is *non-P-selfdual*, i.e., that $\overline{A} \not\leq_m^P A$. Given a **P**-$m$-reduction g, define an F-n^2-extension function f by $f(X|x) = (x, (X|x)(g(x)))$ if $g(x) < x$ and by $f(X|x) = (x, 0), (g(x), 0)$ otherwise. Then, for any x such that A extends f on $A|x$, $A(x) = A(g(x))$ whence g fails to reduce $\overline{A}$ to A.

(b) Let $L^A = \{0^n : \exists x\ (|x| = n\ \&\ x \in A)\}$. Then $L^A \in \mathbf{NP}(A)$. For a proof that $L^A \notin \mathbf{P}(A)$, we have to show that $L^A \neq M^A$ for every given polynomial time bounded oracle Turing machine M^X. To do so, define an F-n^2-extension function f as follows. Fix a polynomial time bound p for M^X and fix c such that $p(n) < 2^n$ for all $n \geq c$. Then, for the finitely many strings $X|x$ with $|x| < c$, let $f(X|x) = (x, 1 - A(x))$ so that A does not extend f on any such $X|x$. For $X|x$ with $|x| \geq c$, let $y_0, \ldots, y_n$ be the queries $\geq x$ made by $M^{X \restriction x}(0^{|x|+1})$, where $X \restriction x = \{y : (X|x)(y) = 1\}$. Now, if $M^{X \restriction x}(0^{|x|+1}) = 0$, let $f(X|x) = (y_0, 0), \ldots, (y_n, 0), (w, 1)$ (in the appropriate order) for the least string w of length $|x| + 1$ which is not among the queries $y_0, \ldots, y_n$ (since $n < p(|x| + 1)$ by choice of p, such a string w must exist by choice of c); otherwise, let

$$f(X|x) = (y_0, 0), \ldots, (y_n, 0), (w_0, 0), \ldots, (w_m, 0)$$

(again in the appropriate order), where $w_0, \ldots, w_m$ are all strings of length $|x| + 1$ which are not already among the queries $y_0, \ldots, y_n$. Note that, for any set X extending $f^*(X|x) = (y_0, 0), \ldots, (y_n, 0)$ on $X|x$, $M^X(x) = M^{X \restriction x}(x)$. So, for any set X extending f on $X|x$, $L^X(0^{|x|+1}) = 1 \neq 0 = M^X(0^{|x|+1})$ in the first case, and $L^X(0^{|x|+1}) = 0 \neq M^X(0^{|x|+1})$ in the second case. Since, by F-n^2-genericity, A meets f, this implies $L^A \neq M^A$.

Corollary 4.9 *(Fenner). There is an L-p-generic set $A \in DTIME(2^{n^2})$ (L-n^k-generic set $A \in \mathbf{E}$ (for all $k \geq 1$)) which is not F-n^2-generic.*

Proof. By Corollary 3.13 and Theorem 4.8.

Theorem 4.8 further implies that, in contrast to the L-n^k-generic sets, F-n^k-generic sets cannot be **P**-m-complete for **E**, in fact not even **P**-T-complete.

Corollary 4.10 *(Fenner). Let A be F-n^2-generic. Then A is not **P**-Turing-complete for $\mathbf{E_2}$ (hence not **P**-Turing-complete for **E**).*

Proof. By closure of $\mathbf{E_2}$ under the **NP**-operator, this follows from the inequality $\mathbf{P}(A) \neq \mathbf{NP}(A)$ in Theorem 4.8.

As the above results indicate, F-genericity is a very powerful tool and it seems that F-n^k-generic sets cover all the standard finite extension arguments over $DTIME(2^{kn})$-sets and reductions, i.e., to be more precise, the standard finite extension arguments, where each extension step corresponds to a direct diagonalization step over such a set or reduction. So F-p-genericity captures standard finite extension arguments over the linear exponential time sets and reducibilities. A recent result of Mayordomo , however, namely the existence of F-p-generic sets which are not **P**-bi-immune, points out some limitations of this concept. We obtain Mayordomo's theorem (and some more facts illuminating the F-genericity concept in Section 8) by stating a general existence result for F-n^k-genericity. For this it is crucial to note that in some sense F-genericity is only a quantitative improvement on L-genericity. The following observation should be compared with Proposition 3.9.

Proposition 4.11 *(a) Let f be an F-$t(n)$-extension function. Then, for*

$$f(X|x) = (x_0, i_0), \ldots, (x_m, i_m),$$

$|x_m| \leq t(|X|x|) \leq t(2^{|x|+1})$. *In particular, for $t(n) = n^k$, $|x_m| \leq 2^{k(|x|+1)}$.*
(b) For any F-$t(n)$-extension function f there is a $2^{t(n)}$-extension function f' such that every set which meets f' meets f too. Hence every L-$2^{t(n)}$-generic set is F-$t(n)$-generic too.

Proof. Part (a) is immediate by definition and by (1.1). For a proof of (b), let f be an F-$t(n)$-extension function. For $f(X|x) = (x_0, i_0), \ldots, (x_m, i_m)$ let $f'(X|x) = \widehat{X}|(x_m + 1)$ where $\widehat{X}(y) = X(y)$ for $y < x$, $\widehat{X}(x_j) = i_j$ for $0 \leq j \leq m$, and $\widehat{X}(y) = 0$ otherwise. Note that $|f'(X|x)| \leq 2^{t(X|x)+1}$ by (a), whence $f' \in DTIME(2^{t(n)})$.

The second part of this proposition intuitively says that every diagonalization proof captured by F-genericity is also captured by L-genericity but may require exponentially higher complexity in case of the latter. In particular, for classes with time bounds closed under the exponential functions (like the elementary recursive or primitive recursive sets) L-genericity and F-genericity yield the same category concept. From the first part of Proposition 4.11 we see that the generic parts in an F-n^k-generic set can be considerably longer than in an L-n^k-generic set. Otherwise, however, the side conditions for the construction of such generic sets are the same. In particular, the generic parts can still be made easily recognizable and arbitrarily long (recursively bounded) gaps can be introduced between the generic parts.

Theorem 4.12 *Let B be any set and let $C \subseteq N$ be infinite such that $\{0^n : n \in C\} \in$ **P**. For $k \geq 2$ there is an F-n^k-generic set $A \in DTIME^{(B,\leq)}(2^{(k+2)n})$ such that*

$$(4.1) \quad \forall x(A(x) \neq B(x) \Rightarrow \exists n \in C(x \in [0^n, 0^{2^{(k+1)n}}]))$$

Furthermore, there is an F-p-generic set $A \in DTIME^{(B,\leq)}(2^{n^2})$ such that

$$(4.2) \quad \forall x(A(x) \neq B(x) \Rightarrow \exists n \in C(x \in [0^n, 0^{2^{n^2}}]))$$

We omit the proof, which, by Proposition 4.11(a), parallels the proof of Theorem 3.10.

Recall that an infinite set A is **P**-*immune* if A does not contain any infinite **P**-set; and A is **P**-*bi-immune* if A and $\overline{A}$ are **P**-immune.

Corollary 4.13 *(Mayordomo [Ma94], [Ma94a]). There is an F-p-generic set $A \in DTIME(2^{n^2})$ (F-n^k-generic set $A \in \mathbf{E}$ ($k \geq 1$)) which is not **P**-immune. Hence the class of* **P** *(-bi)-immune sets is not F-comeager in* **E**.

For the proof of the corollary we need the following well known fact.

Proposition 4.14 *For any recursive function $t : N \to N$ there is an infinite set $D \subseteq N$ such that $\{0^n : n \in D\} \in \mathbf{P}$ and*

$(4.3) \quad \forall m, n \in D(m < n \Rightarrow t(m) < n).$

Proof (sketch). Given t, choose a strictly increasing, time constructible function $\hat{t}$ which dominates t (cf. [HU79]) and let $D = \{\hat{t}^n(0) : n \geq 0\}$ where $\hat{t}^n$ denotes the n-th iteration of $\hat{t}$.

Proof of Corollary 4.13. Let $t(n) = 2^{(n+1)^2}$ and choose D as in Proposition 4.14. Apply Theorem 4.12 to $B = 2^{<\omega}$ and $C = \{n + 1 : n \in D\}$. Since $B \in \mathbf{P}$, this yields an F-p-generic set $A \in DTIME(2^{n^2})$. Moreover, by choice of t, D and C, it follows from (4.2) and (4.3) that $\{0^n : n \in D\} \subseteq A$, whence A is not **P**-immune.

In fact, as observed by Mayordomo too, also by increasing the time-bound of the genericity concept we cannot capture **P**-immunity.

Corollary 4.15 *(to Theorem 3.10). Let t be recursive. There is a recursive L-$t(n)$-generic and F-$t(n)$-generic set A such that A is not **P**-immune.*

Proof. By Proposition 4.11(b) it suffices to make A L-$t(n)$-generic. Apply Proposition 4.14 to $t'(n) = \log t(2^{n+2}) + 1$ and, for the corresponding set D, apply Theorem 3.10 to $B = 2^{<\omega}$ and $C = \{n + 1 : n \in D\}$.

To motivate the new genericity concept introduced in the next section, we have a closer look at the construction of **P**-(bi)-immune sets. **P**-bi-immunity can be ensured by diagonalization and there are **P**-bi-immune sets in **E**. A standard finite extension argument for constructing a **P**-bi-immune set, however, is not effective, whence it does not yield a recursive set. To obtain recursive **P**-immune sets we have to use a so called *slow diagonalization*. In the following we shortly review the standard finite extension and the slow construction of a **P**-bi-immune set (cf. [BS85]).

Let $\{P_n : n \geq 0\}$ be a recursive enumeration of **P** such that $P_n(x)$ can be uniformly computed in exponential time. To make a set A **P**-bi-immune, we have to ensure that A meets the requirements

$R_{2n} : \quad P_n$ infinite $\Rightarrow \exists x \in P_n(A(x) = 0) \quad$ *(Immunity)*
$R_{2n+1} : \quad P_n$ infinite $\Rightarrow \exists x \in P_n(\overline{A}(x) = 0) \quad$ *(Co-Immunity)*

In a *standard finite extension* construction, the s-th requirement is satisfied at stage s:

> *Stage* 2s. {Satisfy requirement R_{2s}}
> Given $A|x_{2s-1}$, distinguish the following cases. If P_s is infinite, then pick $x \geq x_{2s-1}$ minimal with $x \in P_s$; let $x_{2s} = x + 1$ and $A|x_{2s} = (A|x_{2s-1})0^{x_{2s}-x_{2s-1}}$. Otherwise, let $x_{2s} = x_{2s-1}$.

(Stage $2s+1$ is symmetric.) Then the constructed set A is nonrecursive, since it is well known that in general it cannot be decided whether a given set P_s is infinite or not. A problem related to this observation (which actually underlies the proof of Corollary 4.15) is, that there is no uniform recursive bound on the distance between any two consecutive elements of polynomial time computable sets (see Proposition 4.14). So, even if we could decide whether or not a polynomial time set is infinite, there will be no recursive bound on the length of the extensions necessary for satisfying all of the requirements.

These problems are overcome in the slow diagonalization construction of a **P**-bi-immune set A. Here we do not satisfy the requirements in the given order but for any requirement (which is not yet satisfied) we wait for a stage at which an appropriate extension of admissible size is sufficient to meet the requirement. In the case of a **P**-bi-immune set, extensions by one string suffice: Namely, requirement R_{2s} has to wait for a stage such that the next string belongs to P_s. Since the requirements do not act in order, now conflicts may arise whenever two requirements want to act at the same stage. In this case the requirement with lesser index is given higher priority. So the *slow construction* of a **P**-bi-immune set is as follows:

> *Stage* s. {Define $A(s)$; satisfy the least requirement which has not yet been satisfied and which can be satisfied.}
>
> Requirement R_{2n} *requires attention* if $2n < s$, $P_n \upharpoonright s \subseteq A \upharpoonright s$ (i.e. R_{2n} has not yet acted), and $s \in P_n$; requirement R_{2n+1} *requires attention* if $2n + 1 < s$, $P_n \upharpoonright s \subseteq \overline{A} \upharpoonright s$ (i.e. R_{2n+1} has not yet acted), and $s \in P_n$.
>
> Fix m minimal such that R_m requires attention. If m is even, let $A(s) = 0$. If m is odd, let $A(s) = 1$. In either case say that R_m *acts*. {Note that if R_m acts, then R_m is met.}

Obviously, the constructed set A is in **E**. To show that every requirement is met, i.e., that A is **P**-bi-immune, we have to show that the action of a requirement which has to act will not be prevented by other requirements forever. To see this, note that R_{2n} has only to act if P_n is infinite. Since every requirement acts at most once and since there are only finitely many requirements with lesser index, R_{2n} will be prevented from acting only finitely often. By infinity of P_n, however, R_{2n} is prepared to act infinitely often, so that it will eventually act.

The above situation is typical for slow diagonalizations and priority arguments in general: a requirement has to be (actively) satisfied (by some finite extension) only if there are infinitely many chances to do so. In the following section we will capture this type of diagonalization argument by a genericity concept based on *partial* extension functions. The fact, that a requirement has to be met only if there are infinitely many chances to do so, will be mirrored by requiring that a generic set will meet all *densely* defined extension functions.

5 General Genericity and Slow Diagonalizations

Here we extend Fenner's genericity concept by allowing partial F-extension functions. We will define two concepts, general genericity (or shortly G-genericity) and weak general genericity (or shortly G_w-genericity). On the recursive sets both concepts coincide whence, in particular, both genericity notions give rise to the same category concept on $\mathbf{E}$. On the nonrecursive sets, however, G-genericity is strictly stronger than G_w-genericity. As pointed out before, both concepts are designed to extend Fenner-genericity from standard finite extension arguments to slow diagonalization arguments. The motivation for these concepts stems from both, Fenner's concept and the genericity concepts of Ambos-Spies, Fleischhack and Huwig ([AFH84], [AFH87], [AFH88]). The latter captured slow diagonalizations but only such with extensions of constant length. We will come back to this concept in the next section when we discuss the power of different types of constant-length extension arguments in detail.

Definition 5.1 *(a) A set B (of initial segments) is* dense along *a set A if there are infinitely many strings x such that $A|x \in B$, and B is* dense *if B is dense along all sets.*
(b) Let f be a partial F-extension-function. f is dense *(along A) if the domain of f is dense (along A). Let $\widehat{\mathbf{D}}_f = \{A : f$ is not dense along A or A meets $f\}$.*

Remark. Our definition of a dense set is not canonical. In general, corresponding to the notion of a dense class (cf. Definition 2.1), a set A (of initial segments) is called dense if all strings have an extension in A. A set G, however, which meets all F-n^k-extension functions with dense domains (in this more general sense) will not be recursive. (Namely, given any recursive set B, define an F-n-extension function f by letting $f(X|x) = (x,0)$ if, by comparing $X(y)$ and $B(y)$ for the strings $y < x$ in the canonical ordering, some y with $X(y) \neq B(y)$ can be found within the first $O(|x|)$ steps of this procedure, and by letting $f(X|x)$ being undefined, otherwise. Then the domain of f is dense and a set G meets f if and only if $G \neq B$.) So genericity concepts based on this more general density notion will not pertain to $\mathbf{E}$ in the sense of Definition 2.6.

Our new category concepts are based on the following observations.

Proposition 5.2 *Let f be a partial F-extension function. Then $\widehat{\mathbf{D}}_f$ is open-dense. If, moreover, f is dense then $\mathbf{D}_f = \widehat{\mathbf{D}}_f$.*

Proof. For a proof of the first part, fix an initial segment x. We have to show that $\mathbf{B}_y \subseteq \widehat{\mathbf{D}}_f$ for some $y \supseteq x$. W.l.o.g. assume that $\mathbf{B}_x$ is not contained in $\widehat{\mathbf{D}}_f$. Then, by definition of $\widehat{\mathbf{D}}_f$ there is a set $A \in \mathbf{B}_x$ such that f is dense along A. So we may choose z with $x \subseteq z \subseteq A$ such that $f(z)$ is defined. Then, for any y which extends f on z, $x \subseteq y$ and $\mathbf{B}_y \subseteq \widehat{\mathbf{D}}_f$. The second part is immediate by definition.

Note that, for total or dense f, $\mathbf{D}_f$ is open and dense. The classes $\widehat{\mathbf{D}}_f$, however, are, as just shown, open-dense but in general not open. E.g. for f defined by letting $f(X|x) = (x,0)$ if $X \restriction x \neq \emptyset$ and $f(X|x)$ undefined otherwise, f is not dense along $\emptyset$, whence $\emptyset \in \widehat{\mathbf{D}}_f$, but no open neighbourhood of the empty set is completely contained in $\widehat{\mathbf{D}}_f$.

Definition 5.3 *(a) General category (or G-category for short) is the category concept $\mathcal{D} = < \mathcal{D}(t) : t \text{ recursive} >$ where*

$$\mathcal{D}(t) = \{\widehat{\mathbf{D}}_f : f \text{ partial } F\text{-}t(n)\text{-extension function}\}.$$

In particular, a set A is G-$t(n)$-generic if A meets all partial F-$t(n)$-extension functions which are dense along A.
(b) Weak general category (or G_w-category for short) is the category concept $\mathcal{D}$ where $\mathcal{D}(t) = \{\mathbf{D}_f : F \text{ dense } F\text{-}t(n)\text{-extension-function}\}$. In particular, a set A is G_w-$t(n)$-generic if A meets all dense F-$t(n)$-extension functions.

It easily follows from Proposition 5.2 that G-category and G_w-category are category concepts in the sense of Definition 2.4.

Proposition 5.4 *G-category and G_w-category are abstract category concepts.*

Moreover, G- and G_w-category both refine F-category, and G-category refines G_w-category.

Proposition 5.5 *For any set A,*

$$A \; G\text{-}t(n)\text{-generic} \Rightarrow A \; G_w\text{-}t(n)\text{-generic} \Rightarrow A \; F\text{-}t(n)\text{-generic}$$

Proof. The first implication holds by the second part of Proposition 5.2. For a proof of the second implication it suffices to note that any total F-extension function is dense.

So the new concepts are at least as powerful as F-genericity. Hence, to show that the polynomial time bounded versions pertain to **E** it suffices to establish the existence of the corresponding generic sets in exponential time.

Theorem 5.6 *There is a G-n^k-generic (G-p-generic) set in $DTIME(2^{(k+3)n})$* ($DTIME(2^{n^2})$).

Proof. We prove the first part. Given k, we construct a G-n^k-generic set A by a slow diagonalization argument. A *slow* diagonalization is necessary since there is no recursive enumeration of the F-n^k-extension functions which will be *dense along* A (nor is there such an enumeration of the *dense* F-n^k-extension functions). So we have to work with a recursive enumeration of *all* partial F-n^k-extension functions. As one can easily check, there is such an enumeration $\{f_e : e \geq 0\}$ such that, for x with $|x| \geq e$, $f_e(x)$ can be computed in $|x|^{k+1}$ steps (uniformly in e and x). In the following fix such an enumeration.

Then, for every e, we have to ensure that A meets f_e provided that f_e is dense along A. By definition, this means that A has to meet the requirements

$$R_e \quad : \quad \text{If there are infinitely many strings } x \text{ such that } f_e(A|x) \text{ is}$$
$$\text{defined, then } A \text{ extends } f_e \text{ on } A|x \text{ for some } x.$$

for all $e \geq 0$. At stage x of the construction of A we will define $A(x)$. So, for meeting a single requirement, one stage will not suffice. We may begin to work on satisfying the requirement R_e at any stage x_0 such that, for some $x \leq x_0$, $f_e(A|x)$ is defined and equals $(x_0, i_0), \ldots, (x_n, i_n)$. Then at the $n+1$ stages $x_0, \ldots, x_n$ we have to ensure that $A(x_0) = i_0, \ldots, A(x_n) = i_n$. To resolve conflicts among the strategies for meeting the individual requirements, this action may be *injured* at any point (no matter how much progress on making A extend f_e on $A|x$ we have already made) by starting or continuing an attack on a requirement with lesser index. In this case we have to resume working on R_e with some other initial segment $A|x'$.

For the formal description of stage x of the construction we need the following notation: Let $f_e(A|y) = (y_0, i_0), \ldots, (y_n, i_n)$. We say that x is i-*critical for f_e at y* if $x = y_m$ for some $m \leq n$ with $i_m = i$. Then stage x of the construction of A (i.e. the definition of $A(x)$) is as follows:

> Requirement R_e *requires attention via y* if $e \leq |x|$, there is no $z \leq x$ such that $A|x$ extends f_e on $A|z$, $y \leq x$, $A|x$ is consistent with f_e on $A|y$ and y is minimal with this property, and x is i-critical for f_e at y for some $i \leq 1$. Then choose the least e such

that R_e requires attention, say via y, choose i such that x is i-critical for f_e at y, and let $A(x) = i$. Furthermore say that R_e *receives attention*. If there is no such e, let $A(x) = 0$.

Then, by (1.1), it is not difficult to check that $A \in DTIME(2^{(k+3)n})$. To show that the requirements are met, by induction on e show that every requirement requires attention only finitely often and is met. For the inductive step fix e and, by inductive hypothesis, choose a stage x after which no requirement $R_{e'}$ with $e' < e$ requires attention. So R_e will receive attention whenever it requires attention after stage x. Now if R_e does not require attention after stage x, then either $f_e(A|y)$ is undefined for all $y > x$ or A extends f_e on $A|z$ for some $z \leq x$. In either case R_e is met. So w.l.o.g. assume that R_e requires attention at some later stage. Then for the first such stage and for the y via which R_e requires attention at this stage, R_e will continue to require (and will receive) attention via y at all critical stages for f_e at y until the first stage z will be reached such that $A|z$ extends f_e on $A|y$. Then the requirement is met and it will cease to require attention.

To show that these new concepts actually capture slow diagonalizations, we show that G_w-generic sets (hence G-generic sets) are bi-immune.

Theorem 5.7 *Let A be G_w-n^k-generic ($k \geq 1$). Then A is $DTIME(2^{kn})$-bi-immune. Hence every G_w-p-generic set is $\mathbf{E}$-bi-immune.*

Proof. By symmetry it suffices to show that no infinite $DTIME(2^{kn})$-set is contained in A. So fix such a set B and define the partial F-n^k-extension function f by $f(X|x) = (x, 0)$ if $x \in B$, and $f(X|x)$ undefined otherwise. Then, by infinity of B, f is dense, whence A meets f. Obviously this implies that $A(x) = 0$ for some $x \in B$.

The next, more sophisticated, example will allow us to distinguish G_w-genericity and G-genericity.

Definition 5.8 *A set A is $DTIME(t(n))$-predictable if there is a $t(n)$-time bounded deterministic partial oracle Turing machine M such that*

(5.1) $\quad \exists^\infty x (M^{A \upharpoonright x}(x) \downarrow)$ *and*

(5.2) $\quad \forall x (M^{A \upharpoonright x}(x) \downarrow \Rightarrow M^{A \upharpoonright x}(x) = A(x)).$

Otherwise A is $DTIME(t(n))$-unpredictable.

Note that predictability is a partial selfreducibility phenomenon, namely it gives a selfreduction on an infinite domain but not necessarily on all strings. In fact, unpredictability entails both the failure of selfreducibility and bi-immunity. $\mathbf{P}$-unpredictability on the tally universe was used in [AFH88] to characterize the p-genericity concept for tally sets introduced in [AFH84]. Recently, Balcazar and Mayordomo [BM95] lifted this to a characterization of the general genericity concept of [AFH88] (see Section 6 below). The following theorem is implicitly proved in [BM95].

Theorem 5.9 *(Balcazar and Mayordomo). Let A be G-n^{k+1}-generic. Then A is not $DTIME(2^{kn})$-predictable. Hence every G-p-generic set is* **E**-*unpredictable.*

Proof. Fix a partial 2^{kn}-time bounded deterministic oracle Turing machine such that (5.1) holds. To show that (5.2) fails, define a partial F-n^{k+1}-extension function f by letting $dom(f) = \{X|x : M^{X\lceil x}(x) \downarrow\}$ and $f(X|x) = (x, 1 - M^{X\lceil x}(x))$. Then, by (5.1), f is dense along A whence, by G-n^{k+1}-genericity of A, A meets f. By definition of f, this implies $M^{A\lceil x}(x) \downarrow \neq A(x)$ for some x.

The F-extension function defined in the proof of Theorem 5.9 is dense along A but not dense. In fact, as the following theorem shows, **E**-unpredictability cannot be enforced by G_w-genericity.

Theorem 5.10 *There is a (nonrecursive) G_w-p-generic set which is $DTIME(2^{2n})$-predictable (hence* **E**-*predictable). Hence there is a (nonrecursive) sets which is G_w-p-generic but not G-n^3-generic (hence not G-p-generic).*

Proof. We construct a set A with the required properties by a (noneffective) standard finite extension argument. Let $\{f_e : e \geq 0\}$ be a (noneffective) listing of all dense F-p-extension functions. By Proposition 4.11 we may assume that, for $|x| \geq e$,

$$(5.3) \quad f_e(X|x) = (y_0, i_0), \ldots, (y_n, i_n) \Rightarrow |y_n| < 2^{2^{|x|}}.$$

Then, to make A G_w-p-generic it suffices to meet the requirements

$\quad R_e :$ A meets f_e

for $e \geq 0$. To make A **E**-predictable we put all strings into A which are not restrained from A by some requirement. Moreover, we introduce large gaps between the actions for meeting the individual requirements. Together, these precautions will enable us to predict infinitely often that a string belongs to A.

The formal construction is as follows. Given the finite initial segment $A|x_e$ ensuring that A meets the requirements $R_{e'}$ for $e' < e$ ($x_0 = \lambda$), define $x_{e+1} > x_e$ and $A|x_{e+1}$ to meet R_e as follows. Fix m minimal such that

$x_e < 0^{\delta(m)}$, where δ is the iterated double exponential function, and let x be the least string greater than $0^{\delta(m+1)}$ such that $f_e(\widehat{A}|x)$ is defined, say $f_e(\widehat{A}|x) = (y_0, i_0), \dots, (y_n, i_n)$, where $\widehat{A}|x$ extends $A|x_e$ and $(\widehat{A}|x)(z) = 1$ for all z with $x_e \le z < x$. Note that by density of f_e such a string x must exist. Then let $x_{e+1} = y_n + 1$ and define $A|x_{e+1}$ by letting $A(y_p) = i_p$ for $0 \le p \le n$ and $A(y) = 1$ for all other strings y with $x_e \le y < x_{e+1}$.

Obviously, the thus constructed set A meets all requirements R_e, whence A is G_w-p-generic. To show that A is E-predictable, it suffices to prove that, for $I_m = [0^{\delta(m)}, 0^{\delta(m+1)})$,

$$(5.4) \quad \exists^\infty m \ (I_m \cap \overline{A} \neq \emptyset \ \& \ I_{m+1} \cap \overline{A} = \emptyset)$$

and that

$$(5.5) \quad (I_m \cap \overline{A} \neq \emptyset \ \& \ I_{m+1} \cap \overline{A} = \emptyset) \ \Rightarrow 0^{\delta(m+2)} \in A$$

for all m. Namely, for a string y we can decide in polynomial time whether or not it is of the form $0^{\delta(m+2)}$, and the hypothesis of (5.5) can be checked in time exponential in $\delta(m + 2)$ with oracle $A|0^{\delta(m+2)}$. For a proof of (5.5) fix m satisfying the hypothesis. Then, by construction and by (5.3) there is a number e and a string $x \in I_{m-1} \cup I_m$ such that $f_e(\widehat{A}|x) = (y_0, i_0), \dots (y_n, i_n)$, $y_p \in I_m$ for the greatest $p \le n$ such that $i_p = 0$, $A|x_{e+1} = A|(y_n + 1)$, this extension is chosen to extend f_e on $\widehat{A}|x$ in order to meet requirement R_e, and $A(z) = 1$ for all z with $y_p < z \le y_n$. Moreover, again by (5.3), $0^{\delta(m)} < x_{e+1} < 0^{\delta(m+2)}$. By the latter, the definition of $A|x_{e+2}$, ensures that $A(z) = 1$ for all z with $x_{e+1} < z \le 0^{\delta(m+2)}$. So, in particular, $0^{\delta(m+2)} \in A$. The proof of (5.4) is similar: By genericity of A, A is coinfinite. Hence there are infinitely many requirements R_e so that the strategy for meeting these requirements forces a string $y \in [x_e, x_{e+1}]$ in the complement of A. As one can easily check, for any such y and for the least m with $y < 0^{\delta(m+1)}$, m is a witness for (5.4).

In contrast to the second part of Theorem 5.10, for recursive A, G-$t(n)$-genericity and G_w-$t(n)$-genericity coincide. So both notions induce the same category concept on **E**.

Theorem 5.11 *Let A be recursive. Then A is G-$t(n)$-generic iff A is G_w-$t(n)$-generic.*

Proof. For a proof of the nontrivial direction assume that A is G_w-$t(n)$-generic and that f is an F-$t(n)$-extension function which is dense along A. Then it suffices to define a dense F-$t(n)$-extension function f' such that A meets f iff A meets f': For $X|x$ in the domain of f let $f'(X|x) = f(X|x)$. If $f(X|x)$ is undefined, then, for a total of $|x|$ steps, try to compute $A(0), A(1), \dots$ (by simulating some arbitrary but fixed algorithm for A)

and compare each computed value with the corresponding value of X. If this way some y with $A(y) \neq X(y)$ is found, then let $f'(X|x) = (x, 0)$. Otherwise, let $f'(X|x)$ be undefined. Then, for all x, $f(A|x)$ is defined iff $f'(A|x)$ is defined and if so, both agree. So f' is dense along A and A meets f iff A meets f'. To show that f' is dense, it remains to show that f' is dense along all sets X different from A. But for such an X, $f'(X|y) = (y, 0)$ for all sufficiently large y (namely for those y which allow to find the first difference between A and X in at most $|y|$ steps).

We will discuss the limitations of general genericity in Section 7. Next, however, we will somewhat disgress and we will analyse fragments of the so far introduced genericity concepts based on extension functions with constant bounds.

6 Constant Length Extensions: The Genericity Concept of Ambos-Spies, Fleischhack and Huwig

The full power of the above genericity concepts is only required for diagonalizations over truth-table or Turing reductions. For diagonalizations over sets, many-one reductions or, more generally, bounded truth-table reductions it suffices to consider extension functions which determine the values of the next string, of two strings or of a constant number of strings, respectively. In this section we will introduce bounded genericity concepts related to these tasks. For Fenner's genericity notion this will yield a proper hierarchy of bounded genericity concepts, whereas the bounds of different length on general genericity all will yield the same notion. Moreover, this very robust concept coincides with the resource-bounded genericity of Ambos-Spies, Fleischhack and Huwig [AFH88].

The bounded genericity concepts are of further interest since, in contrast to the unbounded case, they are compatible with randomness. So bounded genericity is a useful tool for obtaining simultaneous largeness results in both, resource-bounded category and resource-bounded measure.

Definition 6.1 *A (partial) $F\text{-}t(n)$-extension function f is a (partial)*
   $[F, S]\text{-}t(n)$-extension function
   $[F, k]\text{-}t(n)$-extension function $(k \geq 1)$
if, for every $X|x$ on which f is defined,
   $f(X|x) = (x, i)$ *(some $i \leq 1$)*
   $f(X|x) = (y_0, i_0), \ldots, (y_{k-1}, i_{k-1})$
   (some $y_0, \ldots, y_{k-1} \in \Sigma^*$ and $i_0, \ldots, i_{k-1} \leq 1$),

respectively. f is a (partial) $[F, \omega]$*-t(n)-extension function if f is a (partial)* $[F, k]$*-t(n)-extension-function for some* $k \geq 1$.

As before we say f is a *-p-*extension function* if f is a *-n^k-extension function for some $k \geq 1$. We next introduce the corresponding genericity concepts and leave it to the reader to extend this to category concepts (according to Definition 2.4).

Definition 6.2 *(a) A set A is* $[F, *]$*-t(n)-generic if A meets every total* $[F, *]$*-t(n)-extension function; A is* $[G_w, *]$*-t(n)-generic if A meets every dense* $[F, *]$*-t(n)-extension function; and A is* $[G, *]$*-t(n)-generic if A meets every* $[F, *]$*-t(n)-extension function which is dense along A* $(* \in \{S, \omega\} \cup \{k : k \geq 1\})$.

   (b) A is $[+, *]$*-p-generic if A is* $[+, *]$*-n^c-generic for all* $c \geq 1$ $(+ \in \{F, G_w, G\}, * \in \{S, \omega\} \cup \{k : k \geq 1\})$.

The following implications among these bounded genericity concepts are immediate by definition and by Proposition 5.5.

Proposition 6.3 *For any set A,* $+ \in \{F, G_w, G\}$, $* \in \{S, \omega\} \cup \{k : k \geq 1\}$, *and* $k \geq 1$:

(1) A +-*t(n)-generic* $\Rightarrow$ A $[+, \omega]$*-t(n)-generic*

                            $\Rightarrow$ A $[+, k + 1]$*-t(n)-generic*

                            $\Rightarrow$ A $[+, k]$*-t(n)-generic*

                            $\Rightarrow$ A $[+, S]$*-t(n)-generic*

(2) A $[G, *]$*-t(n)-generic* $\Rightarrow$ A $[G_w, *]$*-t(n)-generic*

                                $\Rightarrow$ A $[F, *]$*-t(n)-generic*

We first discuss the length-bounded F-genericity concepts and we will show that the restrictions of different strength give a proper hierarchy for these concepts and the correspondent category concepts on **E**.

Theorem 6.4 *For* $k \geq 1$ *and* $c \geq 2$, *the following implications are proper:*

(6.1) $[F, \omega]$*-p-generic* $\Rightarrow$ $[F, k + 1]$*-p-generic* $\Rightarrow$ $[F, k]$*-p-generic* $\Rightarrow$ $[F, S]$*-p-generic*

(6.2) $[F, \omega]$*-n^c-generic* $\Rightarrow$ $[F, k + 1]$*-n^c-generic* $\Rightarrow$ $[F, k]$*-n^c-generic* $\Rightarrow$ $[F, S]$*-n^c-generic*

Moreover there are sets in $DTIME(2^{n^2})$ *and* **E** *witnessing the failure of the reverse implications in (6.1) and (6.2), respectively. In fact, in the latter case there are sets* G_S *and* G_k *in* **E** *such that* G_S *is* $[F, S]$*-n^c-generic but not* $[F, 1]$*-n^2-generic and* G_k *is* $[F, k]$*-n^c-generic but not* $[F, k + 1]$*-n^2-generic.*

Note that, by Proposition 2.7, the part on n^c-genericity in Theorem 6.4 implies
a hierarchy theorem for the category notions on **E** induced by these genericity
concepts. I.e., for $[F, *]$-$MEAGER_E$ denoting the family of the classes which
are $[F, *]$-meager in **E**, we obtain:

Corollary 6.5 *For $k \geq 1$*: $[F, S]$-$MEAGER_E \subset [F, k]$-$MEAGER_E \subset$
$[F, k+1]$-$MEAGER_E \subset [F, \omega]$-$MEAGER_E$.

For a proof of the hierarchy theorem 6.4 we will specify the types of polynomial-
time-bounded bounded-truth-table reductions captured by the different genericity notions. These results (which can be extended to exponential-time reductions) are stated as Lemmas 6.6 - 6.8 below. Together they immediately
imply Theorem 6.4.

We first observe that $[F, S]$-genericity captures diagonalizations over exponential time sets but not more.

Lemma 6.6 *If A is $[F, S]$-n^k-generic then $A \notin DTIME(2^{kn})$. Conversely,
if A is not $[F, S]$-n^k-generic then $A \in DTIME(2^{(k+1)n})$. Hence A is $[F, S]$-
p-generic iff $A \notin$ **E**.*

Proof. For a proof of the first part let A be $[F, S]$-n^k-generic and
$B \in DTIME(2^{kn})$. It suffices to show that there is a total $[F, S]$-n^k-extension
function f such that every set meeting f is different from B. Obviously, f
definded by $f(X|x) = (x, 1 - B(x))$ has this property. For a proof of the
second part fix A such that A is not $[F, S]$-n^k-generic. Then there is a total
$[F, S]$-n^k-extension function f such that A does not meet f. Hence, for all x,
$f(A|x) = (x, 1 - A(x))$ so that $A(x)$ can be computed from $A|x$ via f. This
gives an inductive computation of A in time $2^n 2^{kn} = 2^{(k+1)n}$.

Compared with $[F, S]$-genericity, $[F, 1]$-genericity in addition captures diagonalizations over increasing (or decreasing) exponential-time many-one reductions. The next lemma gives an example for this.

Lemma 6.7 *Let A be $[F, 1]$-n^2-generic. Then*

(6.3) $A \not\leq_{m-i}^P A$, *i.e., there is no function $g \in$ **P** such that $x < g(x)$
          and $A(x) = A(g(x))$ for all x.*

(6.4)   *A is not a p-cylinder.*

(6.5)   *A is not **P**-m-complete for $\mathbf{E_2}$.*

Proof. Note that, for any p-cylinder A, $A \leq_{m-i}^P A$ by a padding function for
A, and that **P**-m-completeness and **P**-m-i-completeness coincide for $\mathbf{E_2}$ (see
[Be77]). Hence it suffices to prove (6.3). Fix $g \in$ **P** such that $x < g(x)$ for all
x. To show that $A \leq_m^P A$ via g fails, define the $[F, 1]$-n^2-extension function

f by $f(X|\lambda) = (\lambda, 1 - A(\lambda))$ and $f(X|(x+1)) = (g(x), 1 - X(x))$. Since, by genericity, A meets f, it follows that $A(x) \neq A(g(x))$ for some x.

For capturing diagonalizations over arbitrary many-one reductions of the corresponding time bound, however, in general $[F, 1]$-genericity does not suffice but this requires $[F, 2]$-genericity. For example, the proof of Theorem 4.8 actually shows that no $[F, 2]$-p-generic set is **P**-m-selfdual, whereas it is not difficult to construct a **P**-m-selfdual $[F, 1]$-p-generic set. Here, in order to distinguish $[F, k]$-genericity from $[F, k + 1]$-genericity for any $k \geq 1$, we exploit the power of the latter to capture diagonalizations over k-query truth-table reductions.

For any set A and $k \geq 1$, let $A_k = \{x : \{x, ..., x+k\} \cap A \neq \emptyset\}$. Note that $A_k \leq^P_{(k+1)-tt} A$.

Lemma 6.8 *(a) For any $[F, k + 1]$-n^2-generic set A, $A_k \not\leq^P_{k-tt} A$.*
(b) There is an $[F, k]$-p-generic ($[F, k]$-n^c-generic) set $A \in DTIME(2^{n^2})$ ($DTIME(2^{(c+2)n})$) such that $A_k = \Sigma^*$ (whence $A_k \leq^P_{k-tt} A$).

Proof. For a proof of part (a), let A be $[F, k + 1]$-n^2-generic and fix a **P**-k-tt-reduction $h(g_0, \ldots, g_{k-1})$, where h is a polynomial-time evaluator function and the functions g_m are polynomial-time computable selectors. We have to show that

$$(6.6) \quad A_k(x) \neq h(x, A(g_0(x)), \ldots, A(g_{k-1}(x)))$$

for some string x. By genericity of A, it suffices to define a total $[F, k+1]$-n^2-extension function f such that every set A which meets f will satisfy (6.6). Given $X|x$, for the definition of $f(X|x)$ distinguish the following two cases: If there is a set Y extending $X|x$ such that $h(x, Y(g_0(x)), \ldots, Y(g_{k-1}(x))) = 0$, then for the least such Y and for the least string $z \in \{x, x+1, \ldots, x+k\} - \{g_0(x), \ldots, g_{k-1}(x)\}$ let

$$f(X|x) = (z, 1), (g_0(x), Y(g_0(x))), \ldots, (g_{k-1}(x), Y(g_{k-1}(x))).$$

(Note that, in this case, if A extends f on $X|x$ then $A_k(x) = 1$ (since $A(z) = 1$ and $z \in \{x, x+1, \ldots, x+k\}$) whereas $h(x, A(g_0(x)), ..., A(g_{k-1}(x))) = 0$ (by choice of Y), whence (6.6) holds.) Otherwise let

$$f(X|x) = (x, 0), \ldots, (x+k, 0).$$

(Note that, in this case, if A extends f on $X|x$ then $A_k(x) = 0$ whereas, by case assumption, $h(x, A(g_0(x)), ..., A(g_{k-1}(x))) = 1$ whence (6.6) holds again.)

The proof idea for part (b) is as follows: Note that for ensuring that A meets an $[F, k]$-extension function it suffices to determine A on k strings. Since, moreover, the work on the individual genericity requirements can be arbitrarily spread out (cf. Theorem 4.12), the construction of an $[F, k]$-generic set A is compatible with ensuring that A contains at least one of the $k + 1$ consecutive strings $x, \dots, x + k$ for all x, whence $A_k = \Sigma^*$.

Lemma 6.8 (a) implies that $[F, \omega]$-n^2-genericity suffices to distinguish between the standard bounded-truth-table reducibilities:

Corollary 6.9 *Let A be $[F, \omega]$-n^2-generic. Then, for all $k \geq 1$,*

$$\mathbf{P}_m(A) \subset \mathbf{P}_{k\text{-}tt}(A) \subset \mathbf{P}_{(k+1)\text{-}tt}(A) \subset \mathbf{P}_{btt}(A) \subset \mathbf{P}_{tt}(A).$$

Hence no $[F, \omega]$-n^2-generic set is $\mathbf{P}$-btt-complete for $\mathbf{E}_2$.

Proof. It suffices to prove that the inclusions are proper. As shown in the proof of Theorem 4.8, A is not $\mathbf{P}$-m-selfdual whence $\overline{A} \in \mathbf{P}_{1\text{-}tt}(A) - \mathbf{P}_m(A)$. By Lemma 6.8,

$$(6.7) \quad A_k \in \mathbf{P}_{(k+1)\text{-}tt}(A) - \mathbf{P}_{k\text{-}tt}(A)$$

whence $\mathbf{P}_{k\text{-}tt}(A) \subset \mathbf{P}_{(k+1)\text{-}tt}(A) \subset \mathbf{P}_{btt}(A)$. Finally, to show that $\mathbf{P}_{btt}(A) \subset \mathbf{P}_{tt}(A)$, let $A_\omega = \{0^k 1x : k \geq 1 \ \& \ x \in A_k\}$. Then, obviously, $A_\omega \in \mathbf{P}_{tt}(A)$. Since $A_k \leq^P_m A_\omega$ via $g(x) = 0^k 1x$, however, it follows from (6.7) that $A_\omega \notin \mathbf{P}_{k\text{-}tt}(A)$ for all k, whence $A_\omega \notin \mathbf{P}_{btt}(A)$.

We now turn to the constant length bounded variants of (weak) general genericity. We first observe that the proof of Theorem 5.7 actually shows that every $[G_w, S]$-n^k-generic (hence every $[G, S]$-n^k-generic) set is $DTIME(2^{kn})$-bi-immune ($k \geq 1$). So, by Corollary 4.13, none of the polynomial time bounded F-genericity concepts implies any of the constant-size bounded polynomial G_w- or G-genericity concepts. Moreover, we can adapt the proof of Theorem 5.11 to show that, as in the unbounded case, the corresponding constant-size G_w- and G-genericity concepts coincide on the recursive sets, whence they yield the same category notions for $\mathbf{E}$.

Lemma 6.10 *Let A be recursive. Then A is $[G_w, *]$-$t(n)$-generic if and only if A is $[G, *]$-$t(n)$-generic ($* \in \{S, \omega\} \cup \{k : k \geq 1\}$).*

In contrast to the hierarchy theorem (Theorem 6.4) for the bounded F-genericity notions, for general genericity all of the constant-length bounded concepts coincide. To show this we need the following simple observation, which likewise applies to all of the other genericity concepts discussed in this paper.

Proposition 6.11 *Let A be $[G, S]$-n^c-generic and let f be a partial $[F, S]$-n^c-extension function which is dense along A. There are infinitely many x such that A extends f on $A|x$.*

Proof. Given y, it suffices to show that A extends f on $A|x$ for some $x > y$. Define the $[F, S]$-n^c-extension function f_y by letting $f_y(X|x) = f(X|x)$ for all $X|x$ with $x > y$ and by letting $f_y(X|x)$ be undefined otherwise. Then f_y is dense along A whence A meets f_y, i.e., A extends f_y on $A|x$ for some x. By definition of f_y, however, this implies that $x > y$ and A extends f on $A|x$.

Lemma 6.12 *Let A be $[G, S]$-n^{c+1}-generic. Then A is $[G, \omega]$-n^c-generic.*

Proof. Fix $k \geq 1$ and let f be a partial $[F, k]$-n^c-extension function which is dense along A. It suffices to show that A meets f. Define partial F-extension functions $f_m (m < k)$ as follows: Given $X|y$, $f_m(X|y)$ is undefined if there is no $x \leq y$ such that $f(X|x) = (y_0, i_0), \ldots, (y_{k-1}, i_{k-1})$, $y = y_m$, and $(X|y)(y_r) = i_r$ for $r < m$; and $f_m(X|y) = (y_m, i_m)$ for the least such x otherwise. Then the functions f_m are partial $[F, S]$-n^{c+1}-extension functions. Moreover, if A meets f_{k-1} then A meets f too. So, by the genericity assumption on A, it suffices to show that the functions f_m are dense along A. Density of f_0 along A directly follows from the corresponding property of f. For the inductive step, fix $m > 0$ and assume that f_{m-1} is dense along A. Then, by Proposition 6.11, there are infinitely many y such that A extends f_{m-1} on $A|y$. By definition of f_m this is equivalent to f_m being dense along A.

The following two theorems are immediate by Proposition 6.3 and by Lemmas 6.10 and 6.12.

Theorem 6.13 *For any set A, the following are equivalent:*

1. *A is $[G, S]$-p-generic*

2. *A is $[G, k]$-p-generic ($k \geq 1$)*

3. *A is $[G, \omega]$-p-generic*

If, moreover, A is recursive then the following are equivalent to the above too:

4. *A is $[G_w, S]$-p-generic*

5. *A is $[G_w, k]$-p-generic ($k \geq 1$)*

6. *A is $[G_w, \omega]$-p-generic.*

Theorem 6.14 *For $k \geq 1$:*

$$[G_w, S]\text{-}MEAGER_E = [G_w, k]\text{-}MEAGER_E = [G_w, \omega]\text{-}MEAGER_E =$$
$$[G, S]\text{-}MEAGER_E = [G, k]\text{-}MEAGER_E = [G, \omega]\text{-}MEAGER_E$$

We should point out that the second part of Theorem 6.13 fails if we consider nonrecursive sets. Note that the proof of Theorem 5.9 actually shows that every $[G, S]$-n^{k+1}-generic set is $DTIME(2^{kn})$-unpredictable, whence, by Theorem 5.10, $[G_w, {}^*]$-p-genericity and $[G, {}^*]$-p-genericity (${}^* \in \{S, k, \omega\}$) differ on the nonrecursive sets. Moreover, by combining the ideas of the proofs of Theorem 5.10 and Lemma 6.8 (b), we can prove the following weak hierarchy theorem for constant-size G_w-genericity: For $k \geq 1$, there is a (nonrecursive) $[G_w, k]$-p-generic set A which is not $[G_w, k+1]$-p-generic. Even for nonrecursive A, however, every $[G_w, S]$-n^{k+1}-generic set A is $[G_w, 1]$-n^k-generic, whence $[G_w, S]$-p-genericity and $[G_w, 1]$-p-genericity coincide in general. This is shown as the base step in the inductive argument in the proof of Lemma 6.12.

By Theorem 6.13, $[G, S]$-genericity is a very robust and powerful concept. As we will show next, this concept coincides with the main genericity concept of Ambos-Spies, Fleischhack and Huwig [AFH88].

Definition 6.15 *([AFH88]) A condition is a set C (of initial segments). A set A meets the condition C if, for some string x, $A|x \in C$. C is dense along A if*

$$\exists^{\infty} x \in \Sigma^* \, \exists i \leq 1 ((A|x)i \in C);$$

and C is dense if C is dense along all sets. For any complexity class $\mathbf{C}$, a set A is AFH-$\mathbf{C}$-generic if A meets every condition $C \in \mathbf{C}$ which is dense along A.

In [AFH88], *AFH-$\mathbf{C}$*-generic sets were called $\mathbf{C}$-2-generic sets. Here we abbreviate *AFH-DTIME(n^k)*-genericity and *AFH-P*-genericity by *AFH-n^k-genericity* and *AFH-p-genericity*, respectively, and we call a condition $C \in DTIME(t(n))$ a $t(n)$-condition. As the following theorem shows, conditions can be interpreted as $[F, S]$-extension functions and vice versa.

Theorem 6.16 *A set A is AFH-n^k-generic iff A is $[G, S]$-n^k-generic. Hence A is AFH-p-generic iff A is $[G, S]$-p-generic.*

Proof. First assume that A is *AFH-n^k*-generic and let f be an $[F, S]$-n^k-extension function which is dense along A. We have to show that A meets f. Define the n^k-condition set C_f by $C_f = \{(X|x)i : f(X|x) = (x, i)\}$. Then, for any set X, f is dense along X iff C_f is dense along X, and X meets f iff X meets C_f. So, by *AFH-n^k*-genericity, A meets C_f and therefore f.

Now assume that A is $[G, S]$-n^k-generic and let C be an n^k-condition which is dense along A. We have to show that A meets C. Define the $[F, S]$-n^k-extension function f_C by $f_C(X|x) = (x, i)$ for the least $i \leq 1$ such that $(X|x)i \in C$ if such an i exists, and let $f_C(X|x)$ be undefined otherwise. Then, for any set X, C is dense along X iff f_C is dense along X, and if X meets f_C

then X meets C too. So, by $[G, S]$-n^k-genericity, A meets f_C and therefore C.

Balcazar and Mayordomo ([BM95]) have recently shown that **E**-unpredictability actually coincides with AFH-p-genericity. More results on AFH-genericity, hence on bounded G-genericity can be found in [AFH88] and [ANT94]. For instance, Ambos-Spies, Neis and Terwijn have shown that there are **P**-tt-complete AFH-p-generic sets for $\mathbf{E_2}$, whence this concept in general does not allow diagonalizations over resource-bounded reductions of the unbounded truth-table or Turing type.

A particularly interesting feature of AFH-genericity, hence of the bounded genericity concepts in general, is its compatibility with measure and randomness. Fleischhack [Fl85] has shown that the class of the AFH-p-generic sets has (classical) measure 1. This result was recently extended to resource-bounded measure by Ambos-Spies, Neis, Terwijn and Zheng ([ANT94], [ATZ94]) who have shown that every n^{k+2}-random set is AFH-n^k-generic ($k \geq 1$), whence any p-random set is AFH-p-generic. In [ANT94] this observation has been used to prove largeness results for p-measure via AFH-genericity. There and in [ATZ94] the limitations of this approach have been pointed out too: While n^k-random sets are exponentially dense, there are AFH-p-generic sets of arbitrary density, e.g. sparse ones. Hence there are AFH-p-generic sets which are not p-random.

7 Extended Genericity and Delayed Diagonalizations

The genericity concepts for **E** discussed so far have some limitations not inherent to diagonalizations over **E**. By depending on classical extension functions or F-extension functions, the relevant substeps of the whole diagonalization step are determined at its beginning already. With respect to the resource-bounds, this implies that the complexity of the diagonalization step is bounded by the resources available at the *beginning* of the step, and not - as it would be sufficient - by the resources *locally* available when performing a substep of the diagonalization. The latter option, however, is crucial for the so called *delayed diagonalization* or *looking back* arguments, where a single digonalization step may require the simulation of a given diagonal on a very long interval.

Before we will propose a hierarchy of genericity notions for capturing delayed diagonalizations, we will shortly review the basic ideas underlying such diagonalization proofs.

In a typical delayed diagonalization argument, a diagonal B for a recursively presentable (r.p.) and closed under finite variants (c.f.v.) class **C** is

constructed in such a way that the diagonal has certain additional properties. The fact that $B \notin \mathbf{C}$ is not achieved by a direct diagonalization over
$\mathbf{C}$, however, but B is made sufficiently similar to a given diagonal $A \notin \mathbf{C}$ to
ensure that B is a diagonal too. On the other hand, B can differ from A on
sufficiently many arguments, so that there will be enough room to ensure the
additional properties.

To be a bit more precise, let $\{C_e : e \geq 0\}$ be a recursive listing of the
elements of $\mathbf{C}$. Since $\mathbf{C}$ is c.f.v., the diagonal A will differ from each C_e on
infinitely many strings. So B can be inductively defined in stages, where the
stages to ensure $B \notin \mathbf{C}$ and the stages to ensure the other properties of B
may alternate. E.g. at stage $2e$, $B \neq C_e$ is ensured as follows: Given $B|x_{2e-1}$,
let $B(x) = A(x)$ until the first $x \geq x_{2e-1}$ is reached such that $A(x) \neq C_e(x)$
(since $A \neq^* C_e$ such an x must exist), and complete the diagonalization step
by letting $x_{2e} = x + 1$. In such a construction the complexity of B depends
on the additional tasks achieved at the odd stages, the complexity of A (since
B simulates A), and the complexity of the sets C_e which are used to tell us
when the simulation phases of A are completed. The latter can be avoided
by extending stage $2e$ until a string y is reached so that by *looking back* a
diagonalization witness x can be found in linear time (relative to $|y|$), thereby
delaying the next diagonalization step. This crucial modification allows us to
recognize the intervals ensuring that $B \notin \mathbf{C}$ in polynomial time, whence this
part of B is $\mathbf{P}$-m-reducible to the given diagonal A.

The delayed diagonalization technique has many interesting applications
(see e.g. [La75], [LLR81], [Sch82], [Am87], [BDG95]). E.g. for $r = m, T$, any
nonempty $\mathbf{P}$-r-interval $[A, B]_r^P = \{C : A \leq_r^P C \leq_r^P B\}$, hence any $\mathbf{P}$-r-degree
$deg_r^P(A) = [A, A]_r^P$, is r.p. and c.f.v., so that the density of these reducibilities
can be shown by delayed diagonalization ([La75]).

The basic ideas underlying delayed diagonalizations are formalized by the
following definition and lemma.

Definition 7.1 *Let A and B be sets and let $g: N \to N$ be a recursive function such that $g(n) > n$ for all n. Then $[0^n, 0^{g(n)})$ is the n-th g-interval, and
A and B are g-similar if they agree on infinitely many g-intervals.*

Lemma 7.2 *Let $\mathbf{C}$ be recursively presentable and closed under finite variants
and let A be a recursive set such that $A \notin \mathbf{C}$. There is a nondecreasing time-
constructible function g such that $B \notin \mathbf{C}$ for all sets B which are g-similar
to A.*

Proof. Fix an enumeration $\{C_e : e \geq 0\}$ of $\mathbf{C}$ such that the sets C_e are
uniformly recursive and let $g'(n)$ be the least number $m > n$ such that for
all $e \leq n$ there is a string x with $n \leq |x| < m$ and $C_e(x) \neq A(x)$. Then g' is
recursive and, for B g'-similar to A, $B \notin \mathbf{C}$. Since every recursive function

is dominated by a nondecreasing time constructible function and since, for g' and g with $g' \leq g$, g-similarity implies g'-similarity, this implies the claim.

Note that the replacement of the recursive function g' by the larger, but time constructible function g in the proof of Lemma 7.2 corresponds to the above described "delay" in the delayed diagonalization arguments: If we choose the base points 0^{n_e}, $e \geq 0$, of the g-intervals selected to make B g-similar to A so that $\{0^{n_e} : e \geq 0\} \in \mathbf{P}$ then time constructibility of g ensures that $\bigcup_{e \geq 0}[0^{n_e}, 0^{g(n_e)}) \in \mathbf{P}$ too.

In the following we introduce a hierarchy of genericity notions for $\mathbf{E}$ which depend on a recursive function s as an additional parameter and which are designed to simulate the delayed diagonalization arguments over $\mathbf{E}$ in which the lengths of the diagonalization intervals are bounded by s. For the definition of our concept we use a generalization of the condition-set concept used in the definition of AFH-genericity (see Definition 6.15).

Definition 7.3 *(a) Let $s: N \to N$ be any function. An s-bounded condition set C is a set of (coded) pairs $<x, y>$ of strings x and y such that the following hold.*

(7.1) $\quad \forall x, y(<x, y> \in C \Rightarrow x \subseteq y)$

(7.2) $\quad \forall x, y, z(<x, y> \in C \, \& \, x \subseteq z \subseteq y \Rightarrow <x, z> \in C)$

(7.3) $\quad \forall x, y(<x, y> \in C \Rightarrow |y| \leq s(|x|))$

(b) Let C be an s-bounded condition set. A string y is a C-extension of a string x if $<x, y> \in C$ and, for no string y' properly extending y, $<x, y'> \in C$.
(c) An s-bounded $t(n)$-condition C is an s-bounded condition set $C \in DTIME(t(n))$.

Here the pairing function $< ., . >$ is defined by $< x, y > = 0^{|x|}1xy$, whence, by (7.3), $<x, y> \in C$ implies that $|< x, y >| \leq s(|x|) + 2|x| + 1$.

Note that an s-bounded condition set C describes a partial extension strategy, where the length of all extensions are bounded by s: The strategy is defined on the initial segment x if and only if $< x, x > \in C$, and in this case, every C-extension y of x is an admissible extension of the strategy. Moreover, by (7.2), we can use C as a "map" to produce a C-extension bit by bit. So if C is a $t(n)$-condition, every step of the extension can be (locally) computed in $t(n)$ steps, n the length of the current initial segment (modulo the costs for the pairing function $< ., . >$).

These observations lead to the following definition.

Definition 7.4 *Let C be an s-bounded condition set. C is total if $< X|x, X|x > \in C$ for all initial segments $X|x$. C is dense along a set A, if $< A|x, A|x > \in C$*

for infinitely many strings x; and C is dense *if C is dense along all sets. A set A* meets *the condition C if there are strings x and y such that A|y is a C-extension of A|x; and A* avoids *C otherwise.*

As in the proof of Proposition 5.2 it can be shown that, for any s-bounded condition C, the class

$$\mathbf{D}_C = \{A : C \text{ is not dense along } A \text{ or } A \text{ meets } C\}$$

is open-dense. So we can define a category concept as follows.

Definition 7.5 Extended s-bounded category *(or (E,s)-category for short) is the category concept $\mathcal{D} = <\mathcal{D}(t) : t \text{ recursive}>$ where*

$$\mathcal{D}(t) = \{\mathbf{D}_C : C \text{ is an } s\text{-bounded } t(n)\text{-condition}\}.$$

In particular, A is (E,s)-$t(n)$-generic if A meets all s-bounded $t(n)$-conditions which are dense along A.

Proposition 7.6 *For any function s, (E,s)-category is an abstract category concept in the sense of Definition 2.4.*

Fenner [Fe95] independently introduced an category concept based on similar ideas. His concept is defined in terms of (a new type of) extension functions, not on condition sets, but it can be easily reformulated in terms of the latter (as, conversely, our concept can be rephrased in the terms of Fenner's new extension function concept). The main difference between both concepts is, that Fenner admits only totally defined extension functions, whence his concept relates to E-category like F-category relates to G-category. In our terms his concept can be defined by considering only total conditions:

Definition 7.7 Extended s-bounded total category *(or (E_t,s)-category for short) is the category concept $\mathcal{D} = <\mathcal{D}(t) : t \text{ recursive}>$ where*

$$\mathcal{D}(t) = \{\mathbf{D}_C : C \text{ is a total } s\text{-bounded } t(n)\text{-condition}\}.$$

In particular, A is (E_t,s)-$t(n)$-generic if A meets all total s-bounded $t(n)$-conditions.

Obviously (E,s)-category refines (E_t,s)-category, so that the latter is a category concept in the sense of Definition 2.4 too. In the following three propositions we give some relations among the new genericity concepts, which are immediate by definition, give some trivial relations between the new concepts and F- and G-genericity, and compare L-genericity with E_t-genericity. From now on we will only consider the case of recursive nondecreasing length bounds s.

Proposition 7.8 *Let s, s', t, t' be any nondecreasing recursive functions.*
(a) Any (E, s)-$t(n)$-generic set is (E_t, s)-$t(n)$-generic.
(b) If $s(n) \leq s'(n)$ and $t(n) \leq t'(n)$ almost everywhere, then every (E, s')-$t'(n)$-generic $[(E_t, s')$-$t'(n)$-generic$]$ set is (E, s)-$t(n)$-generic $[(E_t, s)$-$t(n)$-generic$]$.

It is natural to extend the E-genericity concept to *classes* of functions: For function classes **S** and **T** say that A is $(E_*, \mathbf{S})$-**T**-*generic*, if A is (E_*, s)-$t(n)$-generic for all $s \in \mathbf{S}$ and $t \in \mathbf{T}$ $(E_* \in \{E, E_t\})$. Note that for recursive classes **S**, **T** of recursive functions there are recursive functions s and t which majorize all functions in **S** and **T**, respectively, so that, by Proposition 7.8, (E_*, s)-$t(n)$-genericity implies $(E_*, \mathbf{S})$-**T**-genericity. Hence in the following it suffices to discuss the case of single functions s and t.

Proposition 7.9 *Let s be any recursive function such that $s(n) > 2^{2^{n^2}}$ almost everywhere. Then every (E, s)-p-generic $[(E, s)$-n^k-generic$]$ set is G-p-generic $[G$-n^k-generic$]$ and every (E_t, s)-p-generic $[(E_t, s)$-n^k-generic$]$ set is F-p-generic $[F$-n^k-generic$]$.*

Proof. It suffices to express partial (total) F-n^k-extension functions by partial (total) s-bounded n^k-conditions. Given such a function f define a condition C_f by letting $< X|x, X|y > \in C_f$ iff $f(X|x)$ is defined, say $f(X|x) = (x_0, i_0), \ldots, (x_n, i_n)$, and $x \leq y \leq x_n + 1$ and $(X|y)(x_j) = i_j$ for all $j \leq n$ with $x_j < y$. Then C_f is an n^k-condition which, by (1.1) and by Proposition 4.11, is s-bounded. Moreover if f is dense along A then so is C_f and if C_f meets A then so does f.

Proposition 7.10 *For any nondecreasing recursive functions s and t with $s(n)$, $t(n) > n$ there is a recursive function u such that every L-$u(n)$-generic set is (E_t, s)-$t(n)$-generic.*

Proof. Let C be a total s-bounded $t(n)$-condition. Then $f(X|x) = X|y$, where $X|y$ is the least C-extension of $X|x$, is an L-$u(n)$-extension function for $u(n) = s(n) \cdot t(s(n) + 2n + 1)$ and any set which meets f meets C too.

The latter shows that - as F-category - E_t-category is in some sense only a quantitative improvement on L-genericity. We will come back to this later. Next, however, we want to show that the polynomial sections of the new category concepts pertain to **E** (in the sense of Definition 2.6).

Theorem 7.11 *For every recursive function s, there is an (E, s)-n^k-generic $((E, s)$-p-generic$)$ set in $DTIME(2^{(k+3)n})$ $(DTIME(2^{n^2}))$.*

Proof. Since the proof is similar to the proof of Theorem 5.6 we only give a sketch. Given a recursive function s and $k \geq 1$, we construct an (E, s)-n^k-generic set $A \in DTIME(2^{(k+3)n})$ by a slow diagonalization argument. Since, as already observed above, every recursive function is dominated by a nonde-creasing time constructible function, by Proposition 7.8, without loss of generality we may assume that s is time constructible. By time constructibility of s, we may choose a recursive enumeration $\{C_e : e \geq 0\}$ of the s-bounded n^k-conditions such that, for $|x| \geq e$, $C_e(x)$ can be uniformly computed in $|x|^{k+2}$ steps, as follows: Given an enumeration $\{Q_e : e \geq 0\}$ of $DTIME(n^k)$ in $DTIME(n^{k+1})$ define C_e from Q_e as follows: Given $z = <x, y>$, let $C_e(<x,y>) = 1$ if $<x,y> \in Q_e$ and (7.1) - (7.3) hold for this choice of x and y with Q_e in place of C; and let $C_e(<x,y>) = 0$ otherwise. Then it suffices that A meets the requirements

R_e : If C_e is dense along A then A meets C_e.

Requirement R_e *requires attention* at stage x of the construction (at which, given $A|x$, the value of $A(x)$ is determined) if $e < |x|$; R_e is *not yet satisfied* at stage x, i.e. there are no strings $u \leq v \leq x$ such that $A|v$ is a C_e-extension of $A|u$; but, for some string $y \leq x$, $<A|y, A|x> \in C_e$.

Then *stage x of the construction* is as follows: If no requirement requires attention, let $A(x) = 0$. Otherwise, choose e minimal such that R_e requires attention and fix y minimal such that $<A|y, A|x> \in C_e$. Since R_e is not yet satisfied, $A|x$ is not a C_e-extension of $A|y$, whence there is a number $i \leq 1$ such that $<A|y, (A|x)i> \in C_e$. Let $A(x) = i$ for the least such i. (Moreover, say that R_e *receives attention*.)

As in the proof of Theorem 5.6 we may argue that every requirement receives attention only finitely often. So, if C_e is dense along A, there will be a least stage x such that $<A|x, A|x> \in C_e$ and no requirement $R_{e'}$ with $e' < e$ will require attention at stage x or at any later stage. Then either R_e is already satisfied or R_e will receive attention from stage x until the first stage z such that, for the least y with $<A|y, A|x> \in C_e$, $A|z$ will be an C_e-extension of $A|y$.

Lemma 7.12 *Let s be a recursive function such that $s(n) \geq n + 1$ for all n and let A be (E_t, s)-n^k-generic $(k \geq 1)$. Then $A \notin DTIME(2^{kn})$.*

Proof. For any set $B \in DTIME(2^{kn})$,

$$C = \{<X|x, X|x>, <X|x, (X|x)(1 - B(x))>: X|x \in 2^{<\omega}\}$$

is a total s-bounded n^k-condition and B does not meet C.

Corollary 7.13 *For any recursive function s with $s(n) > n$, (E_t, s)-p-category and (E, s)-p-category pertain to $\mathbf{E}$ in the sense of Definition 2.6. In particular, $\mathbf{E}$ is not (E, s)-p-meager, hence not (E_t, s)-p-meager.*

Proof. By Theorem 7.11 and Lemma 7.12.

As we shall show next, the (E, s)-p-category concepts yield a proper hierarchy of category notions for growing s. We will observe first, that similar to the case of the constant size bounded G-genericity concepts (cf. Lemma 6.12), a constant number of iterations of a given length bound will not increase the power of the concept.

Lemma 7.14 *Let s be a nondecreasing time-constructible function, let c, $k \geq 1$, and let A be (E, s)-n^{c+1}-generic. Then A is (E, s^k)-n^c-generic. Hence a set is (E, s)-p-generic if it is (E, s^k)-p-generic.*

Proof. Since the proof is similar to the proof of Lemma 6.12, we give only a sketch. Let C be an s^k-bounded n^c-condition set which is dense along A. We have to show that A meets C. W.l.o.g we may assume that k is minimal such that, for infinitely many x, $A|x$ has an C-extension of length $\leq s^k(|A|x|)$. Define s-bounded n^{c+1}-condition sets C_m $(m < k)$ by letting

$$C_m = \{<z,y>: \exists x \subseteq z\ (|z| = s^m(|x|)\ \&\ |y| \leq s(|z|)\ \&\ <x,y> \in C\}.$$

Then, C_0 is dense along A; C_{m+1} is dense along A if A meets C_m infinitely often $(m < k - 1)$; and A meets C if A meets C_{k-1}. Since (as for the other genericity concepts; cf. Proposition 6.11) an (E, s)-n^{c+1}-generic set meets every s-bounded n^{c+1}-condition which is dense along it not just once but infinitely often, it follows that A meets all the conditions C_m, $m < k$, hence the condition C.

By Lemma 7.14, (E, s')-p-genericity can be more powerful than (E, s)-p-genericity only if s' dominates all finite iterations of s. As the following hierarchy theorem shows, this will also be sufficient if the domination property holds uniformly.

Theorem 7.15 *Let s, s', ind be nondecreasing unbounded recursive functions such that $ind \in \mathbf{P}$, s and s' are time-constructible, $s(n) > n$ almost everywhere, and $s^m(n) < s'(n)$ for $m \leq ind(n)$ and $n \geq 1$. There is an (E, s)-n^k-generic set $A \in DTIME(2^{(k+3)n})$ $[(E, s)$-p-generic set $A \in DTIME(2^{n^2})]$ which is not (E_t, s')-n-generic, hence not (E, s')-n-generic.*

The proof of Theorem 7.15 is based on the following observation.

Lemma 7.16 *Let s be a nondecreasing, time constructible function, let A be (E_t, s)-n^k-generic, and let $B \in DTIME(2^{kn})$. Then there are infinitely many strings (numbers) m such that $A(x) = B(x)$ for all x with $m \leq x < s(m)$.*

Proof. For $m \geq 0$ define total s-bounded n^k-conditions C_m by letting

$$C_m = \{<X|x, X|y> : m \leq x \leq y < s(x)\ \&\ X(z) = B(z)\ \text{for}\ x \leq z < y\}.$$

Then, for any set A which meets C_m, there is a number $m' \geq m$ with $A(x) = B(x)$ for x with $m' \leq x < s(m')$.

Proof of Theorem 7.15. It suffices to construct an (E, s)-n^k-generic set $A \in DTIME(2^{(k+3)n})$ such that

(7.4) $\forall m\, \exists x(m \leq x < s'(m)\ \&\ x \notin A)$.

Then, by (7.4), Lemma 7.16 applied to $B = 2^{<\omega}$ and $k = 1$ shows that A is not (E_t, s')-n-generic. To construct A, proceed as in the proof of Theorem 7.11 with the following modification: Requirement R_e now only requires attention at stage x if $e + 1 < \min(x, ind(x))$ and no requirement $R_{e'}$ with $e' < e$ required attention at stage $x - 1$. This prevents R_e from requiring attention only at finitely many stages, whence R_e will be met as before. This additional restraint, however, ensures that (7.4) is satisfied: Fix m. Since we let $A(x) = 0$ if no requirement requires attention at stage x, it suffices to show that there is a stage x with $m \leq x < s'(m)$ such that no requirement requires attention at stage x. For a contradiction assume that at any such stage some requirement requires attention and let $R_{e(x)}$ be the requirement which receives attention at stage x. Then, by our additional limitations on requiring attention, $e(m) + 1 < ind(m)$ and $e(x)$ is nonincreasing with growing x. Moreover, since the conditions C_e are s-bounded, a requirement which, beginning at stage y, receives attention at $s(y)$-y consecutive stages is satisfied forever. So, if $e(x) = e(y)$ for $m \leq x < y < s'(m)$, then $y < s(x)$. Together this implies that $s'(m) \leq s^{e(m)+1}(m)$. But this contradicts $e(m) + 1 < ind(m)$.

Note that Theorem 7.15 gives a hierarchy for the (E, s)-p-category concepts on **E**. For E_t-category the hierarchy can be refined. There Lemma 7.14 fails. The situation can be compared to the one of the constant length extensions in Section 6, where for F-genericity (corresponding to E_t-genericity) a very fine hierarchy could be proved, whereas for G-genericity (corresponding to E-genericity) the notions differring only by finite iterations collapsed.

Another consequence of the hierarchy theorem is, that there is no recursive set which is (E_t, s)-n-generic for all recursive bounds s. Together with Corollary 7.13 this implies that there is no strongest category concept pertaining to **E**.

Corollary 7.17 *Let A be (E_t, s)-n-generic for all recursive s. Then A is not recursive.*

Corollary 7.17 was independently proved by Fenner [Fe95].

Proof. For a contradiction assume that A is recursive. By Lemma 7.12, $A \notin \mathbf{P}$, whence A is infinite. Hence the function s which maps n to the least element of A greater than n is total and recursive. Moreover, for any n, there is a number m with $n \leq m \leq s(n)$ such that $A(m) \neq \emptyset(m)$. By Lemma 7.16 this implies that A is not $(E_t, s(n) + 1)$-n-generic. A contradiction.

We now show that the E_t-p-category concept, hence the E-p-category concept, captures delayed diagonalizations over **E**, i.e., diagonalizations over r.p. and c.f.v. classes with diagonals in **E**.

The key observation is that g-similarity to a set in $\mathbf{E}$ can be forced by (E_t, s)-n^k-genericity for sufficiently large s and k.

Lemma 7.18 *Let g be a nondecreasing, time constructible function and let $B \in DTIME(2^{kn})$. Then every $(E_t, 2^{g(n+1)})$-n^k-generic set is g-similar to B.*

Proof. This follows from Lemma 7.16 immediately.

We now apply Lemma 7.18 to show that E_t-category captures diagonalizations over r.p. and c.f.v. classes with diagonals in $\mathbf{E}$. The latter was proved first by Fenner [Fe95].

Theorem 7.19 *(Fenner [Fe95]). Let $\mathbf{C}$ be an r.p. and c.f.v. class such that $\mathbf{E}$ is not contained in $\mathbf{C}$. There is a recursive function s and a $k \geq 1$ such that no (E_t, s)-n^k-generic set is in $\mathbf{C}$. So $\mathbf{C}$ is (E_t, s)-p-meager, hence (E_t, s)-meager in $\mathbf{E}$.*

Proof. By assumption, there is a $k \geq 1$ and a set $B \in DTIME(2^{kn})$ such that $B \notin \mathbf{C}$. So, by Lemma 7.2 , there is a nondecreasing, time constructible function g such that every set which is g-similar to B does not belong to $\mathbf{C}$. By Lemma 7.18 this implies that, for $s(n) = 2^{g(n+1)}$, no (E_t, s)-n^k-generic set is in $\mathbf{C}$.

Theorem 7.19 fails - even for the stronger E-category concept - if we fix the recursive length function s. We will demonstrate this by considering the lower and upper P-m-spans of sets in $\mathbf{E}$ in the next section. So, for fixed recursive s, there are r.p. and c.f.v. proper subclasses of $\mathbf{E}$ which are not (E, s)-meager in $\mathbf{E}$, hence not (E_t, s)-meager in $\mathbf{E}$. As observed by Fenner [Fe95], however, all the (E_t, s)-category concepts can be merged to one powerful concept. The same observation applies to E-category.

Definition 7.20 *(Fenner) For $E_* \in \{E, E_t\}$, a class $\mathbf{C}$ is E_*-p-(co)meager if there is a recursive function s such that $\mathbf{C}$ is (E_*, s)-p-(co)meager. $\mathbf{C}$ is E_*-meager in $\mathbf{E}$ if $\mathbf{C} \cap \mathbf{E}$ is E_*-p-meager, and $\mathbf{C}$ is E_*-comeager in $\mathbf{E}$ if $\overline{\mathbf{C}}$ is E-meager in $\mathbf{E}$. A set A is E_*-$t(n)$-generic if A is (E_*, s)-$t(n)$-generic for all recursive s, and A is E_*-p-generic if A is E_*-n^k-generic for all $k \geq 1$.*

Though, by Corollary 7.17, E_t-$t(n)$-generic and E-$t(n)$-generic sets are nonrecursive for $t(n) \geq n$, the E_t-category and E-category concepts on $\mathbf{E}$ are consistent. By Corollary 7.13, $\mathbf{E}$ is not E-meager in $\mathbf{E}$ hence not E_t-meager. Moreover, roughly speaking, the E-p-meager classes are closed under p-uniform unions with uniform recursive length bound, so that E_*-(and E_t-)category on $\mathbf{E}$ resemble the classical Baire category concept (see Fenner [Fe95] for details). The uniformity requirement for the length-bound also becomes apparent, if we characterize E-meagerness in terms of genericity (cf. Proposition 2.7): As one can easily check,

(7.5)   C is $E(E_t)$-meager in **E** iff, for some $k \geq 1$ and some recursive s, C does not contain any (E, s)-n^k-generic $((E_t, s)$-n^k-generic) set.

Next we will show that Theorem 7.19 can be used to completely characterize the classes which are E_t-meager in **E**.

Lemma 7.21 *Let* **C** *be E_t-meager in* **E**. *There is an r.p. and c.f.v. class* **D** *such that* **C** $\cap$ **E** $\subseteq$ **D** $\subset$ **E**.

Proof. Fix $k \geq 1$ and s recursive such that **C**$\cap$**E** is (E_t, s)-n^k-meager and let $\{C_e : e \geq 0\}$ be a recursive enumeration of the total s-bounded n^k-conditions. Then, by (7.5), no element of **C**$\cap$**E** is (E_t, s)-n^k-generic, whence **C**$\cap$**E** $\subseteq$ **D** for

$$\mathbf{D} = \{A \in \mathbf{E} : A \text{ is not } (E_t, s)\text{-}n^k\text{-generic}\} = \{A \in \mathbf{E} : \exists e \, (A \text{ avoids } C_e)\}.$$

By Theorem 7.11, **D** $\subset$ **E**. So it suffices to show that **D** is r.p. and c.f.v. For a proof of the former let $\{E_e : e \geq 0\}$ be a recursive enumeration of **E**, and define $D_{<e,n>}$, $e, n \geq 0$, by letting $D_{<e,n>}(x) = 1$ iff $x \in E_e$ and, for all $y, z \leq x$, $E_e|y$ is not a C_n-extension of $E_e|z$. Then $D_{<e,n>} = E_e$ if E_e avoids C_n and $D_{<e,n>}$ is finite, hence not (E_t, s)-n^k-generic, otherwise. So $\{D_{<e,n>} : e, n \geq 0\}$ is a recursive enumeration of **D**. The closure of the class of the (E_t, s)-n^k-generic sets under finite variants is straightforward.

Corollary 7.22 *For any class* **C**, **C** *is E_t-meager in* **E** *if and only if there is an r.p. and c.f.v. class* **D** *such that* **E** $\cap$ **C** $\subseteq$ **D** $\subset$ **E**.

Proof. By Theorem 7.19 and Lemma 7.21.

Corollary 7.22 can be extended to a characterization of the E_t-p-meager class. Above we presented the delayed diagonalization method as a tool to diagonalize over r.p. and c.f.v. classes (as it is in most cases used in complexity theory). This method, however, can be applied to unbounded classes too. Recall that a class **C** is a Σ_2^0-*class* if

$$\mathbf{C} = \{A : \exists x \, \forall y (M^A(<x, y>) = 1)\},$$

where M is an oracle Turing machine which is total for all oracles. Fenner [Fe95] actually showed that every c.f.v. Σ_2^0-class which does not contain **E** is E_t-p-meager. By extending Lemma 7.21 correspondingly, this gives the following characterization of E_t-p-meagerness.

Theorem 7.23 *For any class* **C**, **C** *is E_t-p-meager if and only if there is a c.f.v. Σ_2^0 class* **D** *such that* **C** $\subseteq$ **D** *but* **E** $\not\subseteq$ **D**.

We omit the proof which is similar to the proof of Corollary 7.22. We only want to mention here the following well known relations between Σ_2^0-classes and r.p. classes: A c.f.v. class which is r.p. is Σ_2^0, while for any c.f.v. classes $\mathbf{A}$ and $\mathbf{B}$ such that $\mathbf{A}$ is Σ_2^0 and $\mathbf{B}$ is r.p., $\mathbf{A} \cap \mathbf{B}$ is r.p. too. So, for a recursively bounded class $\mathbf{A}$ of recursive sets which is c.f.v., $\mathbf{A}$ is Σ_2^0 iff $\mathbf{A}$ is r.p.

We close this section by the observation that Mayordomo's result that the class of the $\mathbf{P}$-immune sets is not F-comeager in $\mathbf{E}$ extends to E_t-category.

Theorem 7.24 *Let s, t be recursive. There is an (E_t, s)-$t(n)$-generic set which is not $\mathbf{P}$-immune. Moreover there is an (E_t, s)-p-generic $[(E_t, s)$-n^k-generic] set A in $DTIME(2^{n^2})$ $[DTIME(2^{(k+3)n})]$ which is not $\mathbf{P}$-immune.*

Proof. The first part is immediate by Corollary 4.15 and Proposition 7.10. The second part can be shown like the correspondent facts for F-genericity in Section 6.

In contrast to Theorem 7.24, already very low levels of E-genericity force $\mathbf{P}$-immunity.

Lemma 7.25 *Let A be $(E, n+1)$-n-generic. Then A is $\mathbf{P}$-immune.*

Proof. For infinite $B \in \mathbf{P}$, the $(n+1)$-bounded n-condition

$$C = \{<X|x, X|x>, <X|x, (X|x)0> : x \in B\}$$

is dense and any set which meets C does not contain B.

Corollary 7.26 *The class of the $\mathbf{P}$-immune sets is E-comeager in $\mathbf{E}$ but not E_t-comeager in $\mathbf{E}$. Hence E-category is a proper refinement of E_t-category.*

Another consequence of Theorems 7.23 and 7.24 is that the class of $\mathbf{P}$-(bi)-immune sets is not a Π_2^0-class. Finally observe that Theorem 7.24 does not extend to E_t-genericity. It can be easily shown that for any infinite set $B \in \mathbf{P}$ there is a recursive function s such that no (E_t, s)-n-generic set contains B (it suffices that $s(n)$ is greater than the least element $> n$ of B). So any E_t-n-generic set is $\mathbf{P}$-bi-immune. In fact an E_t-$t(n)$-generic set meets all recursively bounded $t(n)$-conditions which are dense.

8 Weakly Complete and Weakly Tractable Problems

In classical Baire Category, for any polynomial time reducibility r, all (non-trivial) upper $\mathbf{P}$-r-spans and all lower $\mathbf{P}$-r-spans are meager (see Theorem

2.3). In this section we address the question whether - or to what extent - analogous results can be obtained for the category notions on **E** induced by the different types of resource-bounded genericity concepts discussed in this paper. For simplicity, we will only consider the case of **P**-m-reductions.

As we will see, a full analogue of the classical result can be only obtained for the nonuniform E-category (or E_t-category) concept. For the uniform concepts, the analogy fails, but the extent of the failure heavily depends on the individual concepts. We hope that these investigations further illustrate the power of the different genericity concepts.

As one can easily see, for any set $A \in$ **E**, the lower and upper **P**-m-spans of A, $\mathbf{P}^{\leq}_m(A)$ and $\mathbf{P}^{\geq}_m(A)$, respectively, intersected with **E** are recursively presentable and closed under finite variants. (To avoid trivialities, we ignore $\emptyset$ and $2^{<\omega}$ in this context.) So, Fenner's Theorem (Theorem 7.19) implies that nontrivial **P**-m-spans in **E** are E_t-meager, hence E-meager.

Theorem 8.1 *(Fenner) Let $A \in$ **E** be given. Then*

(8.1) A *not* **P**-m-*complete for* **E** $\Rightarrow$ $\mathbf{P}^{\leq}_m(A) \cap$ **E** E_t-*meager*

(8.2) $A \notin \mathbf{P}$ $\Rightarrow$ $\mathbf{P}^{\geq}_m(A) \cap$ **E** E_t-*meager*

In the following we want to address the question, to what extend analogues of this theorem can be obtained for the stronger, uniform category concepts on **E**. We will consider the case of L-category, F-category, AFH-category, G-category, and $(E, \mathbf{S})$-category, where (for the remainder of this section) **S** is an arbitrary but fixed recursive class of recursive functions such that, for any $s \in \mathbf{S}$ and any polynomial p, **S** contains a nondecreasing, time-constructible function s' with $s'(n) > p(s(n))$ a.e. Moreover, for the sake of (8.3) below, we assume that **S** contains the function $2^{2^{n^2}}$ (cf. Proposition 7.9), but this will not be crucial for the proofs.

Note that, by the relations among the genericity concepts analyzed in the previous sections, for any class **C** the following hold

$$\mathbf{C} \; L\text{-(co)meager in } \mathbf{E} \quad \Rightarrow \quad \mathbf{C} \; F\text{-(co)meager in } \mathbf{E}$$
$$\Rightarrow \quad \mathbf{C} \; G\text{-(co)meager in } \mathbf{E}$$
$$\Rightarrow \quad \mathbf{C} \; (E, \mathbf{S})\text{-(co)meager in } \mathbf{E}$$
(8.3)
$$\Rightarrow \quad \mathbf{C} \; (E)\text{-(co)meager in } \mathbf{E}$$
and
$$\mathbf{C} \; AFH\text{-(co)meager in } \mathbf{E} \quad \Rightarrow \quad \mathbf{C} \; G\text{-(co)meager in } \mathbf{E}$$

In the following we let $* \in \{ L, F, AFH, G, (E, \mathbf{S}) \}$ and we order the prefixes of the category concepts according to their strength, i.e., $L < F < G < (E, \mathbf{S})$ and $AFH < G$. Also recall that we have shown that the polynomial case of all of these category concepts pertains to **E**, whence, by Proposition 2.7, (co)meagerness in **E** can be expressed in terms of n^k-genericity:

(8.4)
$$\begin{aligned} &\textbf{C} \text{ is } *\text{-meager in } \textbf{E} \\ &\quad \Leftrightarrow \quad \exists\, k\, \forall\, G \in \textbf{E}\, (G \;*\text{-}n^k\text{-generic} \;\Rightarrow\; G \notin \textbf{C}) \end{aligned}$$

(8.5)
$$\begin{aligned} &\textbf{C} \text{ is not } *\text{-meager in } \textbf{E} \\ &\quad \Leftrightarrow \quad \forall k\, \exists\, G \in \textbf{E}\, (G \;*\text{-}n^k\text{-generic} \;\&\; G \in \textbf{C}) \end{aligned}$$

(8.6)
$$\begin{aligned} &\textbf{C} \text{ is } *\text{-comeager in } \textbf{E} \\ &\quad \Leftrightarrow \quad \exists\, k\, \forall\, G \in \textbf{E}\, (G \;*\text{-}n^k\text{-generic} \;\Rightarrow\; G \in \textbf{C}) \end{aligned}$$

Note that, by Theorem 8.1 and by (8.3), for any $*$-category concept, $\mathbf{P}_m^{\leq}(A) \cap \mathbf{E}$ and $\mathbf{P}_m^{\geq}(A) \cap \mathbf{E}$ are $*$-comeager only in the trivial cases, i.e. if A is $\mathbf{P}$-m-complete (whence $\mathbf{P}_m^{\leq}(A) \cap \mathbf{E} = \mathbf{E}$) and if $A \in \mathbf{P}$ (whence $\mathbf{P}_m^{\geq}(A) \cap \mathbf{E} = \mathbf{E}$), respectively. This leaves, however, the possibility that also in nontrivial cases, the spans are not $*$-measurable in $\mathbf{E}$, i.e., neither $*$-meager nor $*$-comeager in $\mathbf{E}$.

Definition 8.2 *(a) A set A is $*$-weakly hard if $\mathbf{P}_m^{\leq}(A)$ is not $*$-meager in $\mathbf{E}$. If, moreover, $A \in \mathbf{E}$ then A is $*$-weakly complete.*
(b) A set A is $*$-weakly tractable if $\mathbf{P}_m^{\geq}(A)$ is not $*$-meager in $\mathbf{E}$.

Obviously, any $\mathbf{P}$-m-complete set for $\mathbf{E}$ is $*$-weakly complete, and any $\mathbf{P}$-set is $*$-weakly tractable. So, by (8.3), the relations among the weak completeness and weak tractability concepts are as follows (A any set):

(8.7)
$$\begin{aligned} A \text{ } \mathbf{P}\text{-}m\text{-complete for } \mathbf{E} \;&\Rightarrow\; A \text{ } (E, \mathbf{S})\text{-weakly complete} \\ &\Rightarrow\; A \text{ } G\text{-weakly complete} \\ &\Rightarrow\; A \text{ } F\text{-weakly complete} \\ &\Rightarrow\; A \text{ } L\text{-weakly complete} \end{aligned}$$

and

$$A \text{ } G\text{-weakly complete} \;\Rightarrow\; A \text{ } AFH\text{-weakly complete}$$

(8.8)
$$\begin{aligned} A \in \mathbf{P} \;&\Rightarrow\; A \text{ } (E, \mathbf{S})\text{-weakly tractable} \\ &\Rightarrow\; A \text{ } G\text{-weakly tractable} \\ &\Rightarrow\; A \text{ } F\text{-weakly tractable} \\ &\Rightarrow\; A \text{ } L\text{-weakly tractable} \end{aligned}$$

and

$$A \text{ } G\text{-weakly tractable} \;\Rightarrow\; A \text{ } AFH\text{-weakly tractable}$$

We will first show that, for any $*$-category concept, the $*$-weakly complete sets are abundant. Joseph, Pruim and Young [JPY94] have shown that there are L-weakly complete sets which are not $\mathbf{P}$-m-complete. They obtained their result by a direct, quite involved construction. Here we use a quite different and considerably simpler approach. We show that the recent proof of Ambos-Spies, Terwijn and Zheng [ATZ94] that the weakly complete sets in the sense of Lutz's measure theory on $\mathbf{E}$ have measure 1 in $\mathbf{E}$ can be adopted to most of the $*$-category concepts.

Lemma 8.3 *Let $* \in \{F, AFH, G, (E, \mathbf{S})\}$ and let $A \in \mathbf{E}$ be $*$-n^2-generic. For every $k \geq 1$ there is a $*$-n^k-generic set $A_k \in \mathbf{P}_m^{\leq}(A) \cap \mathbf{E}$.*

Proof. We sketch the proof of the claim for $* = F$ and leave the similar proofs for the other concepts to the reader. Fix c such that $A \in DTIME(2^{cn})$ and let $A_k = \{x : g(x) \in A\}$ for $g(x) = 0^{k|x|}x$. Then $A_k \leq_m^P A$ via g and $A_k \in DTIME(2^{c(k+1)n})$. It remains to show that A_k is F-n^k-generic. Let f be any F-n^k-extension function. We have to show that A_k meets f. Since A is F-n^2-generic this can be established by defining an F-n^2-extension function f' such that

> (8.9) A meets $f' \Rightarrow A_k$ meets f.

For the definition of f' we need the following notation. For a string $X|g(x)$ let $\widehat{X}|x$ be defined by $\widehat{X}(y) = X(g(y))$ for $y < x$. Note that $\widehat{A}|x = A_k|x$. Now given $X|x$ fix y minimal such that $x \leq g(y)$ and let

$$f'(X|x) = (g(z_0), i_0), \ldots, (g(z_m), i_m)$$

where $f(\widehat{X}|y) = (z_0, i_0), \ldots, (z_m, i_m)$.

Theorem 8.4 *Let $* \in \{F, AFH, G, (E, \mathbf{S})\}$. Then a set A is $*$-weakly complete if and only if $\mathbf{P}_m^{\leq}(A) \cap \mathbf{E}$ contains an $*$-n^2-generic set. In particular, every $*$-n^2-generic set $A \in \mathbf{E}$ is $*$-weakly complete, whence the class of $*$-weakly complete sets is $*$-comeager in $\mathbf{E}$.*

Proof. By Lemma 8.3 and by (8.6).

Note that, by Corollary 4.10 and by (8.3), for $* \in \{F, AFH, G, (E, \mathbf{S})\}$, the class of $\mathbf{P}$-m-complete problems for $\mathbf{E}$ is $*$-meager in $\mathbf{E}$. So, by Theorem 8.4, the class of sets which are $*$-weakly complete but not $\mathbf{P}$-m-complete for $\mathbf{E}$ is $*$-comeager in $\mathbf{E}$.

Our proof of Lemma 8.3 (and hence of Theorem 8.4) does not work for L-category and we do not know whether the class of L-weakly complete sets is L-comeager in $\mathbf{E}$. This does not mean that there might be less L-weakly complete sets than weakly complete sets for the other category notions: By (8.7) every $*$-weakly complete set, $* \in \{F, G, (E, \mathbf{S})\}$, is L-weakly complete too, whence, by Theorem 8.4, the class of L-weakly complete sets is $*$-comeager in $\mathbf{E}$. So, by (8.3), the class of L-weakly complete sets is not L-meager, but still L-category might be so weak that it cannot measure the size of this class. Since L-category fails to measure the size of the class of the $\mathbf{P}$-m-complete problems in $\mathbf{E}$ (Corollary 3.14), this leaves the possibility that L-category cannot distinguish the sizes of the classes of the complete and the

L-weakly complete problems, though for the stronger category concepts on **E**, the former is small and the latter is large.

Another consequence of Theorem 8.4 (and (8.3) and (8.7)) is that, for $* \in \{L, F, AFH, G, (E, \mathbf{S})\}$, the class of the $*$-weakly complete sets is E-comeager in **E**, whence, by Theorem 7.19, for any r.p. and c.f.v. class **C** which does not contain **E**, there is a $*$-weakly complete set $A \notin \mathbf{C}$.

We now turn to the weakly tractable problems. We first observe that for the stronger category concepts these sets are much rarer than the weakly complete ones.

Theorem 8.5 *(Ambos-Spies, Neis, Terwijn [ANT94]). Let $* \in \{AFH, G, (E, \mathbf{S})\}$. There is no set which is both, $*$-weakly complete and $*$-weakly tractable.*

Theorem 8.5 is called a *small span theorem* for $*$-category in **E**, since it shows that, for any set $A \in \mathbf{E}$, the upper span or the lower span is small in **E** in the sense of $*$-category. The first small span theorem was proved by Juedes and Lutz for Lutz's measure on **E** (see [JL95]). Ambos-Spies et al. [ANT94] extended this theorem to AFH-genericity. The proof is based on the observations that AFH-n^k-generic sets cannot be compressed by **P**-m-reductions and that such incompressible sets have AFH-p-meager upper spans.

Definition 8.6 *A set A is **P**-m-incompressible if, for any set B and any **P**-m-reduction g from A to B, g is almost one-to-one.*

Lemma 8.7 *Let $A \in DTIME(2^{kn})$ $(k \geq 2)$ be **P**-m-incompressible and assume that $A \leq_m^P B$. Then B is not $[F, 1]$-n^k-generic (hence neither AFH-n^k-generic nor F-n^k-generic).*

Proof. Fix g such that $A \leq_m^P B$ via g. Since g is almost one-to-one and since there are more strings of any length n than strings of any length less than n, for all sufficiently large n there is a string x of length n such that $|g(x)| \geq |x|$. Hence, if we let y_x be the least string $y \geq x$ such that $|g(y)| \geq |y|$ then $|y_x| \leq |x| + 1$ and y_x can be computed in $2^{2|x|}$ steps. It follows that f defined by $f(X|x) = (g(y_x), 1 - A(y_x))$ is an $[F, 1]$-n^k-extension-function, and, since $A \leq_m^P B$ via g, B does not meet f. So B is not $[F, 1]$-n^k-generic.

Lemma 8.8 *Let A be AFH-n^2-generic. Then A is **P**-m-incompressible.*

Proof. Let g be a **P**-m-reduction which is not almost one-to-one, i.e., $I_g = \{x : \exists y < x \ (g(y) = g(x))\}$ is infinite. We have to show that there are strings x and y such that $g(x) = g(y)$ but $A(x) \neq A(y)$ so that g does not reduce A to any set. Define the partial $[F, 1]$-n^2-extension function f by letting $f(X|x) = (x, 1 - X(y_x))$ if $x \in I_g$ and y_x is the least string $y < x$ with $g(y) = g(x)$, and let $f(X|x)$ be undefined otherwise. Then, by infinity of I_g,

f is dense. So, since A is AFH-n^2-generic, hence $[G, 1]$-n^2-generic, A meets f. By definition of f this implies the desired inconsistency of g with A.

Proof of Theorem 8.5. By (8.3) it suffices to consider $* = AFH$. Fix $A \in \mathbf{E}$ and w.l.o.g. assume that A is AFH-weakly complete. Then, by (8.5), there is an AFH-n^2-generic set $G \in \mathbf{P}_m^{\le}(A) \cap \mathbf{E}$. Obviously, $\mathbf{P}_m^{\ge}(A) \subseteq \mathbf{P}_m^{\ge}(G)$ and, by Lemmas 8.7 and 8.8, there is a number $k \ge 1$ such that $\mathbf{P}_m^{\ge}(G)$ does not contain any AFH-n^k-generic set. It follows with (8.4) that $\mathbf{P}_m^{\ge}(G)$, hence $\mathbf{P}_m^{\ge}(A)$, is AFH-meager in $\mathbf{E}$. So A is not AFH-weakly tractable.

Though, by Theorems 8.4 and 8.5, for $* \in \{AFH, G, (E, \mathbf{S})\}$ the class of $*$-weakly tractable problems is $*$-meager, there exist such problems which are not in $\mathbf{P}$.

Theorem 8.9 *Let* $* \in \{AFH, G, (E, \mathbf{S})\}$ *and let* $A \in \mathbf{E}$ *be* $*$-n^2-generic. *There is a* $*$-weakly tractable problem B with $B \in (\mathbf{P}_m^{\le}(A) \cap \mathbf{E})$-$\mathbf{P}$. *In particular, there are* $*$-weakly tractable problems in $\mathbf{E}$-$\mathbf{P}$.

Proof. Define A_k and g as in the proof of Lemma 8.3 and let $B = \{0^{n!} : n \ge 0\} \cap A$. Since $\{0^{n!} : n \ge 0\} \in \mathbf{P}$, $B \in \mathbf{P}_m^{\le}(A) \cap \mathbf{E}$ is immediate. Moreover, since A is $*$-n^2-generic, hence $\mathbf{P}$-bi-immune, $B \notin \mathbf{P}$. Finally, to show that B is $*$-weakly tractable, by choice of the sets A_k and by (8.5), it suffices to show that $B \le_m^P A_k$ for all $k \ge 1$. Note that $0^m \in A_k$ iff $0^{(k+1)m} \in A$. So a reduction function g from B to A_k can be defined as follows: If $x = 0^{n!}$ and $k + 1 \le n$ (i.e. $k + 1$ devides $n!$) then let $g(x) = 0^{n!/(k+1)}$; if x is not of the form $0^{n!}$ then let $g(x)$ be a fixed element of $\overline{A_k}$; and if $x = 0^{n!}$ for $n < k + 1$ then choose $g(x)$ so that $B(x) = A_k(x)$.

For L-category and F-category, weakly tractable problems occur more frequently. In case of L-category, the class of $\mathbf{P}$-m-complete problems for $\mathbf{E}$ is not meager in $\mathbf{E}$ (see Corollary 3.14), which immediately implies that *every* set in $\mathbf{E}$ is L-weakly tractable. For F-category, the class of F-weakly tractable sets is neither F-meager nor F-comeager.

Theorem 8.10 *(a) The class of the F-weakly tractable problems is not F-meager in* $\mathbf{E}$.
(b) The class of the F-weakly tractable problems is AFH-meager in $\mathbf{E}$, *hence not F-comeager in* $\mathbf{E}$.

Proof. We first prove part (b) of the theorem. By Lemmas 8.7 and 8.8, for any AFH-n^2-generic set A in $\mathbf{E}$ there is a number k such that A is not $\mathbf{P}$-m-reducible to any F-n^k-generic set. So no AFH-n^2-generic set in $\mathbf{E}$ is F-weakly tractable.
For a proof of part (a), by (8.5), it suffices to prove the following lemma.

Lemma 8.11 *Let* $c \ge 1$. *There is an F-n^c-generic set* $A \in \mathbf{E}$ *such that, for every* $k \ge 1$, *there is an F-n^k-generic set A_k such that* $A \le_m^P A_k$ *and* $A_k \in \mathbf{E}$.

Proof. For the definition of A, apply the first part of Theorem 4.12 to $B = \emptyset$ and $C = \{\delta(2n) : n \geq m(c)\}$, where δ is the iterated double exponential function and $m(c)$ is chosen so that $2^{(c+1)\delta(n)} < \delta(n+1)$ for all $n \geq m(c)$. Then A is F-n^c-generic, $A \in \mathbf{E}$, and, by (4.1), $A \subseteq EVEN_\delta$ for the **P**-set $EVEN_\delta = \bigcup_{n \geq 0} [0^{\delta(2n)}, 0^{\delta(2n+1)})$. The required sets A_k are obtained by a second application of Theorem 4.12: Given k, apply the first part of the theorem to $B' = A$ and $C' = \{\delta(2n+1) : n \geq m(k)\}$. Then A_k is F-n^k-generic, $A_k \in DTIME^{(A, \leq)}(2^{(k+2)n})$, which, by $A \in \mathbf{E}$, is contained in $\mathbf{E}$, and, by (4.1) and by choice of C' and $m(k)$, $A_k \cap EVEN_\delta = A$. Since $EVEN_\delta \in \mathbf{P}$, the latter implies that $A \leq_m^P A_k$.

By Theorems 8.4 and 8.10, there are sets in $\mathbf{E}$ which are both, F-weakly complete and F-weakly tractable. Hence a small span theorem fails for F-category. In fact, by a more sophisticated, direct construction we can show that there is a set $A \in \mathbf{E}$ such that the **P**-m-degree of A is not F-meager in $\mathbf{E}$. (Note that by Corollary 4.10 such a set cannot be **P**-T-complete for $\mathbf{E}$.)

9 Conclusion

We have discussed three types of resource-bounded genericity concepts related to three of the most fundamental diagonalization techniques in structural complexity theory, namely F-genericity due to Lutz [Lu90] and Fenner [Fe91], which captures standard finite extension arguments; the new G-genericity concept which, in the spirit of the genericity concept of Ambos-Spies et al. [AFH88], extends F-genericity to a concept handling slow diagonalizations too; and the new E_t- and E-genericity concepts, the former - independently introduced by Fenner [Fe95] - designed to capture delayed diagonalizations, and the latter unifying all of these concepts.

These genericity notions can be applied to all sufficiently closed deterministic time and space classes extending $\mathbf{E}$, though in this paper we only considered the case of the latter class. As observed already in the introduction, however, an extension of these concepts to subexponential time classes might be nontrivial. Moreover, it might be interesting to introduce and study corresponding genericity concepts for other fundamental complexity classes like the nondeterministic or probabilistic time classes.

Further research might be directed to the analysis of other important diagonalization techniques in computational complexity, like e.g. the speed-up diagonalizations (see e.g. [Am87a]) which played an important role in the study of the polynomial degrees.

Finally, a further study of the properties of the genericity concepts discussed in this paper might clarify the question whether they actually meet the tasks they have been defined for and might give more insight in the power and limitations of the fundamental diagonalization concepts in complexity theory.

Acknowledgements. I thank Wolfgang Merkle, Yongge Wang and Xizhong Zheng for several useful suggestions and comments on a draft of this paper.

References

[AS94]   E. Allender and M. Strauss, Measure on small complexity classes, in: Proc. 35th Symp. on Foundations of Comput. Sci., 1994, IEEE Comput. Soc. Press.

[Am87]   K. Ambos-Spies, Polynomial time degrees of NP-sets, in: "Trends in Theoretical Computer Science" (E. Börger, Ed.), 95-142, Computer Science Press, 1987.

[Am87a]   K. Ambos-Spies, Minimal pairs for polynomial time reducibilities, in: "Computation Theory and Logic", Lect. Notes Comput. Sci. 270 (1987) 1-13, Springer Verlag.

[AFH84]   K. Ambos-Spies, H. Fleischhack, and H. Huwig, P-generic sets, in: Proc. ICALP 1984, Lect. Notes Comput. Sci. 172 (1984) 58-68, Springer Verlag.

[AFH87]   K. Ambos-Spies, H. Fleischhack, and H. Huwig, Diagonalizations over polynomial time computable sets, Theor. Comput. Sci. 51 (1987) 177-204.

[AFH88]   K. Ambos-Spies, H. Fleischhack, and H. Huwig, Diagonalizing over deterministic polynomial time, in: Proc. CSL '87, Lect. Notes Comput. Sci. 329 (1988) 1-16, Springer Verlag.

[ANT94]   K. Ambos-Spies, H.-C. Neis and S. A. Terwijn, Genericity and measure for exponential time, Theor. Comput. Sci. (to appear) [Extended abstract in: Proc. MFCS 1994, Lect. Notes Comput. Sci. 841 (1994) 221-232, Springer Verlag].

[ATZ94]   K. Ambos-Spies, S. A. Terwijn and X. Zheng, Resource bounded randomness and weakly complete problems, Theor. Comput. Sci. (to appear) [Extended abstract in: Proc. ISAAC'94, Lect. Notes Comput. Sci. 834 (1994) 369-377, Springer Verlag].

[BDG90]   J. L. Balcazar, J. Diaz, J.Gabarro, Structural Complexity II, Springer Verlag, 1990.

[BDG95] J. L. Balcazar, J. Diaz, J.Gabarro, Structural Complexity I (Second Ed.), Springer Verlag, 1995.

[BGS75] T. Baker, J. Gill and R. Solovay, Relativizations of the P=?NP question, SIAM J. Comput. 5 (1975) 431-442.

[BM95] J. L. Balcazar and E. Mayordomo, A note on genericity and bi-immunity, in: Proc. 10th Structure in Complexity Theory Conference, 1995, 193-196, IEEE Comput. Soc. Press.

[BS85] J. L. Balcazar and U. Schöning, Bi-immune sets for complexity classes, Mathematical Systems Theory 18 (1985) 1-10.

[Be77] L. Berman, Polynomial reducibilities and Complete Sets, PhD thesis, Cornell University, 1977.

[Fe65] S. Feferman, Some applications of the notions of forcing and generic sets, Fund. Math. 56 (1965) 325-245.

[Fe91] S. A. Fenner, Notions of resource-bounded category and genericity, in: Proc. 6th Structure in Complexity Theory Conference, 1991, 196-212, IEEE Comput. Soc. Press.

[Fe95] S. A. Fenner, Resource-bounded Baire category: a stronger approach, in: Proc. 10th Structure in Complexity Theory Conference, 1995, 182-192, IEEE Comput. Soc. Press.

[Fl85] H. Fleischhack, On Diagonalizations over Complexity Classes, Dissertation, Universität Dortmund, Dept. Comput. Sci. Tech. Rep. 210, 1985.

[Fl86] H. Fleischhack, P-genericity and strong P-genericity, in: Proc. MFCS 1986, Lect. Notes Comput. Sci. 233 (1986) 341-349, Springer Verlag.

[GHS87] J. Geske, D. Huyn and A. Selman, A hierarchy theorem for almost everywhere complex sets with applications to polynomial complexity degrees, in: Proc. 4th Symp. on Theor. Aspects of Comput. Sci., Lect. Notes Comput. Sci. 247 (1987) 125-135, Springer Verlag.

[Hi69] P. G. Hinman, Some applications of forcing to hierarchy problems in arithmetic, Z. Math. Logik Grundlagen Math. 15 (1969) 341-352.

[HU79] J. E. Hopcroft and J. D. Ullman, Introduction to Automata Theory, Languages, and Computation, Addison-Wesley, 1979.

[Jo80]   C. G. Jockusch, Degrees of generic sets, in: Recursion Theory: its Generalisations and Applications, London Math. Soc. Lect. Notes Series 45 (1980) 110-139, Cambridge University Press.

[Jo85]   C. G. Jockusch, Genericity for recursively enumerable sets, in: Proc. Recursion Theory Week 1984, Lect. Notes Math. 1141 (1985) 203-232, Springer Verlag.

[JL95]   D. W. Juedes and J. H. Lutz, The complexity and distribution of hard problems, SIAM J. Comput. 24 (1995) 279-295.

[JPY94]   D. Joseph, R. Pruim and P. Young, Weakly hard problems in resource-bounded category, Manuscript, 1994.

[Ku95]   M. Kumabe, Degrees of generic sets, this volume.

[La75]   R. E. Ladner, On the structure of polynomial time reducibility, Journal of the ACM 22 (1975) 155-171.

[LLR81]   L. H. Landweber, R. J. Lipton and E. L. Robertson, On the structure of sets in NP and other complexity classes, Theor. Comput. Sci. 15 (1981) 181-200.

[LLS75]   R. E. Ladner, N. A. Lynch and A. L. Selman, A comparison of polynomial time reduciblities, Theor. Comput. Sci. 1 (1975) 103-123.

[Le83]   M. Lerman, The Degrees of Unsolvability, 1983, Springer Verlag.

[Lu90]   J. H. Lutz, Category and measure in complexity classes, SIAM J. Comput. 19 (1990) 1100-1131.

[Lu92]   J. H. Lutz, Almost everywhere high nonuniform complexity, J. Comput. System Sci. 44 (1992) 220-258.

[Lu93]   J. H. Lutz, The quantitative structure of exponential time, in: Proc. 8th Structure in Complexity Theory Conference, 1993, 158-175, IEEE Comput. Soc. Press.

[Lu94]   J. H. Lutz, Weakly hard problems, in: Proc. 9th Structure in Complexity Theory Conference, 1994, 146-161, IEEE Comput. Soc. Press.

[Ma82]   W. Maass, recursively enumerable generic sets, J. Symbolic Logic 47 (1982) 809-823.

[Ma94] E. Mayordomo, Contributions to the Study of Resource-bounded Measure, Doctoral thesis, Universitat Politecnica de Catalunya, Barcelona, 1994.

[Ma94a] E. Mayordomo, Almost every set in exponential time is P-bi-immune, Theor. Comput. Sci. 136 (1994) 487-506.

[Ma94b] E. Mayordomo, Measuring in PSPACE, in: Proc. IMYCS '92, Topics in Comput. Sci. 6 (1994) 93-100, Gordon and Breach.

[Od89] P. Odifreddi, Classical Recursion Theory, 1989, North-Holland.

[Ox80] J. C. Oxtoby, Measure and Category, 1980, Springer Verlag.

[Ro67] H. Rogers, Theory of Recursive Functions and Effective Computability, 1967, McGraw Hill.

[Sch82] U. Schöning, A uniform approach to obtain diagonal sets in complexity classes, Theor. Comp. Sci. 18 (1982) 95-103.

[So87] R. I. Soare, Recursively Enumerable Sets and Degress, 1987, Springer Verlag.

On Isolating r.e. and Isolated d-r.e. Degrees

Marat M. Arslanov[*]
Kazan University, Kazan, Russia

Steffen Lempp[†]
University of Wisconsin, Madison WI 53706-1388 USA

Richard A. Shore[§]
Cornell University, Ithaca NY 14853 USA

1. Introduction

The notion of a *recursively enumerable (r.e.) set*, i.e. a set of integers whose members can be effectively listed, is a fundamental one. Another way of approaching this definition is via an approximating function $\{A_s\}_{s\in\omega}$ to the set A in the following sense: We begin by guessing $x \notin A$ at stage 0 (i.e. $A_0(x) = 0$); when x later enters A at a stage $s+1$, we change our approximation from $A_s(x) = 0$ to $A_{s+1}(x) = 1$. Note that this approximation (for fixed) x may change at most once as s increases, namely when x enters A. An obvious variation on this definition is to allow more than one change: A set A is 2-r.e. (or d-r.e.) if for each x, $A_s(x)$ change at most twice as s increases. This is equivalent to requiring the set A to be the difference of two r.e. sets $A_1 - A_2$. (Similarly, one can define n-r.e. sets by allowing at most n changes for each x.)

The notion of d-r.e. and n-r.e. sets goes back to Putnam [1965] and Gold [1965] and was investigated (and generalized) by Ershov [1968a, b, 1970]. Cooper showed that even in the Turing degrees, the notions of r.e. and d-r.e. differ:

---

[*]Partially supported by Russia Foundation of Fundamental Investigations Grant 93-011-16004 and a Fulbright Fellowship held at Cornell University and the University of Wisconsin.

[†]Partially supported by NSF Grant DMS-9100114.

[§]Partially supported by NSF Grant DMS-9204308 and ARO through MSI, Cornell University, DAAL-03-C-0027.

Theorem 1.1. (Cooper [1971]) *There is a properly d-r.e. degree, i.e. a Turing degree containing a d-r.e. but no r.e. set.*

In the eighties, various structural differences between the r.e. and the d-r.e. degrees were exhibited by Arslanov [1985], Downey [1989], and others. The most striking difference is probably the following result which stands in contrast with the well-known Sacks Density Theorem for the r.e. degrees:

Theorem 1.2. (Cooper, Harrington, Lachlan, Lempp, Soare [1991]) *There is a maximal incomplete d-r.e. degree below $\mathbf{0}'$; thus the d-r.e. degrees are not densely ordered.*

The distribution of r.e. degrees within the structure of the d-r.e. degrees has also been investigated, starting with Lachlan's observation (unpublished) that any noncomputable d-r.e. degree bounds a noncomputable r.e. degree.

Cooper and Yi [1995] defined the notion of an *isolated d-r.e. degree* $\mathbf{d}$ as a Turing degree such that the r.e. degrees strictly below $\mathbf{d}$ contain a greatest r.e. degree $\mathbf{a}$, say. ($\mathbf{a}$ is then said to *isolate* $\mathbf{d}$.) They established the following results about this notion:

Theorem 1.3. (Cooper, Yi [1995]) *(i) There exists an isolated d-r.e. degree.*
  (ii) There exists a non-isolated properly d-r.e. degree.
  (iii) Given any r.e. degree $\mathbf{a}$ and d-r.e. degree $\mathbf{d} > \mathbf{a}$, there is a d-r.e. degree $\mathbf{e}$ between $\mathbf{a}$ and $\mathbf{d}$.

They raise the question of whether the phenomena in (i) and (ii) above occur densely relative to the r.e. degrees (i.e. whether we can find such degrees between any two comparable r.e. degrees), and whether every noncomputable incomplete r.e. degree isolates some d-r.e. degree. LaForte answered the first of these questions positively:

Theorem 1.4. (LaForte [1995]) *Given any two comparable r.e. degrees $\mathbf{v} < \mathbf{u}$, there exists an isolated d-r.e. degree $\mathbf{d}$ between them.*

(Ding and Qian [1995] independently obtained a partial answer to the above by showing that there is an isolated d-r.e. degree below any noncomputable r.e. degree.)

We answer the other two questions in the present paper:

Theorem 2.1. *Given any two comparable r.e. degrees $\mathbf{v} < \mathbf{u}$, there exists a non-isolated d-r.e. degree $\mathbf{d}$ between them.*

Before stating the answer to the last question, we state the following proposition, connecting d-r.e. and REA in $\mathbf{a}$ degrees:

Proposition 3.1. *If* $\mathbf{d} > \mathbf{a}$ *is d-r.e. (or n-r.e. for any $n \in \omega$) then there is a* $\mathbf{c} \leq \mathbf{d}$ *which is r.e. in* $\mathbf{a}$ *and strictly above* $\mathbf{a}$. *So, in particular, if* $\mathbf{a}$ *isolates* $\mathbf{d}$ *then* $\mathbf{a}$ *isolates* $\mathbf{c}$.

The following is then a negative answer to the last question of Cooper and Yi mentioned above:

Theorem 3.2. *There is a noncomputable r.e. degree* $\mathbf{a}$ *which isolates no degree REA in it.*

We extend this result by showing that the non-isolating degrees are downward dense in the r.e. degrees and that they occur in any jump class:

Theorem 3.7. *For every noncomputable r.e. degree* $\mathbf{c}$, *there is a noncomputable r.e. degree* $\mathbf{a} \leq \mathbf{c}$ *which isolates no degree REA in it.*

Theorem 3.8. *If* $\mathbf{c}$ *is REA in* $\mathbf{0}'$ *then there is a noncomputable r.e. degree* $\mathbf{a}$ *with* $\mathbf{a}' = \mathbf{c}$ *which isolates no degree REA in it.*

We close with another result relating the d-r.e. degrees to the notion of relative enumerability.

Theorem 4.2. *Given r.e. degrees* $\mathbf{v} < \mathbf{u}$, *there is a d-r.e. degree* $\mathbf{d}$ *between them which is not r.e. in* $\mathbf{v}$.

We generally follow the notation of Soare [1987]. Familiarity with the proof of the weak density result of Cooper, Lempp, Watson [1989] is frequently assumed throughout the paper.

2. Non-isolated d-r.e. degrees

In this section we show that between any two r.e. degrees there is a properly d-r.e. degree which is not isolated by any r.e. degree. The proof of this theorem uses an infinite injury argument and is essentially the same as in Cooper, Lempp, Watson [1989] where, given r.e. sets $U >_T V$, a d-r.e. set C of properly d-r.e. degree such that $U >_T C >_T V$ is constructed.

Theorem 2.1. *Given r.e. sets* $U >_T V$ *there is a d-r.e. set* C *of properly d-r.e. degree such that* $U >_T C >_T V$, *and, for any r.e. set* B, *if* $B <_T C$ *then* $B <_T W <_T C$ *for some r.e. set* W.

Proof. We construct r.e. sets $A_1, A_2 \leq_T U$. If $A = A_1 - A_2$ then $C = V \oplus A$ will be the desired set. To ensure that $V \oplus A$ is not of r.e. degree we satisfy for every e the requirement

$$R_e : A \neq \Theta_e^{W_e} \quad \vee \quad W_e \neq \Phi_e^{V \oplus A}.$$

To ensure that the degree of $V \oplus A$ is not isolated we satisfy for every e the requirement

$$S_e : W_e = \Psi_e^{V \oplus A} \Rightarrow (\exists \text{ r.e. } U_e \leq_T V \oplus A)(\forall i)(U_e \neq \Omega_i^{W_e}).$$

Here $\{(W_e, \Theta_e, \Phi_e, \Psi_e, \Omega_e)\}_{e \in \omega}$ is some enumeration of all possible five-tuples of r.e. sets W and partial recursive functionals Θ, Φ, Ψ and Ω.

Since we handle the requirements $\{R_e\}_{e \in \omega}$ in the same way as in Cooper, Lempp, Watson [1989] we will consider here only the requirements $\{S_e\}_{e \in \omega}$. In satisfying S_e we shall construct a r.e. set U_e with the intention that if $W_e = \Psi_e^{V \oplus A}$ then $U_e \neq \Omega_i^{W_e}$ for all i and $U_e \leq_T V \oplus A$ through a modified permitting argument. We break S_e up into subrequirements $S_{e,i}$:

$$S_{e,i} : W_e = \Psi_e^{V \oplus A} \Rightarrow U_e \neq \Omega_i^{W_e}.$$

Basic module. Let us first consider requirements $S_{e,i}$ without the claim that $A \leq_T U$ and in the absence of any V-changes. (This is just the proof that there is a non-isolated d.r.e. degree.) The strategy proceeds as follows:

(1) Choose an unused candidate x for $S_{e,i}$ greater than any number mentioned in the construction thus far.

(2) Wait for a stage s such that

$$\Omega_{i,s}^{W_{e,s}}(x) \downarrow = 0,$$

and for some least u such that

$$W_{e,s} \upharpoonright \omega_{i,s}(x) = \Psi_{e,s}^{(V \oplus A)_s \upharpoonright u} \upharpoonright \omega_{i,s}(x).$$

(If this never happens then x is a witness to the success of $S_{e,i}$).

(3) Protect $A \upharpoonright u$ from other strategies from now on.

(4) Put x into U_e and A.

(5) Wait for a stage s' such that

$$\Omega_{i,s'}^{W_{e,s'}}(x) \downarrow = 1,$$

and for some least u' such that

$$W_{e,s'} \upharpoonright \omega_{i,s'}(x) = \Psi_{e,s'}^{(V \oplus A)_{s'} \upharpoonright u'} \upharpoonright \omega_{i,s'}(x).$$

(If this never happens then again x is a witness to the success of $S_{e,i}$. If it does happen then the change in $\Omega_i^{W_e}(x)$ between stages s and s' can only be brought about by a change in $W_e \upharpoonright \omega_{i,s}(x)$, which is irreversible since W_e is a r.e. set.)

(6) Remove x from A and protect $A \upharpoonright u'$ from other strategies from now on.

(Now x is a permanent witness to the success of $S_{e,i}$ because

$$\Psi_e^{V \oplus A} \upharpoonright \omega_{i,s}(x) = \Psi_{e,s}^{(V \oplus A)_s} \upharpoonright \omega_{i,s}(x) = W_{e,s} \upharpoonright \omega_{i,s}(x) \neq W_e \upharpoonright \omega_{i,s}(x).)$$

We see that the $S_{e,i}$-strategy in isolation and without the claims $A \leq_T U$ and $V \leq_T A$ is essentially the same as the R_e-strategy under similar assumptions. (Note that since we have refuted the overall hypothesis of S_e we no longer need to maintain the reduction $U_e \leq_T A$.) It allows us to meet all requirements $\{S_{e,i}\}_{e,i \in \omega}$ and $\{R_e\}_{e \in \omega}$ together in the same way as in the similar theorem from Cooper, Lempp, Watson [1989].

As in Cooper, Lempp, Watson [1989], we handle the condition $V \leq_T A$ by imposing "indirect" restraints to protect V, threatening $U \leq_T V$ via a functional Γ. We make infinitely many attempts to satisfy $S_{e,i}$ as above by an ω-sequence of "cycles", each cycle k proceeding as above with its own witness and with the following step inserted after step 3:

$(3\frac{1}{2})$ Set $\Gamma_e^V(k) = U_s(k)$ with *use* $\gamma(k) = u$, start cycle $k+1$ simultaneously, wait for $U(k)$ to change, then stop cycles $k' > k$ and proceed.

Finally, we ensure that $A \leq_T U$ through a permitting argument. So x has to be permitted to enter A by U at step (4) and to leave A at step (6). The former permission is already given by the $U(k)$-change, the latter we build into the strategy as in Cooper, Lempp, Watson [1989].

Now the basic module for the $S_{e,i}$-strategy repeats the module for the R_e-strategy from Cooper, Lempp, Watson [1989]. It consists of an $(\omega \times \omega)$-sequence of cycles $(j, k), j, k \in \omega$. Cycle $(0, 0)$ starts first, and each cycle (j, k) can start cycles $(j, k+1)$ or $(j+1, 0)$ and stop, or cancel, cycles $(j', k') > (j, k)$ (in the lexicographical ordering). Each cycle (j, k) can define $\Gamma_j^V(k)$ and $\Delta^V(j)$.

A cycle (j, k) now proceeds as follows:

(1) Choose an unused candidate x such that $x - 1$ is greater than any number mentioned thus far in the construction.

(2) Wait for a stage s_1 such that

$$\Omega_{i,s_1}^{W_{e,s_1}}(x) \downarrow = 0,$$

and for some least u such that

$$W_{e,s_1} \restriction \omega_{i,s_1}(x) = \Psi_{e,s_1}^{(V \oplus A)_{s_1} \restriction u} \restriction \omega_{i,s_1}(x).$$

(3) Protect $A \restriction u$ from other strategies from now on.

(4) Set $\Gamma_j^V(k) = U_{s_1}(k)$ with *use* $\gamma_j(k) = u$, and start cycle $(j, k+1)$ to run simultaneously.

(5) Wait for $V \restriction u$ or $U(k)$ to change.

      If $V \restriction u$ changes first then cancel cycles $(j', k') > (j, k)$, drop the A-protection of cycle (j, k) to 0, and go back to step (2).

      If $U(k)$ changes first then stop cycles $(j', k') > (j, k)$ and proceed to step (6).

(6) Put x into A and U_e.

(7) Wait for a stage s_2 such that

$$\Omega_{i,s_2}^{W_{e,s_2}}(x) \downarrow = 1,$$

and for some least u' such that

$$W_{e,s_2} \restriction \omega_{i,s_2}(x) = \Psi_{e,s_2}^{(V \oplus A)_{s_2} \restriction u'} \restriction \omega_{i,s_2}(x).$$

(8) Protect $A \restriction u'$ from other strategies from now on.

(9) Set $\Delta^V(j) = U_{s_2}(j)$ with *use* $\delta(j) = u'$ and start cycle $(j+1, 0)$ simultaneously.

(10) Wait for $V \restriction u'$ or $U(j)$ to change.

      If $V \restriction u'$ changes first then cancel cycles $(j', k') \geq (j+1, 0)$, drop the A-protection of cycle (j, k) to u, and go back to step (7).

      If $U(j)$ changes first then stop cycles $(j', k') \geq (j+1, 0)$ and proceed to step (11).

(11) Remove x from A.

(12) Wait for $V \restriction u \neq V_{s_1} \restriction u$.

(13) Reset $\Gamma_j^V(k) = U(k)$, put $x + 1$ into A, cancel cycles $(j', k') > (j, k)$, start cycle $(j, k+1)$, and halt cycle (j, k).

Whenever a cycle (j, k) is started, any previous version of it has been cancelled and its functionals have become undefined through V-changes and, therefore, Γ_j and Δ are defined consistently.

The basic module has four possible outcomes similar to those of the basic module of the R_e-strategy.

(A) There is a stage s after which no cycle acts. Then some cycle (j_0, k_0) eventually waits at step 2, 7 or 12 forever. Thus we win requirement $S_{e,i}$ through the cycle (j_0, k_0).

(B) Some cycle (j_0, k_0) acts infinitely often but no cycle $< (j_0, k_0)$ does so. Then it goes from step 5 to step 2, or from step 10 to step 7, infinitely often. Thus Ψ_e or Ω_e is partial. Notice that the overall restraint of all cycles has finite liminf.

(C) There is a (least) j_0 such that every cycle $(j_0, k), k \in \omega$, eventually waits at step 5 or 13 forever. ("Row j_0 acts infinitely".) This means that $U \leq_T V$ via Γ_{j_0} contrary to hypothesis.

(D) For every j there is a cycle (j, k_j) that eventually waits at step 10 forever. ("Every row acts finitely"). This means that $U \leq_T V$ via Δ contrary to hypothesis.

The verification now proceeds as in Cooper, Lempp, Watson [1989], and we leave the details to the reader, except for the following item: When we remove x from A, we also lose the U_e-permission for x (which must, of course, remain in U_e). But note that the win on S_e is global (and so U_e is no longer needed) unless $V \restriction u$ changes later. In that case, however, $x+1$ is enumerated into A, and so $V \oplus A$ can recognize this. $\square$

3. Nonisolating r.e. degrees

Following Cooper and Yi [1995] we say that a r.e. degree $\mathbf{a}$ *isolates* the degree $\mathbf{d} > \mathbf{a}$ if, for every r.e. $\mathbf{b} \leq \mathbf{d}$, we have $\mathbf{b} \leq \mathbf{a}$. Cooper and Yi ask (**Q 4.3**) if every r.e. degree $\mathbf{a}$ isolates some d-r.e. degree $\mathbf{d}$. In this section we supply a strong negative answer to this question. Our basic construction shows (Theorem 3.2) that there is a noncomputable r.e. degree $\mathbf{a}$ which does not isolate any $\mathbf{d}$ which is REA in it. The answer to **Q 4.3** then follows immediately from the next proposition.

Proposition 3.1. *If* $\mathbf{d} > \mathbf{a}$ *is d-r.e. (or n-r. e. for any $n \in \omega$) and $\mathbf{a}$ is r.e. then there is a degree $\mathbf{c} \leq \mathbf{d}$ which is r.e. in $\mathbf{a}$ and strictly above it. So, in particular, if $\mathbf{a}$ isolates $\mathbf{d}$ then $\mathbf{a}$ isolates $\mathbf{c}$.*

Proof. By Jockusch and Shore [1984], **d** is 2-REA, i.e. there is a r.e. degree **e** such that **d** is REA in **e**. Now if **e** $\leq$ **a** then **d** itself is REA in **a** and so the degree **c** required in the Proposition. If not, then **a** < **e** $\vee$ **a** $\leq$ **d** and so **e** $\vee$ **a** is the degree **c** required by the Proposition. (Essentially the same argument now works for **d** n-r.e. by induction on n.) The assertion about **a** isolating **c** follows by definition.

We also supply two variations on this basic construction that show that the degrees **a** not isolating any **d** which is REA in **a** are widely distributed in the r.e. degrees. Theorem 3.7 shows that such degrees exist below every nonrecursive r.e. **c** and Theorem 3.8 shows that they exist in every jump class, i.e. for every **c** REA in **0**′ there is such a r.e. degree **a** with **a**′ = **c**.

We begin with the basic construction.

Theorem 3.2. *There is a nonrecursive r.e. set A such that its degree **a** isolates no degree REA in it, i. e.* $\forall e(A <_T W_e^A \rightarrow \exists B(B$ *is r.e.* & $B \leq_T W_e^A$ & $B \not\leq_T A))$.

There are two types of basic requirements:

$P_e : \Phi_e \neq A$ (for each partial recursive function Φ_e).

$N_e : A <_T W_e^A \rightarrow B_e \leq W_e^A$ & $B_e \not\leq A$ (for each e we construct an appropriate r.e. set B_e).

The requirements N_e are divided up into subrequirements:

$N_{e,i} : \Phi_i^A \neq B_e$ (for each partial recursive functional Φ_i).

We order the requirements $P_e, N_{e,i}$ in an ω type list $\langle R_n \rangle$. The procedures for satisfying the individual requirements are fairly standard. We will diagonalize against Φ_e by putting some witness x into A at a stage s to satisfy P_e when $\Phi_e(x) = 0[s]$. For $N_{e,i}$ we will wait until some $x \in \omega^{[\langle e,i \rangle]}$ with $\Phi_i(A; x) = 0[s]$ is permitted by W_e^A at s and then put x into B_e. To implement the permitting we first approximate W_e^A in the usual way: $x \in W_e^A[s]$ iff $\Phi_e(A; x) \downarrow [s]$. We then say that x is *permitted by* W_e^A at s if it looks as if some $y < x$ is in W_e^A at s but it does not, at s, look as if it was in at $s - 1$ by an A-correct computation, i.e.

$$\exists y < x\{y \in W_e^A[s] \,\&\, (y \notin W_e^A[s-1] \lor \exists z < \varphi_e(y, s-1)[z \in A_s - A_{s-1}])\}.$$

The restraint necessary to preserve the A-use relevant to this computation will be imposed automatically by our procedure for choosing potential witnesses for the P_e. We now present the formal construction and verifications.

Construction:

At stage s we find the first requirement R_n in our list such that one of the following two cases holds:

1) $R_n = P_e$; there is no z such that $\Phi_e(z) = 0\,[s]$ and $z \in A$; $\Phi_e(x) = 0\,[s]$ for the least $x \in \omega^{[e]}$ which is larger than any stage at which we have acted for any requirement of higher priority than P_e. We call this x the *current potential witness for P_e*.

2) $R_n = N_{e,i}$; there is no z such that $\Phi_i(A; z) = 0\,[s]$ and $z \in B_e$; there is a least $x \in \omega^{[\langle e, i \rangle]}$ larger than any stage at which we have acted for any requirement of higher priority than $N_{e,i}$ such that $\Phi_i(A; x) = 0\,[s]$ and larger than any current potential witness for any higher priority P-requirement; and x is permitted by W_e^A at s.

If there is no such n, we go on to stage $s + 1$. Otherwise, we now act for requirement R_n according to which of the above two cases applies:

1) If $R_n = P_e$ then we put x into A.
2) If $R_n = N_{e,i}$ we put x into B_e.

Verifications:

Lemma 3.3. *We act for each requirement only finitely often.*

Proof. We proceed by induction through the priority ordering. Suppose we never act for any R_m with $m < n$ after stage s. If $R_n = P_e$ and we act for this requirement at $t > s$ by putting x into A then it is clear that we never act for it again as $\Phi_e(x) = 0\,[t]$ and $x \in A\,[t + 1]$. If $R_n = N_{e,i}$ and we act for this requirement at $t > s$ by putting x into B_e, we never put any number less than t into A at any later stage since no P_j of lower priority can do so by construction and none of higher priority can act by our choice of s. Thus, by the usual conventions that the Φ_i use at t is at most t, no number less than the use $\varphi_i(x, t)$ can ever enter A after t. In particular, $\Phi_i(A; x)\,[t] = \Phi_i(A; x)\,[t'] = 0$ and $x \in B_e$ for every $t' > t$. So we never act again for $N_{e,i}$. $\square$

Lemma 3.4. *Each requirement P_e is satisfied, i.e. $\Phi_e \neq A$.*

Proof. Let s be the last stage at which we act for a requirement of higher priority than P_e and let x be the least element of $\omega^{[e]}$ larger than s. If we ever act for P_e after s we have $\Phi_e(x) = 0$ and we put x into A to satisfy P_e. If we never act for P_e after s, then either there is some other z such that $\Phi_e(z) = 0$ and $z \in A$ or $x \notin A$ and $\neg(\Phi_e(x) = 0)$. In either case $\Phi_e \neq A$ as required. $\square$

Lemma 3.5. *If $A <_T W_e^A$ then we satisfy each requirement $N_{e,i}$, i.e. $\Phi_i^A \neq B_e$ for each i.*

Proof. Consider any requirement $N_{e,i}$ and let s be a stage after which we never act for any requirement of higher priority than $N_{e,i}$. If we ever act for $N_{e,i}$ at a stage $t > s$ by putting some x into B_e then the argument for Lemma 3.3 shows that $\Phi_i(A; x) = 0$ and so $\Phi_i^A \neq B_e$ as required. If we never act for $N_{e,i}$ after stage s then $x \notin B_e$ for each $x \in \omega^{[\langle e,i \rangle]}$ which is larger than some fixed $s' > s$. Unless $\Phi_i(A; x) = 0$ for each such x, we have also shown that $\Phi_i^A \neq B_e$.

If neither of these situations satisfying $N_{e,i}$ occurs, we show that $W_e^A \leq_T A$ for a contradiction. To compute $W_e^A(x)$ for $x > s'$, then find a $z \in \omega^{[e]}$ such that $z > x$ and a stage $t > z$ such that $\Phi_i(A; z) = 0\,[t]$ by an A-correct computation, i. e. $A \restriction \varphi_i(z, t) = A_t \restriction \varphi_i(z, t)$. We claim that $x \notin W_e^A$ unless $x \in W_e^A[t]$ by an A-correct computation, i. e. $\Phi_e(A; x) \downarrow [t]$ and $A \restriction \varphi_e(x, t) = A_t \restriction \varphi_e(x, t)$. Of course, if $x \in W_e^A[t]$ by an A-correct computation, then $x \in W_e^A$. On the other hand, if $x \in W_e^A$ but not by an A-correct computation at t, then there must be a $v > t$ (the first stage at which we have the A-correct computation of $\Phi_e(A; x)$) at which W_e^A would permit z and so we would act for $N_{e,i}$ at v by putting z into B_e contrary to our assumption. $\square$

Lemma 3.6. $B_e \leq_T W_e^A \oplus A$.

Proof. To determine if $x \in \omega^{[\langle e,i \rangle]}$ is in B_e, wait until a stage s such that $A \restriction x = A_s \restriction x$ and, for every $y < x$ such that $y \in W_e^A$, $y \in W_e^A[s]$ by an A-correct computation. We claim that if $x \notin B_{e,s}$ then $x \notin B_e$. The only way x can enter B_e at some $t > s$ is by our acting for $N_{e,i}$ at t and so, in particular, by W_e^A permitting x at t. Thus some $y < x$ is in $W_e^A[t]$ that was not previously (and so not at s) in W_e^A by an A-correct computation. By construction, no requirement of lower priority than $N_{e,i}$ can injure the computation of $\Phi_e(A; y)[t]$. On the other hand, the current potential witnesses for P_j of higher priority than $N_{e,i}$ must all be less than x by construction.

Thus none of them can enter A by our choice of s. If any of these potential witnesses changes at a later stage $v > t$ (because of some action by a yet higher priority $N_{k,l}$ requirement), it must change to a number grater than $v > t > \varphi_e(y, t)$ and so also cannot injure the computation putting y into W_e^A. Thus $y \in W_e^A$ but is not in W_e^A by an A-correct computation at s for the desired contradiction. $\square$

We may combine this last construction with r.e. permitting to construct the desired A below any given nonrecursive r.e. set C.

Theorem 3.7. *For every nonrecursive r.e. set C there is a nonrecursive r.e. set $A <_T C$ such that $\forall e(A <_T W_e^A \rightarrow \exists B(B \text{ is r.e. } \& \ B \leq_T W_e^A \& \ B \not\leq_T A))$.*

Proof. We adjust the previous construction by possibly appointing many current potential witnesses for each requirement P_e. More specifically, if there is no z such that $\Phi_e(z) = 0\,[s]$ and $z \in A$ and $\Phi_e(y) = 0\,[s]$ for every current potential witness for P_e at s, then we act for P_e by appointing as a potential witness the least $x \in \omega^{[e]}$ which is larger than any stage at which we have acted for any requirement of higher priority than P_e and larger than every current potential witness. We cancel this potential witness at any later stage at which we act for some requirement of higher priority than P_e. If there is now a potential witness x with $\Phi_e(x) = 0$ which is permitted by C (i.e. some $y < x$ enters C at s) then we act for P_e by putting x into A. Otherwise, the construction is the same as before. The verifications now follow the usual pattern of a permitting argument. Assuming we never act for any requirement of higher priority than P_e after stage s, we use the nonrecursiveness of C to show that we act only finitely often for P_e and eventually satisfy it. (If we act infinitely often without putting a number into A (necessarily by appointing more and more potential witnesses) then we calculate C by noting that once $\Phi_e(x) = 0$ at a stage $t > s$ for some potential witness x, we can never later have a number $y < x$ enumerated in C.) The other verifications now proceed as before. $\square$

Finally we show that there is a nonrecursive r.e. set A in every jump class which does not isolate any set D which is REA in A and so (by Proposition 3.1) not any d-r.e. degree above it either.

Theorem 3.8. *If C is REA in $\emptyset'$ then there is a nonrecursive r.e. set A such that $A' \equiv_T C$ and $\forall e(A <_T W_e^A \rightarrow \exists B(B$ is r.e. & $B \leq_T W_e^A$ & $B \not\leq_T A))$.*

We follow the usual proof of the Sacks jump theorem by starting with an r.e. D such that $x \in C$ implies $D^{[x]} = \{y | y < n\}$ for some n and $x \notin C$ implies $D^{[x]} = \omega$. Moreover, for technical convenience we assume that if a number z is enumerated in $D^{[e]}$ at stage s then $z < s$ and every $x < s$ which is in $\omega^{[e]}$ and not already in $D^{[e]}$ is enumerated in $D^{[e]}$ at s. We will make A a thick subset of D, i. e. for every x, $A^{[x]} \subseteq D^{[x]}$ and $D^{[x]} - A^{[x]}$ is finite. As usual this guarantees that $C \leq_T A'$. We now use a typical tree construction to satisfy the following requirements:

$P_e : A^{[e]} \subseteq D^{[e]}$ and $D^{[e]} - A^{[e]}$ is finite.

$N_{e,i} : \Phi_i^A \neq B_e$ (for each e we construct a r.e. set B_e satisfying $N_{e,i}$ for each partial recursive functional Φ_i if $A <_T W_e^A$).

Our priority tree is constructed as usual given that we assign nodes on level $2e$ to P_e and their possible outcomes are, in left to right order, $i < 0 < 1 < \ldots < n < \ldots$ while ones on level $2\langle e, i \rangle + 1$ area assigned to $N_{e,i}$ and their possible outcomes are $w < 0 < 1 < \ldots < s < \ldots$. (The intended meaning of the outcomes for P_e are $i : D^{[e]} = \omega$ and $n : n$ is the last stage at which a number is enumerated in $D^{[e]}$ (and so the first number not in $D^{[e]}$). The intended meaning of the outcomes for $N_{e,i}$ are w : we are waiting for a chance to diagonalize and $s \in \omega$: we succeed in diagonalizing by putting some x into B_e for which $\Phi_i(A; x) = 0$ at stage s. The nodes α assigned to requirements $N_{e,i}$ may impose *restraint* $r(\alpha, s)$ at stage s. We define $R(\alpha, s)$ the *restraint* imposed at s on a requirement α assigned to a requirement P_e as $\max\{r(\beta, s) | \beta < \alpha\}$.

Construction:

At each stage s we define a sequence of length s of *accessible* nodes and act accordingly. We begin with $\emptyset$, the root of our priority tree, as the first accessible node at each stage s. Suppose a node α of length less than s has just been declared accessible. If P_e is assigned to α then we see if there has been a number enumerated in $D^{[e]}$ since the last stage at which α was accessible (since stage 0 if this is the first stage at which α is accessible). If so, then the outcome of P_e is i; we declare $\alpha \hat{~} i$ to be accessible; and we put every $x > R(\alpha, s)$ which is in $D_s^{[e]}$ into A. If not, we declare $\alpha \hat{~} n$ to be accessible where n is the last stage at which some number was enumerated in $D^{[e]}$. If α is assigned to $N_{e,i}$ and we have acted for α at some previous stage t, then the outcome of α is t and we declare $\alpha \hat{~} t$ to be accessible. Otherwise, the outcome of α is w and $\alpha \hat{~} w$ is accessible. When we reach a node β of length s we see if there is any node α for which we have not yet acted which has previously been accessible (but is not necessarily accessible now) and is assigned to a requirement $N_{e,i}$ and any $x \in \omega^{[\alpha]}$ satisfying the following conditions:

(1) x is larger than the first stage $u' > u$ at which α was accessible where u is the last stage at which any $\beta <_L \alpha$ has been accessible.

(2) x is smaller than the last stage at which α was accessible.

(3) x is permitted by W_e^A at s.

(4) $\Phi_i(A; x) = 0\,[s]$ via an α-believable computation. (We say that the computation $\Phi_i(A; x) = 0$ is *α-believable* at s if $\forall z \forall k (\, z \in \omega^{[k]} \ \& \ \alpha(k) = i \ \& \ z > R(\alpha \restriction k, s) \ \& \ z < \varphi_i(x, s) \ \rightarrow \ z \in A_s)$.)

If there is such an α, we act for the highest priority one by putting the smallest such x into B_e and set $r(\alpha, s) = s$. If not, we go on to stage $s + 1$.

Verifications:

As each node that is accessible infinitely often clearly has a leftmost immediate successor which is accessible infinitely often, there is a path TP in the priority tree consisting of the leftmost nodes which are accessible infinitely often.

Lemma 3.9. *If $\alpha \in TP$ and α is first accessible at s then no $\beta <_L \alpha$ is ever accessible at $t > s$. Moreover, there is a stage $t \geq s$ after which we never act for any $\beta < \alpha$ assigned to a requirement $N_{e,i}$. Thus if α is assigned to some P_e then $R(\alpha, v)$ is constant for $v \geq t$.*

Proof. Proceeding by induction along TP the first claim is obvious from the definition of when the various outcomes of each node are accessible. As we can act at most once for each β assigned to a requirement $N_{e,i}$, the other assertions are also immediate. $\square$

Lemma 3.10. *Suppose $\alpha \in TP$ is assigned to requirement P_e and s is the first stage at which α is accessible and $t \geq s$ is the first stage after which we never act for any $\beta < \alpha$ assigned to a requirement $N_{e,i}$ (such a stage exists by Lemma 3.9). If $D^{[e]} = \omega$ then $\alpha\hat{\ }i \in TP$ and for all $x \in \omega^{[e]}$, $x \in A^{[e]}$ iff $x > R(\alpha, t) \vee x \in D_t^{[e]}$. Otherwise, $D^{[e]}$ is finite and if n is the last stage at which some number is enumerated in $D^{[e]}$ then $\alpha\hat{\ }n \in TP$ and no number is put into $A^{[e]}$ after the first stage at which $\alpha\hat{\ }n$ is accessible.*

Proof. Suppose $D^{[e]} = \omega$. It is immediate from the definition of the accessible successor of α that $\alpha\hat{\ }i \in TP$. Now, $R(\alpha, t) = R(\alpha, v)$ for every $v > t$ by our choice of s and Lemma 3.9. Thus if $x < R(\alpha, t)$ and $x \notin D_t^{[e]}$ then $x \notin A^{[e]}$ by construction. On the other hand, if $x > R(\alpha, t)$ then there is a stage $v > t$ after x has entered $D^{[e]}$ at which $\alpha\hat{\ }i$ is accessible. By construction, we put x into $A^{[e]}$ at v.

If $D^{[e]}$ is finite and n is the last stage at which a number is enumerated in $D^{[e]}$ then it is clear from the definition of the accessible successor of α that $\alpha\hat{\ }n \in TP$ and from the construction and Lemma 3.9 that no number is put into $A^{[e]}$ after the first stage at which $\alpha\hat{\ }n$ is accessible. $\square$

Lemma 3.11. *If $\alpha \in TP$ is assigned to $N_{e,i}$ and we act for α at s by putting x into B_e, then $\Phi_i(A; x) = 0$.*

Proof. By construction, α has been accessible before stage s or we could not act for it. Thus by Lemma 3.9 no node to the left of α can ever be accessible again. In particular, no action for a node $\beta <_L \alpha$ can put any number into A after stage s. No node of lower priority can put any number less than $\varphi_i(x, s) < s$ into A after stage s as we set $r(\alpha, s) = s$ and never change it. Finally, we claim no node $\beta \subset \alpha$ will ever put a number less than $\varphi_i(x, s)$ into A after stage s. If $\beta \subset \alpha$ and $\beta\hat{\ }n \subseteq \alpha$ for some n, then β puts no numbers at all into A after s by Lemma 3.10. On the other hand, if $\beta\hat{\ }i \subseteq \alpha$, then note that $R(\beta, t)$ is nondecreasing in t and so we will not put in any number less than $R(\beta, s)$ for β after s while all others that it might ever put into

A less than $\varphi_i(x, s)$ are already in A by the definition of $\Phi_i(A; x)[s]$ being α-believable. $\square$

Lemma 3.12. *We satisfy* N_e, *i.e. if* $A <_T W_e^A$ *then* $\Phi_i^A \neq B_e$ *for each* i.

Proof. Suppose $A <_T W_e^A$ and consider the node $\alpha \in TP$ assigned to $N_{e,i}$. If we ever act for α by putting some $x \in \omega^{[\alpha]}$ into B_e then, by Lemma 3.11, $\Phi_i(A; x) = 0 \neq B_e(x)$ as required. If we never act for α then $B_e \cap \omega^{[\alpha]} = \emptyset$ by construction. In this case, if, contrary to the conclusion of our Lemma, $\Phi_i^A = B_e$ then $\Phi_i(A; x) = 0$ for every sufficiently large $x \in \omega^{[\alpha]}$. Let s be the first stage at which α is accessible. We now argue exactly as in Lemma 3.5 with α replacing e that $W_e^A \leq_T A$ for the desired contradiction. $\square$

Lemma 3.13. *For every* e, $B_e \leq_T W_e^A \oplus A$.

Proof. To determine if $x \in \omega^{[\alpha]}$ with α assigned to $N_{e,i}$ is in B_e assume we have already calculated $B_e \upharpoonright x$ and all numbers in $B_e \upharpoonright x$ have already been enumerated in B_e by stage $u > x$. Now choose a w such that for every e in the domain of α there is a $z \in \omega^{[e]}$ with $w > z > u$ and wait until a stage $s > w$ such that $A \upharpoonright w = A_s \upharpoonright w$ and, for every $y < x$ such that $y \in W_e^A$, $y \in W_e^A[s]$ by an A-correct computation. We claim that if $x \notin B_{e,s}$ then $x \notin B_e$. First, note that if any node $\beta <_L \alpha$ has been accessible since stage x then x cannot later enter B_e by condition (1) on our choice of x. Moreover, until such a β becomes accessible, no number greater than x can be put into any B_j for any $\beta <_L \alpha$ by condition (2) on our choice of x. Thus, by our choice of u, the restraints imposed by such β remain constant after stage u and are less than u until some such β becomes accessible. Now, the only way x can enter B_e at some $t > s$ is by our acting for α at t and so, in particular, by W_e^A permitting x at t. Thus some $y < x$ is in $W_e^A[t]$ that was not previously (and so not at s) in W_e^A by an A-correct computation. By construction no requirement of lower priority than α can injure the computation of $\Phi_e(A; y)[t]$ after t. On the other hand, no action for a node $\beta \subset \alpha$ or $\beta <_L \alpha$ can injure the computation without our moving to the left of α or already having first moved to its left. Suppose then that we move to the left of α at some $v > t$. This can happen only when some $\alpha \upharpoonright e$ is accessible at v and some has been enumerated in $D^{[e]}$ since stage n where $\alpha(e) = n \in \omega$. When this happens, we must enumerate all numbers in $\omega^{[e]}$ which are less than v into $A^{[e]}$ unless they are below $R(\alpha \upharpoonright e, v)$. Our previous remarks, however, show that $R(\alpha \upharpoonright e, v) < u$ and so some number $z \in \omega^{[e]}$ with $w > z > u$ is enumerated into A at v contradicting our choice of s. Thus $\varphi_e(y, t)$ would never be injured contradicting our choice of s once again. $\square$

Lemma 3.14. $A' \equiv_T C$.

Proof. By Lemma 3.10, A is a thick subset of D and so $C \leq_T A'$. We claim that $TP \leq_T C$ and that $A' \leq_T TP$. We first recursively calculate TP from C. Suppose we have $\alpha \in TP$ and want to find the immediate successor of α on TP. If α is assigned to some $N_{e,i}$ then $\alpha\hat{}w \in TP$ unless there is a stage s at which we act for α. In this case, $\alpha\hat{}s \in TP$. Of course, $\emptyset'$ can tell if there is such a stage and $\emptyset' \leq_T C$. If α is assigned to some P_e then $\alpha\hat{}i \in TP$ if $D^{[e]}$ is infinite and otherwise $\alpha\hat{}n \in TP$ where n is the last stage at which a number is enumerated in $D^{[e]}$. As $D^{[e]}$ is infinite if and only if $e \notin C$, C can tell which case applies and so (using $\emptyset'$ again in the second case) find the correct immediate successor of α on TP.

Now, we calculate A' from TP. We begin with a fixed e such that $A <_T W_e^A$. To determine if $j \in A'$ find an i such that, for every z, $\Phi_i(A; z) = 0$ iff $\Phi_j(A; j) \downarrow$. It is now clear from the proof of Lemma 3.12 that $j \in A'$ iff $\alpha\hat{}w \notin TP$ for the node $\alpha \in TP$ assigned to $N_{e,i}$. $\square$

Proof (of Theorem 3.8). If $C \equiv_T \emptyset'$ then Theorem 3.7 provides the required A. Otherwise, our last construction supplies the desired set by Lemmas 3.12, 3.13 and 3.14. $\square$

4. D-r.e. degrees and REA degrees

Theorem 4.1. *Let* $\mathbf{v}$ *be a r.e. degree such that* $\mathbf{0}' > \mathbf{v}$. *Then there is a d-r.e. degree* $\mathbf{d} > \mathbf{v}$ *which is not r.e. in* $\mathbf{v}$.

Proof. Let $K \in \mathbf{0}'$ and $V \in \mathbf{v}$ be fixed r.e. sets. We will construct a d-r.e. set A so that $D = V \oplus A$ does not have degree r.e. in $\mathbf{v}$.

To satisfy the last property we meet the following requirements for all e,

$$R_e : A \neq \Phi_e^{W_e^V} \vee W_e^V \neq \Psi_e^{V \oplus A},$$

where $\{(W_e^V, \Phi_e, \Psi_e)\}_{e \in \omega}$ is some enumeration of all possible triples consisting of sets W^V r.e. in V and partial recursive functionals Φ and Ψ.

We use a common convention (see, for example, Rogers [1967]) that W_e^V enumerates an element x by listing in W_e a quadruple $\langle x, 1, u, v \rangle$ with $D_u \subseteq V$ and $D_v \subseteq \bar{V}$.

Obviously, if for some finite set $X \subset \omega$, $X \subseteq W_e^V$ then there is a stage s such that for all $t \geq s$ we have $X \subseteq W_{e,t}^{V_t}$. (Note here that we denote by $X \subseteq W_e^V$ that X is a *subset* and *not* necessarily a substring.) Besides, if for some s and a (least) number θ, $X \subseteq W_{e,s}^{V_s}$, and

$$\forall x(x \in X \rightarrow \langle x, 1, u, v \rangle \in W_{e,s} \wedge (D_u \subset V_s \upharpoonright \theta \wedge D_v \subset \bar{V}_s \upharpoonright \theta)),$$

then $X \not\subseteq W_e^V$ implies $V \restriction \theta \neq V_s \restriction \theta$. We call θ the X-use for $W_{e,s}^{V_s}$.

In satisfying R_e we shall construct functionals $\Gamma_j (j \in \omega)$ and Δ with the intention that if R_e fails then $K \leq_T V$ via some Γ_j, or Δ, contrary to our hypothesis.

Basic module. As usual, we will choose a sequence of candidates (one for each "cycle" of the strategy), one of which will witness the failure of one or both of the statements:

1. $A = \Phi_e^{W_e^V}$,

2. $W_e^V = \Psi_e^{V \oplus A}$.

This will be sufficient for R_e to succeed.

We make infinitely many attempts to satisfy R_e by an $\omega \times \omega$-sequence of "cycles", where each cycle (j, k) proceeds as follows:

(1) Choose an unused candidate $x_{j,k}$ greater than any number mentioned thus far in the construction.

(2) Wait for a stage s at which for some n, $\Psi_{e,s}^{V_s \oplus A_s} \restriction n$ is defined, and for $X \subseteq \{0, \ldots, n\}, X \restriction n = \Psi_{e,s}^{V_s \oplus A_s} \restriction n$,

$$X \subseteq W_{e,s}^{V_s}$$

(with X-use θ), and

$$A(x_{j,k}) = \Phi_e^{X \restriction n}(x_{j,k}).$$

(It is easy to see that if this never happens then $x_{j,k}$ is a witness to the success of R_e.)

(3) Protect $A \restriction \psi_{e,s}\varphi_{e,s}(x_{j,k})$ from other strategies from now on.

(4) Set $\Gamma_j^V(k) = K_s(k)$ with *use* $\gamma_j(k) = \max\{\theta, \psi_{e,s}\varphi_{e,s}(x_{j,k})\}$, and start cycle $(j, k + 1)$ to run simultaneously with cycle (j, k).

(5) Wait for $K(k)$ to change (at a stage s', say).

(If there is a $V \restriction \psi_e\varphi_e(x_{j,k})$-change between stages s and s', we kill the cycles $(j', k') > (j, k)$, drop the A-protection of this cycle (j, k) to 0, and go back to step (2). If there is a $V \restriction \theta$-change between stages s and s', but there is no $V \restriction \psi_e\varphi_e(x_{j,k})$-change, we kill the cycles $(j', k') > (j, k)$, and go back to step (2). In both cases, the parts of the functionals Γ_j, Δ defined by cycles $(j', k') > (j, k)$ become undefined by the V-change.)

(6) Stop cycles $(j', k') > (j, k)$ and put $x_{j,k}$ into A.

(7) Wait for a stage s'' at which, for some n', $\Psi_{e,s''}^{(V \oplus A)_{s''}} \upharpoonright n'$ is defined, and for $X' \subseteq \{0, \ldots, n'\}, X' \upharpoonright n' = \Psi_{e,s''}^{(V \oplus A)_{s''}} \upharpoonright n'$,

$$a) X \underset{\neq}{\subseteq} X',$$

$$b) X' \subseteq W_{e,s''}^{V_{s''}}$$

with X'-use θ', and

$$c) A(x_{j,k}) = \Phi_e^{X' \upharpoonright n'}(x_{j,k}).$$

(Note that if this never happens then $x_{j,k}$ is again a witness to the success of R_e. Indeed, if b) and c) never happen then obviously either $A(x_{j,k}) \neq \Phi_e^{W_e^V}(x_{j,k})$ or $W_e^V \neq \Psi_e^{V \oplus A}$. If b) and c) do happen with some X', but $X \not\subseteq X'$, then while enumerating V we must have seen some $V \upharpoonright \psi_{e,s}\varphi_{e,s}(x_{j,k})$-change or a $V \upharpoonright \theta$-change and would go back to step 2, otherwise we would win R_e by $x_{j,k}$: we have $X \upharpoonright n' \underset{\neq}{\subseteq} W_e^V \upharpoonright n'$ and $\Psi_e^{V \oplus A} \upharpoonright n' = X' \upharpoonright n'$, therefore $\Psi_e^{V \oplus A} \upharpoonright n' \neq W_e^V \upharpoonright n'$).
Notice also that if we now remove $x_{j,k}$ from A, we would have (in the absence of a $V \upharpoonright \theta'$-change or $V \upharpoonright \psi_e\phi_e(x_{j,k})$-change)

$$\Psi_e^{V \oplus A} \upharpoonright \varphi_e(x_{j,k}) = \Psi_{e,s}^{(V \oplus A)_s} \upharpoonright \varphi_e(x_{j,k}) = X \upharpoonright \varphi_e(x_{j,k}) \underset{\neq}{\subseteq} X' \upharpoonright \varphi_e(x_{j,k}).$$

So if $X' \upharpoonright \varphi_e(x_{j,k}) \subseteq W_e^V \upharpoonright \varphi_e(x_{j,k})$ then this is enough for the success of R_e. But, unfortunately, W_e^V is reversible through a $V \upharpoonright \theta'$-change (even if $V \upharpoonright \theta$ does not change) and we may again have $X \upharpoonright \varphi_e(x_{j,k}) = W_e^V \upharpoonright \varphi_e(x_{j,k})$.

To avoid this difficulty we will use these changes of $V \upharpoonright \theta'$ to threaten $K \leq_T V$ via a new functional Δ.

(8) Set
$$\Delta^V(j) = K(j)$$
with *use* $\delta(j) = \max\{\theta', \psi_{e,s''}(\varphi_{e,s''}(x_{j,k}))\}$, and start cycle $(j + 1, 0)$ to run simultaneously.

(9) Wait for $K(j)$ to change (at stage s^*, say).

(10) Stop all cycles $(j', k') \geq (j + 1, 0)$, remove the number $x_{j,k}$ from A, and preserve $A \upharpoonright \psi_{e,s^*}\varphi_{e,s^*}(x_{j,k})$.

(11) Wait for a $V \restriction \delta(j)$-change.

(12) Drop the A-protection of this cycle to 0, set

$$K(j) = \Delta^V(j)$$

with a new use $\delta(j)$, stop cycle (j, k), cancel all cycles $> (j, k)$, and start cycle $(j + 1, 0)$.

Whenever some cycle sees a $V \restriction \delta(j)$-change between stages s'' and s^*, it will kill the cycles $(j', k') > (j, k)$, make their functionals (including Δ^V) undefined, and go back to step 7.

If some cycle sees a $V \restriction \psi_{e,s}\varphi_{e,s}(x_{j,k})$-change between stages s and s^*, it will again kill the cycles $(j', k') > (j, k)$, make their functionals and $\Delta^V(j)$ undefined, and go back to step 2.

Note that if a cycle (j, k) sees a $V \restriction \delta(j)$-change between stages s'' and s^* but there is no $V \restriction \psi_{e,s}\varphi_{e,s}(x_{j,k})$-change after stage s then it goes back to step 7 and proceeds. If later the cycle again comes to step 8 it redefines $\Delta^V(j)$ (with the same j) with a new *use* $\delta(j)$. So in this case (when there is no $V \restriction \psi_{e,s}\varphi_{e,s}(x_{j,k})$-change), other cycles $(j', k') \neq (j, k)$ cannot define $\Delta^V(j)$.

The module has the following possible outcomes:

(A) There is a stage s after which no cycle acts. Then some cycle (j_0, k_0) eventually waits at step (2), (7) or (11) forever. It means that we were successful in satisfying R_e through the cycle (j_0, k_0).

(B) Some (least) cycle (j_0, k_0) acts infinitely often. Then it goes from step (5) to step (2), or from step (9) to step (7) or (2) infinitely often. Thus Φ_e or Ψ_e is partial. Notice that the overall restraint of all cycles has finite liminf.

(C) Every cycle acts only finitely often but there are infinitely many cycles (j_0, k) (for some least j_0) which collectively act infinitely often. Then $\Gamma^V_{j_0} = K$, contrary to hypothesis.

(D) Otherwise. Then, for each j, the last time some cycle (j, k) acts, it defines $\Delta^V(j)$ permanently and correctly, so $\Delta^V = K$, contrary to hypothesis.

The explicit construction and the remaining parts of the proof are now essentially the same as in Cooper, Lempp and Watson [1989] with only obvious changes. So we will not give them here.

Moreover, adding to the construction a permitting argument in exactly the same way as in Cooper, Lempp, Watson [1989], we can prove the following theorem.

Theorem 4.2. *Let* **u** *and* **v** *be r.e. degrees such that* **v** $<$ **u**. *Then there is a d-r.e. degree* **d** *such that* **v** $<$ **d** $<$ **u** *and* **d** *is not r.e. in* **v**.

5. Bibliography

Arslanov, M. M. [1985], Structural properties of the degrees below 0', *Sov. Math. Dokl. N.S.* **283** no. 2, 270-273.

Arslanov, M. M. [1988], On the upper semilattice of Turing degrees below 0', *Sov. Math.* **7**, 27-33.

Arslanov, M. M. [1990], On the structure of degrees below 0', in *Recursion Theory Week*, K. Ambos-Spies, G. H. Müller and G. E. Sacks eds., *LNMS* **1432**, Springer-Verlag, Berlin, 1990, 23-32.

Arslanov, M. M., Lempp, S. and Shore, R. A. [1995], Interpolating d-r.e. and REA degrees between r.e. degrees, to appear.

Cooper, S. B. [1971], *Degrees of Unsolvability*, Ph. D. Thesis, Leicester University, Leicester, England.

Cooper, S. B. [1990], The jump is definable in the structure of the degrees of unsolvability, *Bull. Am. Math. Soc. (NS)* **23**, 151-158.

Cooper, S. B. [1991], The density of the low$_2$ n-r.e. degrees, *Arch. Math. Logic* **31**, 19-24.

Cooper, S. B. [1992], A splitting theorem for the n-r.e. degrees, *Proc. Am. Math. Soc.* **115**, 461-471.

Cooper, S. B. [1993], Definability and global degree theory, in *Logic Colloquium '90*, J. Oikkonen and J. Väänänen eds., *Lecture Notes in Logic* **2**, Springer-Verlag, Berlin, 25-45.

Cooper, S. B. [1994], Rigidity and definability in the noncomputable universe, in *Proceedings of the 9th International Congress of Logic, Methodology and Philosophy of Science*, D. Prawitz, B. Skyrms and D. Westerstahl eds., North-Holland, Amsterdam, 1994, 209-236.

Cooper, S. B., Harrington, L., Lachlan, A. H., Lempp, S. and Soare, R. I. [1991], The d-r.e. degrees are not dense, *Ann. Pure and Applied Logic* **55**, 125-151.

Cooper, S. B., Lempp, S. and Watson, P. [1989], Weak density and cupping in the d-r.e. degrees, *Israel J. Math.* **67**, 137-152.

Cooper, S. B. and Yi, X. [1995], Isolated d-r.e. degrees, to appear.

Ding, D. and Qian, L. [1995], An r.e. degree not isolating any d-r.e. degree, to appear.

Epstein, R. L., Haas, R. and Kramer, R. L. [1981], Hierarchies of sets and degrees below 0', in *Logic Year 1979-80*, M. Lerman, J. H. Schmerl and R. I. Soare eds., Springer-Verlag, *LNMS* **859**, Berlin, 32-48.

Ershov, Y. [1968a] On a hierarchy of sets I, *Algebra i Logika* **7** no. 1, 47-73.

Ershov, Y. [1968b] On a hierarchy of sets II, *Algebra i Logika* **7** no. 4, 15-47.

Ershov, Y. [1970] On a hierarchy of sets III, *Algebra i Logika* **9** no. 1, 34-51.

Gold, E. M. [1965], Limiting recursion, *J. Symb. Logic* **30**, 28-48.

Ishmukhametov, Sh. T. [1985], On differences of recursively enumerable sets, *Izv. Vyssh. Uchebn. Zaved. Mat.* **279**, 3-12.

Kaddah, D. [1993], Infima in the d-r.e. degrees, *Ann. Pure and Applied Logic* **62**, 207-263.

LaForte, G. [1995], *Phenomena in the n-r.e. and n-REA degrees*, Ph. D. Thesis, University of Michigan.

Putnam, H. [1965], Trial and error predicates and the solution to a problem of Mostowski, *J. Symb. Logic* **30**, 49-57.

Sacks, G. E. [1963], On the degrees less than $0'$, *Ann. Math. (2)* **77**, 211-231.

Sacks, G. E. [1964], The recursively enumerable degrees are dense, *Ann. Math. (2)* **80**, 300-312.

Shore, R. A. [1982], On Homogeneity and Definability in the First Order Theory of the Turing Degrees, *J. Symb. Logic*, **47**, 8-16.

Slaman, T. A. and Woodin, W. H. [1996], *Definability in Degree Structures*, in preparation.

Soare, R. I., [1987], *Recursively Enumerable Sets and Degrees*, Springer-Verlag, Berlin.

Soare, R. I. and Stob, M. [1982], Relative recursive enumerability, in *Proc. Herbrand Symposium, Logic Colloquium 1981*, J. Stern, ed., North-Holland, Amsterdam, 299-324.

A CHARACTERISATION OF THE JUMPS
OF MINIMAL DEGREES BELOW 0'

S. BARRY COOPER

School of Mathematics, University of Leeds,
Leeds LS2 9JT, England

In computability theory, Gödel's incompleteness theorem [1934] finds expression in the definition of the jump operator. Thus, the Friedberg [1957] jump inversion theorem completely characterises the scope of the Gödel undecidability phenomenon within the Kleene-Post [1954] degree structure for classifying unsolvable problems.

Minimal degrees of unsolvability naturally arise as the structural counterpart of decision problems whose solutions are extremely specialised (their solution does not have any other nontrivial applications). Spector [1956] showed that minimal degrees exist, while Sacks [1961] constructed one below $\mathbf{0}'$.

The first result concerning the jumps of minimal degrees was due to Yates [1970], who obtained a low minimal degree as a corollary to his construction of a minimal $\mathbf{m}$ below any given nonzero computably enumerable (or r.e.) degree. A global characterisation of such jumps was provided by the Cooper [1973] jump inversion theorem, while in the same paper it was shown that such a theorem could not hold locally (there are no high minimal degrees). The intuition that minimal degrees are in a sense close to $\mathbf{0}$ (the degree of the computable sets) was reinforced by Jockusch and Posner [1978], who found that all minimal degrees are in fact generalised low$_2$. This, with Sasso's [1974] construction of a non-low minimal degree below $\mathbf{0}'$, supported Jockusch's conjecture (see Yates [1974], p.235) that the jumps of minimal degrees below $\mathbf{0}'$ can be characterised as those $\mathbf{0}'$-REA degrees which are low over $\mathbf{0}'$.

1991 *Mathematics Subject Classification.* Primary 03D25, 03D30; Secondary 03D35.

Preparation of this paper partially supported by E.P.S.R.C. research grant no. GR/H02165, and by EC Human Capital and Mobility network 'Complexity, Logic and Recursion Theory'. This paper originated from joint work, February–March 1992, with David Seetapun. In particular, the proving of Theorem 1.1 below is very much a product of that collaboration.

Typeset by $\mathcal{A}_{\mathcal{M}}\mathcal{S}$-TEX

The purpose of this paper is to characterise the jumps of the Δ_2^0 minimal degrees as those $0'$-REA degrees which are 'almost Δ_2^0', in so doing providing a negative answer to the Jockusch conjecture.[1]

In section 1 we outline a proof of the existence of a Σ_2^0 degree $\mathbf{c}$ low over $0'$ such that no $\mathbf{d} \geq \mathbf{c}$ is the jump of a minimal degree below $0'$. In section 2, in the spirit of Ambos-Spies $et\ al$ [1984], we distil from that construction the essential ingredients of the construction of $C \in \mathbf{c}$ which ensure a jump class $\mathbf{c}^{-1} \subset \Delta_2^0$ free of minimal degrees. This leads to the definition of the notion of $almost\ \Delta_2^0$ for Σ_2^0 sets which Turing compute ϕ'. The remainder of the paper is devoted to verifying that the almost Δ_2^0 degrees comprise the jumps of the degrees below $0'$, and form a proper subclass of the Σ_2^0 degrees low over $0'$.

1. The basic non-inversion theorem

For basic notation and terminology we refer to Lerman [1983] and Soare [1987].

THEOREM 1.1. *There is a degree $\mathbf{c}$ r.e. in and above $0'$ with $\mathbf{c}' = 0''$ which is not the jump of a minimal degree $\leq 0'$. (And in fact, no $\mathbf{d} \geq \mathbf{c}$ is such a jump.)*

PROOF. We construct $C \in \Sigma_2^0$ with standard sequence of computable approximations $\{C^s\}_{s \in \omega}$. In particular, we will define $\{C^s\}_{s \in \omega}$ with infinitely many *thin* stages (stages at which $C^s \subset C$).

We assume (Θ, Φ_X, W_X, X) to be a typical member of a standard listing of quadruples of partially computable (p.c.) functionals Θ, Φ computably enumerable sets W and Σ_2^0 sets X, each with standard computable approximating sequences. We use the usual $[s]$ suffix to indicate the value of an expression at stage s.

DEFINITION 1.2. We say that a stage s is x, Θ-*use thin* if and only if

$$\Theta^{C,K} \downarrow [s]\ \&\ \forall t > s\,[(C^s, K^s) \upharpoonright \theta^s(x) \subseteq (C^t, K^t)],$$

where θ is the standard use function for Θ.

We seek to satisfy for each $(\Theta, \Phi_X, W_X, X, x)$ the following requirements:

$$\mathcal{N}_{x,\Theta}:\ \mathrm{Lim}\,\{\Theta^{C,K}(x)[s] \mid s \text{ is } x, \Theta\text{-use thin}\} \text{ exists,}$$

$$\mathcal{R}_X:\ \Theta(W^X) = C \Rightarrow [\exists \text{ a 1-generic } A = \Gamma^X \leq_T X] \vee X \notin \Delta_2^0,$$

---

[1] Downey, Lempp and Shore [ta] have independently refuted the Jockusch conjecture. See also Cooper [1986].

where the set $A = A_X$ and the p.c. $\Gamma = \Gamma_X$ are to be constructed, and we write Φ, W for Φ_X, W_X respectively.

Let S be a typical member of a standard listing of the computably enumerable sets of binary strings (with each S closed under upward inclusion of strings). Then we break each $\mathcal{R}_X$ into infinitely many subrequirements of the form

$$\mathcal{R}_{X,S}: \ \Phi(W^X) = C \Rightarrow \exists \sigma \subset \Gamma^X \left[\sigma \in S \vee (\forall \tau \supset \sigma)(\tau \notin S) \right] \vee X \notin \Delta_2^0,$$

where we write $\subset$ for string inclusion.

We assume a computable priority ordering of all requirements $\mathcal{N}_{x,\Theta}$, $\mathcal{R}_{X,s}$.

LEMMA 1.3. *If C satisfies all the $\mathcal{N}$-requirements then $\mathbf{c} \cup \mathbf{0}'$ is $\mathbf{0}'$-low.*

PROOF. We check that the relation $\lambda x, \Theta \left[\Theta^{C.K}(x) \downarrow \right]$ is Δ_3^0.

Since K is computably enumerable and we are assuming infinitely many thin stages for $\{C^s\}_{s\in\omega}$, we have

$$\Theta^{C,K}(x) \downarrow$$

> $\Leftrightarrow$ there exists an x, Θ-use thin stage such that the use $\theta(x)$ changes at
>
>   no greater x, Θ-use thin stage

$$\Leftrightarrow [(\exists s)[(\forall t > s)((C,K) \restriction \theta(x)[s] \subseteq (C^s, K^s) \ \& \ (\forall t > s)[(\exists t')$$
$$((C,K) \restriction \theta(x)[t] \not\subseteq (C^{t'}, K^{t'}) \vee ((C,K) \restriction \theta(x)[t] = (C,K) \restriction \theta(x)[s])]]$$
$$\in \Sigma_3^0.$$

Also, since $\mathrm{Lim}\{\Theta^{C,K}(x)[s] \mid s \text{ is } x, \Theta\text{-use thin}\}$ exists and (by the existence of infinitely many thin stages) if $\Theta^{C,K}(x) \downarrow$ there are infinitely many x, Θ-use thin stages, we have

$$\Theta^{C,K}(x) \downarrow$$

> $\Leftrightarrow$ there exist infinitely many x, Θ-use thin stages at which $\Theta^{C,K}(x) \downarrow$

$$\Leftrightarrow [(\forall t)(\exists t > s)[\Theta^{C,K}(x) \downarrow^t \ \& \ (\forall t' > t)((C,K) \restriction \theta(x)[x] \subset (C^{t'}, K^{t'}))]]$$
$$\in \Pi_3^0.$$

The lemma follows.   □

We now give an informal outline of the naive strategy for satisfying an instance of $\mathcal{R}_{X,S}$ on the true path before setting out the detailed module.

Assuming $\Phi(W^X) = C$, we first select some string σ, anticipating $\sigma = \Gamma^X \restriction n$, some n.

If we ever get some $\tau \supset \sigma$ with $\tau \in S$ at a stage $s+1$, say, then we want to end up with $\tau \subset \Gamma^X$. To do this we must move X away from any previously defined beginning $X \restriction \gamma(y)$, say, for X for which an existing axiom for Γ dictates $\Gamma(X \restriction \gamma(y), y) \downarrow \neq \tau(y)$, $y < \ell h(\tau)$. This can be achieved if we obtain $\sigma = \Gamma^X \restriction n$ but with $X \restriction \gamma(n) \not\subset X^t$ at each stage $t < s+1$. To this end, we try to maintain a situation previous to stage $s+1$ in which $\gamma(n) >$ some suitable $\varphi^X(y)$ (φ^X the X-use function for $\Phi(W^X)$) at all stages at which $\gamma(n)$ is defined (although only partial satisfaction of this condition will be possible). To do this we must choose an appropriate $\mathcal{R}$-*agitator* y, and only define $\Gamma^X(n)$ when $\Phi(W^X, y) \downarrow = C(y) = 1$ (that is, at y-*replete*, or just *replete*, stages).

When a suitable $\tau \supset \sigma$ appears, we attempt to move X away from the finite tree ξ_n of beginnings of X used in previously defined Γ-axioms at argument n via $C(y)$-changes in conjunction with the Φ-*equation* $\Phi(W^X) = C$. Making $C(y)$-changes at y-replete stages requires an $\mathcal{R}$-response of a change in $W^X \restriction \varphi(y)$ (φ the standard use function for Φ).

Let $X^* \in \xi_n$ where we define $\Gamma^{X^*}(n)$ for the first time at a y-replete stage $t+1$, say (that is, X^* is Γ-*held* at stage $t+1$). We can assume that there are certain numbers $z \in W^{X^*} \restriction \varphi(y)$ at stage $t+1$, called X^*-*designated* members of W^{X^*}.

The strategy for X avoiding X^* is to use $C(y)$-changes with the overall requirement for y-replete stages to shake W^X changes out of the Φ-equation. If the changes are in X^*-designated numbers, we will get X to move away from X^* (since $X^* = X \restriction \gamma(n)[t+1]$ and we define $\gamma(n) >$ the X-use of all X^*-designated numbers $z \in W^X$ (that is, $\gamma(n) > w(z)$), each such z). Otherwise we use restoration of $C(y) = 1$ to attempt to reinstate $W^X \restriction \varphi(y)[t+1]$ so as to re-initialise the process.

We may still fail with $\tau \subset \Gamma^X$ despite the occurrence of changes in designated numbers $z \in W^X$, but we will then have $X \notin \Delta_2^0$ since there will be infinitely many X-changes over the finite tree ξ_n. We say the occurrence of a designated $z \nearrow W^X$ ('z exiting from W^X') is a *prerequisite* for $\tau \subset \Gamma^X$.

It is quite possible that no such prerequisite is permanently in place for $\tau \subset \Gamma^X$ (there may not even be any designated z for σ), and the various rectifications of the Φ-equation needed over a sequence of y-attacks may be made via undesignated $W^X(z)$-changes. It is necessary therefore to use y-attacks as occasions for selecting further strings σ' (say) as candidates for $\sigma' \subset \Gamma^X$, and looking for $\tau' \supset \sigma'$ with $\tau' \in S$ with a view to initiating a similar cycle of y-attacks. The new strings σ' will take up new numbers $z \in W^X$ as their own designated numbers.

The rectifications of the Φ-equation, we will want to argue, which do not provide prerequisites, will be those involving some z entering W^X but leaving before we get $\tau' \in S$ with $\tau' \supset \sigma'$, z designated for σ'. This situation will be important in ensuring Φ-rectification at infinitely many $W^X \upharpoonright \varphi(y)$-thin stages with $C(y) = 1$. (In the case in which there are many prerequisites but no designated changes we will have $\Phi(W^X, n)\uparrow$.)

We need to follow through the various possibilities arising from infinitely many y-attacks.

If a y-attack (that is $y \nearrow C$) produces no prerequisite for $\tau \subset \Delta^X$, maintenance of $\mathcal{R}$ requires some $z \searrow W^X$ ('z entering W^X') with $z < \varphi(y)$. As outlined previously, we now switch to the Ψ-*equation* $\Psi^{C,K} = W^X$, and wait for rectification up to argument z. Then, in the absence of $K \upharpoonright \psi(z)$-changes, we can reinstate the value of $C \upharpoonright \psi(z)$ which gave $z \notin \Psi^{C,K}$ (in particular, by enumerating y into C), so requiring $z \nearrow W^X$ for $\mathcal{R}$-maintenance, and re-initialising in preparation for a renewed agitation of the Φ-equation.

There are two possible outcomes corresponding to infinitely many y-attacks. There is the possibility that every y-attack results in the extraction from W^X of a z which entered at some stage after the selection of y, so that the resulting X-change is of no help in pursuing the $\tau \subset \Delta^X$ outcome. Again, as previously described, we avoid this by clearing such z from, $W^X \upharpoonright \varphi(y)$ as part of the pre-attack initialisation, using the Ψ-equation and reinstatement of the appropriate $C \upharpoonright \psi(z)$. This means that each y-attack entails some $z \searrow W^X \upharpoonright \varphi(y)$ followed by $z \nearrow W^X$ due to C-reinstatement in the Ψ-equation.

Outcome (i): If a particular z is involved in the y-attack infinitely often, so that $\mathrm{Lim}_s\, W^X(z)[s]$ does not exist, we have $z \notin W^X$ and $y \notin C$. Also, any z involved in a finite number of y-attacks produces the same situation. This gives $y \in \Phi(W^X)$ but $y \notin C$, and $\mathcal{R}$ is trivially satisfied.

Outcome (ii): If infinitely many such numbers z are involved, we obtain $\mathcal{R}$-satisfaction via $\Phi(W^X, y)\uparrow$. This involves a similar argument to (i), but showing that this results in unbounded $\varphi(y)$ over W^X.

It remains to incorporate the possibility of $\Psi^{C,K}(z)$ computations becoming K-*incorrect* (the K-use changes). If this happens, an essential element in the strategy for forcing $\tau \subset \Delta^X$ is lost, and we must select a new σ, σ' say, of greater length, while retaining the $\mathcal{R}$-agitator y.

With infinitely many such K-incorrect computations occuring, there are two possibilities:

(i) We get infinitely many losses of K-computation for some $\Psi^{C,K}(z)$, so that $\Psi^{C,K}$ is not total, or

(ii) We get infinitely many new numbers $z \in W^X$ involved in a

$\Phi(W^X, y)$ computation at some stage, whereby $\Psi^{C,K}(z)$ becomes vulnerable to the K-incorrect computations, which will result in $\Phi(W^X)$ not being total ($\Phi(W^X, y)\uparrow$).

The basic module for $\mathcal{R}$ at stage $s + 1$

Let $\ell^{\Phi,\Psi} = $ the greatest y, s.t. (at stage $s + 1$ $\Phi(W^X) \upharpoonright y = C \upharpoonright y$ & $\Psi^{C<K} \upharpoonright \varphi(y) = W^X \upharpoonright \varphi(y)$.

Overall requirement:

The basic module only operates at stages at which $\ell^{\Phi,\Psi} > $ any existing σ-agitator y for $\mathcal{R}$ (σ as in 2 below), that is, at *replete* stages $s + 1$. We also assume that we routinely define $\Delta \upharpoonright \ell^{\Phi,\Psi}$ at stage $s + 1$, the uses or values only changing when noted below.

1. *Select* an agitator y and *define* $y \in C$.

2. *Select* $\sigma = \Delta^X \upharpoonright n$ (say), and *restrain* $C \upharpoonright \psi(\varphi(y))$. If we get a $K \upharpoonright \psi(\varphi(y))$-change at any subsequent stage (that is the computation $\Psi^{C,K} \upharpoonright \varphi(y)$ becomes K-*incorrect*) we *cancel* σ and its associated C-restraint and *return* to 2.

3. *Test* for the existence of a string $\tau \supset \sigma$ with $\tau \in S$.

4. Test *negative:* For each $\Delta^X(z)\uparrow$, $\ell^{\Phi,\Psi} > z \geq n$, define $\Delta^X(z)$ with the new $\delta(z) > \varphi^X(y)$ (the X-use of $\Phi(W^X)$). *Place* $X \upharpoonright \delta(n)$ in ξ_σ. *Return* to 3.

5. Test *positive: Proceed* to 6.

6. *Test* for a beginning of X in ξ_σ.

7. Test *negative: Define* $\Delta^X(z) = \tau(z)$ for each z, $\min(\ell h(\tau), \ell^{\Phi,\Psi}) > z \geq n$, with $X \upharpoonright \delta(z)$ incompatible with each $\pi \in \xi_\sigma$. *Return* to 6.

8. Test *positive: Test* for $y \in C$.

9. Test (8) *negative: Define* $y \in C$ (that is *reinstate* $C \upharpoonright \psi(\varphi(y))$). *Return* to 6.

10. Test (8) *positive: Extract* y from C (that is, carry out a y-*attack*). *Return* to 6.

The construction of C consists of a standard implementation of the basic module, with verification routinely amplifying the above motivational remarks.   □

In the next section we extract from the above construction the properties of C essential to the satisfaction of the requirements.

2. The almost Δ_2^0 degrees

The motivation for definition 2.1 below can be summarised as follows.

Firstly, it appears from the previous discussion that if we can make $y \in C$ coincide with infinitely many replete stages, then the basic conditions needed for the construction of a 1-generic set below A, with $C \leq_T W_e^A$, are all in place. The particular choice of replete stages is Δ_3^0, depending on the eventual true path.

Moreover, if for every x we can cofinitely often computably predict a latest stage at which $x \in C$ ($C \in \Sigma_2^0$) will be extracted from C (if ever), we can carry out a full-approximation jump inversion construction of a minimal M, $M' \equiv_T C$. We make $C \leq_T \operatorname{dom}\Gamma(M)$, Γ p.c., in such a way that the predictions re $x \notin C$ ensure bounded use in the reduction of C to $\operatorname{dom}\Gamma(M)$. If $x \searrow C$, we need to shift certain $y \in \operatorname{dom}\Gamma(M)$ by choosing a new beginning of M beyond a designated apex in the nest of splitting/nonsplitting trees. If $x \nearrow C$ before certain bad types of higher priority splitting pairs appear, such splittings need not be used, and a bounded number of such y will suffice.

DEFINITION 2.1.

(1) We say s is a *false stage* (*for* $\{C^s\}_{s \in \omega}$, a standard approximating sequence to $C \in \Sigma_2^0$) *at* x if and only if $C^s \restriction x \neq C \restriction x$.

(2) We say a function f is *true for* $\{C^s\}_{s \in \omega}$ *at* x if and only if only finitely many values of f are false stages at x.

(3) We say $\{C_s\}_{s \in \omega}$ is *tame* if and only if there exists a computable $\lambda x, s[f(x,s)]$ with $f \in \Delta_3^0$ defined by $f(x) = \operatorname{Lim\,inf}_s f(x,s)$ each x, such that for almost all $e \in \omega$ $\varphi_{f(x)}$ is a computable function true for $\{C_s\}_{s \in \omega}$ at x, with the range of $\varphi_{f(x)}$ infinite.

(4) We say C or $\deg(C)$ (where $\deg(C)$ is $\mathbf{0}'$-REA) is *almost* Δ_2^0 if and only if C has a tame standard Σ_2^0 approximating sequence $\{C^s\}_{s \in \omega}$.

We notice that:

(1) Trivially, $\mathbf{0}'$ is almost Δ_2^0.

(2) If $\deg(C)$ is almost Δ_2^0, then $\deg(C)$ is $\mathbf{0}'$-low. This is because we can get a Δ_3^0 approximation to $J(C)$ (J p.c. with $\operatorname{dom} J(C) = C'$) by enumerating x into $J(C)$ whenever $J^s(C^s, x) \downarrow$ at a stage $s \in \operatorname{range}(\varphi^s_{f(j^s(x),s)})$ (j the standard use function for J) and extracting x at a stage $t > s$, $t \in \operatorname{range}(\varphi^t_{f(j^t(x),t)})$, at which either (a) $f(j^s(x),s) \neq f(j^t(x),s)$, or (b) $f(j^t(x),s) = f(j^s(x),s)$ but $C^s \restriction j^s(x) \neq C^t \restriction j^t(x)$.

We say that a stage s is *on the true path* if and only if $f(j^s(x),s) = \operatorname{Lim\,inf}_{s'} f(j^s(x),s')$. Then $x \in J(C) \Leftrightarrow x$ is enumerated into $J(C)$ at infinitely many stages on the true path, a Δ_3^0 relation.

THEOREM 2.2. *If* **c** *is almost* Δ_2^0 *then there exists a minimal degree* **m** $< \mathbf{0}'$ *with* $\mathbf{m}' = \mathbf{c}$.

PROOF. Let C be almost Δ_2^0 with corresponding f. We can construct M on a standard nest of p.c. splitting/non-splitting trees by full approximation, getting $M \in \Delta_2^0$ of minimal degree. (See Lerman [1983] or Epstein [1979] for a description of the full-approximation construction of a minimal degree.) We must combine forcing of the jump at x with selection of the apex of the $(x+1)^{\text{th}}$ tree in the usual way to get $\mathbf{m}' \leq \mathbf{c}$ (that is, only changes in guesses concerning membership of C can injure restraints related to forcing of jump). To get $\mathbf{c} \leq \mathbf{m}'$, we must make $C = \Psi(\operatorname{dom}\Gamma(M))$, Ψ, Γ p.c. functionals to be constructed, by defining $\Psi(\{\operatorname{dom}\Gamma(T_{y+1}(\phi))\} \restriction z, x) = 1$, for suitable y, z, when $x \in C$.

To be explicit, we may tie $x \notin C$ to $T_y(0) \subset M$, maintaining $\psi(\operatorname{dom}\Gamma(T_y(0)), x) \uparrow$ at stages at which $x \notin C$. If $x \in C$, we will exercise the option of defining certain $\Gamma(T_y(1), w)$, $w < \psi(x)$, and setting $\Psi(\operatorname{dom}\Gamma(T_y(1)), x) = 1$ with $T_y(1) \subset M$.

There are two immediate problems:

  (a) A boundary on a higher priority $T_{y'}$ may move due to the definition of a new splitting pair $T_{y'}(\tau * 0), (\tau * 1)$, where $T_y(1) \subset T_{y'}(\tau * 0), (\tau * 1)$, so forcing any w already a member of $\operatorname{dom}\Gamma(T_y(1))$ to be in the new $\operatorname{dom}\Gamma(T_y(0))$. So if $x \searrow C$ at some later stage, we need some $w' \neq w$ with $w \searrow \operatorname{dom}\Gamma(T_y(0))$ and $w < \psi(x)$ to enable us to define $\Psi(M, x) = 0$ $(T_y(0) \subset M)$. For one $x \searrow C, x \nearrow C$ cycle, there is no problem in that we can bound the number of such 'bad' splittings emanating from higher priority trees. But:

  (b) $M \in \Delta_2^0$ requires us to shift T_y up the nest of trees. If we have a full complement of bad splittings at each level, we may end up with $\psi(x)$ unbounded on $\operatorname{dom}\Gamma(M)$, and $\Psi(\operatorname{dom}\Gamma(M))$ partial. However, if we can be sure that the $x \searrow C$ part of the cycle precedes the appearance of the bad splitting, we can use the knowledge that we need $T_y(0) \subset M$ to prohibit such bad splittings. We cannot prohibit $T_y(1) \subset$ some $T_{y'}(\tau * i)$ $(i \leq 1)$, but we can verify that in this situation if the only splittings $T_{y'}(\tau * 0), (\tau * 1) \supset T_y(1)$ then $\Phi_{y'-1}^M$ is partial.

This is where the almost Δ_2^0-ness of C is essential – the tame $\{C^s\}_{s \in \omega}$ provides us with a 'true path' with a correct $f(x)$ along which we can (eventually) correctly anticipate $x \nearrow C^s$ and pre-empt the need for unbounded $\psi(x)$ over $\operatorname{dom}\Gamma(M)$. We must use a tree of Γ definitions to ensure that actions off the true path do not interfere with what we do with Γ and Ψ on the true path. Also, we must make sure that actions to build the tree of trees do not spoil the partial computability of the nest along the true path, and facilitate $M \in \Delta_2^0$. □

THEOREM 2.3. *If* $\mathbf{a} \leq \mathbf{0}'$ *and* $C \in \Sigma_2^0$ *is not almost* Δ_2^0, *then* $\mathbf{a}' \geq \deg(C) \Rightarrow \mathbf{a}$ *bounds a* 1-*generic degree.*

PROOF. Unlike the earlier version of the basic module, we can no longer make to-order C-changes. Instead, the lack of tameness of $\{C^s\}_{s \in \omega}$ will give the needed $C(y)$-changes at replete stages.

Assume given p.c. Φ, computably enumerable W and $X \in \Delta_2^0$, such that $\Phi(W^X) = C$. Let $\{S_i\}_{i \in \omega}$ be a standard listing of the computably enumerable sets of strings (with $\sigma \in S_i \Rightarrow$ all $\tau \supset \sigma \in S_i$). We need to construct $A = \Delta^X$ satisfying the following requirements ($i \in \omega$):

$$\mathcal{R}_i : \exists \sigma = \Delta^X \upharpoonright n \, (\text{say}) \, [\sigma \in S_i \vee (\forall \tau \supset \sigma)(\tau \notin S_i)] \vee X \notin \Delta_2^0 \vee \Phi(W^X) \neq C.$$

We outline the basic strategy for dealing with $\mathcal{R}$-requirements (dropping the index i).

We must first select some σ, anticipating $\sigma = \Delta^X \upharpoonright n$, some n.

If we get some $\tau \supset \sigma$ in S at some stage ($s{+}1$ say) we need to ultimately obtain $\tau \subset \Delta^X$. To do this, we must move X away from the X-uses of previously defined axioms for Δ which contradict τ, that is, we need to obtain $\sigma = \Delta^X \upharpoonright n$ with $X \upharpoonright \delta(n) \not\subset X^t$ at each stage $t < s + 1$. To do this, we try to maintain a situation previous to stage $s + 1$ in which $\delta(n) >$ some suitable $\varphi^X(y)$ at all stages at which $\delta(n)$ is defined (although only partial satisfaction of this condition will be possible). To this end we try to obtain a situation in which we only define $\Delta^X(n)$ when some $\Phi(W^X, y) \downarrow = C(y) = 1$ (at that stage in the construction) but in the final outcome $C(y) = 0$, with $Y <$ some y^* selected for $\mathcal{R}$ (where y is an $\mathcal{R}$-*agitator*); that is, we try to only define $\Delta^X(n)$ at y^*-*replete*, or just *replete*, stages s (when $\Phi(W^X) \upharpoonright y^* \downarrow = C \upharpoonright y^*$), which are also false stages for C at y^*. Since C is not computably enumerable, the Σ_2^0 version of the Miller-Martin Lemma provides such a σ via delayed thin stages. When suitable $\tau \supset \sigma$ appears, we rely on the definition of a computable function f appropriate to our level on the tree of strategies, together with the non-tameness of $\{C^s\}_{s \in \omega}$ to provide us with an attempt to move X away from the finite tree ξ_n of old beginnings of X used in previously defined Δ-axioms via certain $C(y)$-changes in conjunction with the Φ-*equation* $\Phi(W^X) = C$. Such a $C(y)$-change subsequent to such a y^*-replete stage requires an $\mathcal{R}$-response of a change in $W^X \upharpoonright \varphi(y^*)$.

Let $X^* \in \xi_n$ where we define $\Delta^{X^*}(n)$ for the first time at a y^*-replete stage $t + 1$, say (that is, X^* is Δ-*held* at stage $t + 1$). We can assume that there are certain numbers $z \in W^{X^*} \upharpoonright \varphi(y^*)$ at stage $t + 1$, called X^*-*designated* members of W^{X^*}.

The strategy for X avoiding X^* is to work with the computable f to produce a false stage for C at y^* at some y^*-replete stage, and then to use the resulting $C(y)$-changes with the overall requirement for y^*-replete stages to shake W^X changes out of the Φ-equation. If the changes are in X^*-designated numbers, we will get X to move away from X^* (since $X^* = X \restriction \delta(n)[t+1]$ and we define $\delta(n) >$ the X-use of all X^*-designated numbers $z \in W^X$, that is $\delta(n) > w(z)$, each such z). Otherwise we rely on a later false stage to restore some $C(y) = 1$ in an attempt to reinstate $W^X \restriction \varphi(y^*)[t+1]$ at a y^*-replete stage so as to re-initialise the process.

Assuming we still fail with $\tau \subset \Delta^X$ despite the occurrence of changes in designated numbers $z \in W^X$, we will then have $X \notin \Delta^0_2$, since there will be infinitely many X-changes over the finite tree ξ_n, a pseudo-outcome. We say the occurrence of a designated $z \nearrow W^X$ is a *prerequisite* for $\tau \subset \Delta^X$.

It is quite possible that no such prerequisite is permanently in place for $\tau \subset \Delta^X$ (there may not even be any designated z for σ), and the various rectifications of the Φ-equation needed over a sequence of y^*-attacks may be made via undesignated $W^X(z)$-changes. It is necessary therefore to use y^*-attacks as occasions for selecting further strings σ' (say) as candidates for $\sigma' \subset \Delta^X$, and looking for $\tau' \supset \sigma'$ with $\tau' \in S$ with a view to initiating a similar cycle of y^*-attacks. The new strings σ' will take up new numbers $z \in W^X$ as their own designated numbers. The rectifications of the Φ-equation, which do not provide prerequisites, will be those involving some z entering W^X but leaving before we get $\tau' \in S$ with $\tau' \supset \sigma'$, z designated for σ'. This situation will be important in ensuring Φ-rectification at infinitely many $W^X \restriction \varphi(y^*)$-thin stages false at y^*.

We need to follow through the various possibilities arising from infinitely many such y^*-attacks. First consider the case in which there are infinitely many replete stages $s+1 \in$ range f which are false stages at y^* and at which $W^X \restriction \varphi(y^*)$ is *thin* (that is $W^X \restriction \varphi(y^*)[s] \subset W^X \restriction \varphi(y^*)$). There are two subcases:

(a) For some such stage we get $W^X \restriction \varphi(y^*)[s] = W^X \restriction \varphi^s(y^*)$. In that case we get $\Phi(W^X, y) = 1 \neq C(y)$ for some $y < y^*$.

(b) Otherwise. But this can only happen if $\Phi(W^X, y) \uparrow$ for some $y < y^*$, as $\Phi(W^X, y) \downarrow = k$, say, means $\Phi(W^X, y) \downarrow = k$ at all sufficiently large $W^X \restriction \varphi(y^*)$-thin stages.

Now say there is some s^* such that at every replete stage $s+1 \in$ range f which is false at y^* if $s+1 > s^*$ then $W^X \restriction \varphi(y^*)$ is not thin. It will now follow that the failure in obtaining $\tau \subset \Delta^X$ will provide designated numbers for some σ' which will help us succeed in choosing some $\tau' \subset \Delta^X$ ($\tau' \supset \sigma'$, $\tau' \in S$). This is because otherwise, even if $W^X \restriction \varphi(y^*)$ is not thin at such stages, for each w there is a stage after which $W^X \upharpoonright w$ is thin at all such stages. And since $\Phi(W^X, y^*) \downarrow$, this gives the required outcome. □

(We notice that in the above proof one can utilise the fact that $\{C^s\}_{s \in \omega}$ is tame if and only if there is a p.c. $\lambda x, s[f(x,s)]$ such that for almost all x $\lambda s[f(x,s)]$ total $\Rightarrow$ range $\lambda s[f(x,s)]$ infinite & $\lambda s[f(x,s)]$ true for $\{C^s\}_{s \in \omega}$ at x.)

THEOREM 2.4. *There exists a* **0′**-*REA degree* **c** *which is* **0′**-*low but not almost* Δ_2^0.

PROOF. Let $\hat{\varphi}_e(x) = \operatorname{Lim\,inf}_s \varphi_e^{(2)}(x,s)$, each e, x, so that all Δ_3^0 functions f are included in the list $\{\hat{\varphi}_e\}_{e \in \omega}$. We set up a tree with the $\langle e, x \rangle + 1^{\text{th}}$ level relating to the requirement

$$\mathcal{R}_{\langle e,x \rangle} : \varphi_{\hat{\varphi}_e(x)} \text{ is not a computable function true for } \{C^s\}_{s \in \omega} \text{ at } x,$$

C a Σ_2^0 set to be constructed with computable approximating sequence $\{C^s\}_{s \in \omega}$. We label nodes with numbers. A branching m on the tree at the $\langle e, x \rangle + 1^{\text{th}}$ level is to the left of node n if and only if $m < n$, and corresponds to a guess that $\hat{\varphi}_e(x) = m$.

We can organise the construction of C so that movements to the left during the construction do not result in numbers x being enumerated into C at any given stage. Then we make C to be Δ_3^0 by making the limits of $x \in K^C$ exist restricting ourselves to stages $s+1$ at which we visit the true path of correct guesses at values $\varphi_e^{(2)}(x,s) = \hat{\varphi}_e(x)$.

At different stages of the construction we may have to take action to prevent $\varphi_{\hat{\varphi}_e(x)}$ being true for $\{C^s\}_{s \in \omega}$ at x. Typically, this will involve attempting to extract some $y < x$ from C, where $y \in$ some $C[\varphi_{\hat{\varphi}_e(x)}(n)]$, where y has been previously designated for this purpose. The restraints in the interests of limits $x' \in K^C$ existing along the true path will be given level x' priority in the tree. This may preclude us from using $y < x$ to prevent $\varphi_{\hat{\varphi}_e(x)}$ being true for $\{C^s\}_{s \in \omega}$ at x, but we can fix on *some* y and use it as if $y < x$ at the priority level $\langle e, x \rangle + 1$, which will then make it available for lower priority $\langle e, x'' \rangle + 1$, in which case $y < x''$ will almost always be satisfied. This must of course be done in such that actions on $\varphi_{\hat{\varphi}_e(x)}$ (different numbers x) maintain their independence. □

It would be interesting to find a natural definition of the jumps of the Δ_2^0 minimal degrees in the **0′**-REA degrees. (Such jumps are of course definable in $\mathcal{D}$, by Cooper [ta].)

QUESTIONS 2.5. (1) Is there a property of computably enumerable sets which relativises to a property equivalent to that of being almost Δ_2^0? And does this property (if it exists) have a natural degree-theoretic characterisation in $\mathcal{R}$?

(2) Does the set of all almost Δ_2^0 degrees form an ideal in $\mathbf{0}'$-REA? And is the set of $\mathbf{0}'$-REA degrees which are not almost Δ_2^0 a filter in $\mathbf{0}'$-REA?

REFERENCES

K. Ambos-Spies, C. G. Jockusch, Jr., R. A. Shore and R. I. Soare [1984], *An algebraic decomposition of the recursively enumerable degrees and the coincidence of several degree classes with the promptly simple degrees*, Trans. Amer. Math. Soc. **281**, 109–128.

S. B. Cooper [1973], *Minimal degrees and the jump operator*, J. Symbolic Logic **38**, 249–271.

S. B. Cooper [1986], *Some negative results on minimal degrees below $\mathbf{0}'$*, item 353 (abstract), Recursive Function Theory Newsletter **34**.

S. B. Cooper [ta], *On a conjecture of Kleene and Post*, to appear.

R. G. Downey, S. Lempp and R. A. Shore [ta], *Jumps of minimal degrees below $\mathbf{0}'$*, to appear.

R. L. Epstein [1979], *Degrees of Unsolvability: Structure and Theory*, Lecture Notes in Mathematics no. 759, Springer-Verlag, Berlin, Heidelberg, New York.

R. M. Friedberg [1957], *A criterion for completeness of degrees of unsolvability*, J. Symbolic Logic **22**, 159–160.

K. Gödel [1934], *On undecidable propositions of formal mathematical systems*, mimeographed notes, The Undecidable. Basic Papers on Undecidable Propositions, Unsolvable Problems, and Computable Functions, (M. Davis, ed.), Raven Press, New York, 1965, pp. 39–71.

C. G. Jockusch, Jr. and D. Posner [1978], *Double jumps of minimal degrees*, J. Symbolic Logic **43**, 715–724.

S. C. Kleene and E. L. Post [1954], *The upper semi-lattice of degrees of recursive unsolvability*, Ann. Math. (2) **59**, 379–407.

M. Lerman [1983], *Degrees of Unsolvability*, Perspectives in Mathematical Logic, Omega Series, Springer-Verlag, Berlin, Heidelberg, London, New York, Tokyo.

G. E. Sacks [1961], *A minimal degree less than $\mathbf{0}'$*, Bull. Amer. Math. Soc. **67**, 416–419.

L. P. Sasso [1974], *A minimal degree not realising least possible jump*, J. Symbolic Logic **39**, 571–574.

R. I. Soare [1987], *Recursively Enumerable Sets and Degrees*, Perspectives in Mathematical Logic, Springer-Verlag, Berlin, Heidelberg, London, New York, Paris, Tokyo.

C. Spector [1956], *On degrees of recursive unsolvability*, Ann. of Math. (2) **64**, 581–592.

C. E. M. Yates [1970], *Initial segments of the degrees of unsolvability, Part II: Minimal degrees*, J. Symbolic Logic **35**, 243–266.

C. E. M. Yates [1974], *Prioric games and minimal degrees below $\mathbf{0}'$*, Fund. Math. **82**, 217–237.

ARRAY NONRECURSIVE DEGREES AND GENERICITY

ROD DOWNEY, CARL G. JOCKUSCH, AND MICHAEL STOB

ABSTRACT. A class of r.e. degrees, called the array nonrecursive degrees, previously studied by the authors in connection with multiple permitting arguments relative to r.e. sets, is extended to the degrees in general. This class contains all degrees which satisfy $\mathbf{a}'' > (\mathbf{a} \cup \mathbf{0}')'$ (i.e. $\mathbf{a} \in \overline{GL_2}$) but in addition there exist low r.e. degrees which are array nonrecursive (a.n.r.). Many results for $\overline{GL_2}$ degrees extend to the a.n.r. degrees and thus to certain low degrees. A new notion of genericity (called pb-genericity) is introduced which is intermediate between 1-genericity and 2-genericity. It is shown that the upward closure of the pb-generic degrees is the set of a.n.r. degrees.

1. INTRODUCTION

This paper is a sequel to [5], which was a study of certain recursively enumerable sets called *array nonrecursive sets*. A number of characterizations of the degrees of these sets were obtained, showing that they are precisely those r.e. degrees relative to which various "multiple permitting constructions" may be performed. (In such a construction an attempt to meet a requirement may require a finite, but recursively bounded, number of permissions from the oracle.) For example, it was shown that, for any increasing recursive function f the array nonrecursive degrees are precisely those r.e. degrees $\mathbf{a}$ such that some function of degree $\mathbf{a}$ is not f-r.e. (A function is f-r.e. if it has a recursive approximation which changes at most $f(n)$ times on argument n for each n.) Subsequently Kummer [10, Theorem 3.2] characterized the array nonrecursive degrees as those r.e. degrees which contain an r.e. set A which is "complex" in the sense of Kolmogorov complexity. (This means that there is a constant c such that, for infinitely many n, the Kolmogorov complexity of the first n bits of the characteristic function of A is at least $2 \log n - c$.)

In the current paper we extend the notion of array nonrecursiveness from r.e. degrees to degrees in general. This is a conservative extension in the sense that the r.e. degrees which are array nonrecursive in the sense of

1991 *Mathematics Subject Classification.* Primary 03D30; Secondary 03D25.
Jockusch was supported by NSF grant DMS 92-02833.

Typeset by $\mathcal{A}\mathcal{M}\mathcal{S}$-TEX

[5] coincide with the r.e. degrees which are array nonrecursive in the sense of the current paper, as we will show. It will be clear from our definition that every degree $\mathbf{a} \in \overline{GL_2}$ ($= \{\mathbf{a} : \mathbf{a}'' > (\mathbf{a} \cup \mathbf{0}')'\}$) is array nonrecursive. However, the array nonrecursive (a.n.r.) degrees properly include the $\overline{GL_2}$ degrees, and in fact there are low r.e. degrees which are a.n.r. by [5, Theorem 2.1]. On the other hand, we will show that a.n.r. degrees have many of the properties of $\overline{GL_2}$ degrees. For example, we will show that every a.n.r. degree $\mathbf{a}$ bounds a 1-generic degree (extending Jockusch-Posner [8]) and that every recursive lattice with distinct least and greatest elements can be embedded in $\mathcal{D}_{\leq}(\mathbf{a})$ preserving the lattice operations and the least and greatest elements (extending Fejer [6]). Downey [4] previously showed that every r.e. degree which is a.n.r. in the sense of [5] is the supremum of a minimal pair of (not necessarily r.e.) degrees, and our result extends this with a much easier proof.

We will define a new sort of genericity, called pb-genericity. This concept will be strictly intermediate in strength between 1-genericity and 2-genericity. The upward closure of the pb-generic degrees will be shown to be the a.n.r. degrees. To compare this notion of genericity with 1-genericity, recall that, roughly speaking, a 1-generic set is one which has all properties which can be guaranteed by a Kleene-Post construction with an effective list of requirements $\{R_n\}$ such that, for each requirement R_n and each string σ, either every set extending σ satisfies R_n, or a certain effective search yields a string $\tau \supseteq \sigma$ such that every set extending τ satisfies R_n. The intuitive idea of pb-genericity is that instead of a single effective search as above, there may be a finite sequence of effective searches. The number of such searches must be bounded in advance by a recursive function of n and σ. The first search simply produces σ, and no subsequent search begins until the previous one has terminated. The final terminating search yields a string $\tau \supseteq \sigma$ such that any set extending τ satisfies the requirement. Thus, this paper concerns multiple search arguments, rather than multiple permitting arguments as in [5]. We use the pb-generic sets as a convenient tool to study the cupping properties of a.n.r. degrees. (For background on genericity see, for example, [7] or [9].)

Our notation is standard and follows that of [14].

We are indebted to Peter Fejer and Martin Kummer for helpful corrections and suggestions.

Our new definition of array nonrecursiveness is based on domination properties of functions. We first recall that $f \leq_{wtt} A$ (for f a function and A a set) means that there is a number e and a recursive function b such that, for all n, $f(n) = \{e\}^A(n)$ and, furthermore, for each n, the use of the computation of $\{e\}^A(n)$ does not exceed $b(n)$. Let K be the usual complete r.e. set. It is easily seen that $f \leq_{wtt} K$ iff there are recursive functions $h(.,.)$ and $p(.)$ such that, for all n, $f(n) = \lim_s h(n, s)$ and $|\{s : h(n, s) \neq h(n, s + 1)\}| \leq p(n)$.

Definition 1.1. *A degree* **a** *is array nonrecursive (a.n.r.) if for each* $f \leq_{wtt} K$ *there is a function* g *recursive in* **a** *such that* $g(n) \geq f(n)$ *for infinitely many* n.

It is stated in [8, Lemma 1] that an arbitrary degree **a** is in $\overline{GL_2}$ iff for each function f recursive in $\mathbf{a} \cup 0'$ there is a function g recursive in **a** such that $g(n) \geq f(n)$ for infinitely many n. From this we immediately obtain the following:

Proposition 1.2. *For any degree* **a**, *if* $\mathbf{a} \in \overline{GL_2}$, *then* **a** *is a.n.r.*

As we will show, many results for $\overline{GL_2}$ degrees extend to a.n.r. degrees. The following result gives some characterizations of the a.n.r. degrees. As in [5], we define a *very strong array* to be a sequence $\{F_n\}$ of nonempty, pairwise disjoint finite sets of strictly increasing cardinality with union ω such that the canonical index of F_n is a recursive function of n. Fix a recursive enumeration $\{K_s\}$ of K, and define $m_K(n)$ to be the least s such that $K \restriction n = K_s \restriction n$, where $A \restriction n = \{i < n : i \in A\}$.

Theorem 1.3. *Let* **a** *be a degree, and let* $\{F_n\}$ *be a very strong array. Then the following three conditions are equivalent:*

(i) **a** *is a.n.r.*

(ii) There is a function h *recursive in* **a** *such that* $h(n) \geq m_K(n)$ *for infinitely many* n.

(iii) There is a function r *recursive in* **a** *such that for all* e *there exists* n *with* $W_e \cap F_n = W_{e,r(n)} \cap F_n$.

Proof. To prove $(ii) \rightarrow (i)$, let f be given with $f \leq_{wtt} K$, and let h satisfy (ii). We must find g recursive in **a** with $g(n) \geq f(n)$ for infinitely many n. Fix e and a recursive function b such that $f(n) = \{e\}^K(n)$ with use at most $b(n)$ for all n. We may assume without loss of generality that h and b are increasing. To compute $g(n)$, let s be minimal such that $s > h(b(n + 1))$ and $\{e\}_s^{K_s}(n) \downarrow$ with use at most $b(n)$, and let $g(n) = \{e\}_s^{K_s}(n)$. Clearly g is recursive in **a**. Let n and k be such that $b(n) \leq k \leq b(n + 1)$ and $h(k) \geq m_K(k)$, and let s be as in the definition of $g(n)$. We have $s \geq h(b(n + 1)) \geq h(k) \geq m_K(k) \geq m_K(b(n))$, so K_s and K are the same below the use of $\{e\}_s^{K_s}(n)$. Hence $g(n) = f(n)$. Since there are infinitely many j with $h(j) \geq m_K(j)$, there are infinitely many n for which k exists as described above, and hence $g(n) = f(n)$ for infinitely many n.

The implication $(i) \rightarrow (iii)$ is obtained by applying (i) to the function $f(n) = (\mu s)(\forall e \leq n)[W_e \cap F_n = W_{e,s} \cap F_n]$.

It remains only to show that $(iii) \rightarrow (ii)$. Let r witness (iii). Then for each e there are infinitely many n with $W_e \cap F_n = W_{e,r(n)} \cap F_n$. (If this fails, one obtains a contradiction by defining an index e' so that there is no n with $W_{e'} \cap F_n = W_{e',r(n)} \cap F_n$. If $x \in F_n$ for one of the finitely many n such that $W_e \cap F_n = W_{e,r(n)} \cap F_n$, let $\phi_{e'}(x)$ converge in strictly more than $r(n)$ steps, and otherwise let $\phi_{e'}(x)$ converge (if ever) in at least the same number of steps as $\phi_e(x)$.) Hence it suffices to show that there is an e such that, for all n, $\mu s[W_{e,s} \cap F_n = W_e \cap F_n] \geq m_K(n)$, since then it follows that r also witnesses (ii). The set W_e will be $V = \cup_s V_s$, where V_s is defined as follows. The idea is to make V change on F_n whenever an element $< n$ enters K. Let $V_0 = \emptyset$. Given V_s, let $c_{n,s}$ be the least element (if any) of $F_n - V_s$, and let

$$V_{s+1} = V_s \cup \{c_{n,s} : (\exists z < n)[z \in K_{s+1} - K_s]\}.$$

Note that $|F_n \cap V| \leq |\{s : (\exists i < n)[i \in K_{s+1} - K_s]\}| \leq n < |F_n|$ so that $c_{n,s}$ is defined for all n and s. It follows that $(\mu s)[V_s \cap F_n = V \cap F_n] \geq m_K(n)$. By the proof of the slowdown lemma [14, page 284] (which requires only that the sets $V_{e,s}$ be recursive uniformly in e and s and not necessarily finite), there exists e such that $W_e = V$ and, for all s, $W_{e,s} \subseteq V_s$. Hence for all n, $\mu s[W_{e,s} \cap F_n = W_e \cap F_n] \geq m_K(n)$, and the proof is complete. $\square$

Since the truth of (i) of Theorem 1.3 does not depend on the choice of the very strong array $\{F_n\}$, it follows that the truth of (iii) is also independent of the choice of $\{F_n\}$. A similar result on independence of array occurs as Theorem 2.5 of [5]. The following result which shows in particular that the notion of array nonrecursiveness in Definition 1.1 is equivalent for r.e. degrees to the definition of array nonrecursiveness in [5].

Proposition 1.4. *Let $\mathbf{a}$ be an r.e. degree and let $\{F_n\}$ be a very strong array. Then the following are equivalent:*

(i) $\mathbf{a}$ is a.n.r.

(ii) There is an r.e. set A of degree $\mathbf{a}$ such that $(\forall e)(\exists n)[W_e \cap F_n = A \cap F_n]$.

Proof. To prove $(ii) \rightarrow (i)$, we assume that (ii) holds and show that (iii) of Theorem 1.3 holds with $r(n) = (\mu s)[A_s \cap F_n = A \cap F_n]$, where $\{A_s\}$ is a recursive enumeration of A. To do this, for each e let $V_e = \{x : (\exists s)[x \in W_{e,s} - A_s]\}$. Then V_e is an r.e. set so by (ii) there exists n such that $A \cap F_n = V_e \cap F_n$. It is then easily seen that $W_e \cap F_n = W_{e,r(n)} \cap F_n$ as needed to prove (iii) of Theorem 1.3.

For the converse, assume that $\mathbf{a}$ is a.n.r. as defined in Definition 1.1. Let $\{F_n\}$ be any very strong array. We shall construct an r.e. set A recursive in $\mathbf{a}$ such that $(\forall e)(\exists n)[W_e \cap F_n = A \cap F_n]$. (This suffices to prove (ii) by [5, Corollary 2.8].) Let $f(n) = (\mu s)(\forall e \leq n)[W_{e,s} \cap F_{\langle e,n \rangle} = W_e \cap F_{\langle e,n \rangle}]$.

Clearly $f \leq_{wtt} K$, so there exists g of degree at most **a** with $g(n) \geq f(n)$ for infinitely many n. By the Modulus Lemma, there is a recursive function $h(n,s)$ and a function p recursive in **a** such that $g(n) = h(n,s)$ for all $s \geq p(n)$. We now define the r.e. set A. It suffices to ensure that if $n \geq e$ and $g(n) \geq f(n)$ then $A \cap F_{\langle e,n \rangle} = W_e \cap F_{\langle e,n \rangle}$. Whenever $h(n,s) \neq h(n,s+1)$ and $e \leq n \leq s$, put all elements of $W_{e,h(n,s+1)} \cap F_{\langle e,n \rangle}$ into A at stage s, and let A be the set of all numbers obtained in this fashion. If $x \in A \cap F_{\langle e,n \rangle}$, then $x \in A_{p(n)}$, so A is recursive in p and hence A has degree at most **a**. Suppose now that $n \geq e$ and $g(n) \geq f(n)$. It follows from the definition of f that $W_{e,g(n)} \cap F_{\langle e,n \rangle} = W_e \cap F_{\langle e,n \rangle}$. Choose s as large as possible so that $h(n,s) \neq h(n,s+1)$. (There is no loss of generality in assuming there is at least one such $s \geq n$.) Then $h(n,s+1) = g(n)$ and so $A_{s+1} \cap F_{\langle e,n \rangle} = W_{e,h(n,s+1)} \cap F_{\langle e,n \rangle} = W_{e,g(n)} \cap F_{\langle e,n \rangle} = W_e \cap F_{\langle e,n \rangle}$. Furthermore, by the maximality of s, no elements of $F_{\langle e,n \rangle}$ enter A after $s+1$, so $A \cap F_{\langle e,n \rangle} = W_e \cap F_{\langle e,n \rangle}$, as needed to complete the proof. $\square$

If references to recursive enumerability are deleted from Proposition 1.4, then $(i) \rightarrow (ii)$ still holds but $(ii) \rightarrow (i)$ fails. In fact, for any very strong array $\{F_n\}_{n \in \omega}$ there is a set A such that $(\forall e)(\exists n)[W_e \cap F_n = A \cap F_n]$ and every function recursive in A is majorized by a recursive function, so that $\deg(A)$ fails badly to be a.n.r. This may be proved by modifying the standard construction of a hyperimmune-free degree so that the recursive perfect trees used are compatible with the condition that $(\forall e)(\exists n)[W_e \cap F_n = A \cap F_n]$. Specifically, one constructs a descending sequence $\{T_e\}_{e \in \omega}$ of recursive perfect trees and obtains A as their unique common branch. (Here a recursive perfect tree is a recursive function T from $2^{<\omega}$ to $2^{<\omega}$ such that, for any string σ, $T(\sigma 0)$ and $T(\sigma 1)$ are incompatible extensions of $T(\sigma)$.) The additional condition imposed on each T_e is that there exist infinitely many n such that if σ is any string of length $\min F_n$ extendible to a string in the range of T_e, then any string τ which extends σ and has length $(\max F_n) + 1$ is also extendible to a string in the range of T_e.

2. WORKING BELOW A.N.R. DEGREES

A number of results about initial segments determined by $\overline{GL_2}$ degrees extend easily to a.n.r. degrees. This is illustrated by the following two results.

Theorem 2.1. *If* **a** *is any a.n.r. degree, then there is a 1-generic degree* **b** $\leq$ **a**. *Hence, no a.n.r. degree is minimal.*

Proof. It is shown in [8 , Lemma 3] that there is a function f recursive in **0**$'$ such that for any g with $g(n) \geq f(n)$ for infinitely many n there is a 1-generic set B recursive in g. It suffices to show that the function f specified in the proof of that lemma satisfies $f \leq_{wtt} K$. This is clear since f has the form $f(n) = \max\{\psi(e,\sigma) : e \leq n, |\sigma| = n, \psi(e,\sigma) \downarrow\}$, for a certain partial recursive function ψ. (Here σ ranges over binary strings.) Thus the natural recursive approximation to $f(n)$ changes at most once for each pair (e,σ), with $e \leq n$ and $|\sigma| = n$. □

Downey [4, Theorem 1.1] proved that every r.e. degree $\mathbf{a}$ which is a.n.r. in the sense of [5] is the sup of a minimal pair of (possibly non-r.e.) degrees. Below, this result is extended to a.n.r. degrees in general and also (applying the construction of Fejer [6]) to a more general class of lattices. (Another extension to a.n.r. degrees in general will be given in Theorem 2.5.)

Theorem 2.2. *Let* $\mathbf{a}$ *be any a.n.r. degree. Then any recursively presented lattice with distinct least and greatest elements can be embedded in* $\mathcal{D}_{\leq}(\mathbf{a})$ *with least and greatest elements preserved.*

Proof. This is established by the proof of Theorem 6 of [6], once it is shown that the function F defined in that proof satisfies $F \leq_{wtt} K$. The function F has the form $F(n) = \max\{f(\sigma,e) : (\sigma,e) \in D_{g(n)}\}$, where g is recursive, D_k is the finite set with canonical index k, and $f(\sigma,e)$ is a certain function. (Here σ ranges over $\omega^{<\omega}$.) The definition of f is quite involved, but makes it clear that f has a natural recursive approximation which changes at most 5 times on each argument. Thus $f \leq_{wtt} K$, so $F \leq_{wtt} K$. □

T. Slaman [12] proved that there is a nonzero r.e. degree $\mathbf{b}$ which is not the sup of a minimal pair of (not necessarily r.e.) degrees. Downey [4, §1] then answered a question raised by Slaman by showing that there is a low r.e. degree $\mathbf{a}$ such that every r.e. degree $\mathbf{b} \geq \mathbf{a}$ is the sup of a minimal pair. (This used the existence of a low r.e. degree which is a.n.r. and the result [4, Theorem 1.1] cited above.) The following corollary removes the restriction that $\mathbf{b}$ be r.e. from Downey's result and also gives a simpler proof.

Corollary 2.3. *There is a low r.e. degree* $\mathbf{a}$ *such that every degree* $\mathbf{b} \geq \mathbf{a}$ *is the sup of a minimal pair.*

Proof. Let $\mathbf{a}$ be any low r.e. a.n.r. degree and apply Theorem 2.2 (or Theorem 2.5, to follow) and the obvious fact that the a.n.r. degrees are upward closed. It is shown in [5, Theorem 2.1] that there is a low r.e. degree which is a.n.r. in the sense of [5], but such degrees are also a.n.r. in the sense of the current paper by Proposition 1.4. □

It is shown in [8, Lemma 6] that if $\mathbf{a} \in \overline{GL_2}$ and $\{S_n\}$ is a sequence of dense sets of strings uniformly recursive in $\mathbf{a} \cup \mathbf{0}'$, then there is a function F recursive in $\mathbf{a}$ which meets each S_n. It is easily seen that the converse holds. (If the conclusion holds for $\mathbf{a}$, then for each f recursive in $\mathbf{a} \cup \mathbf{0}'$, we can let $S_n = \{\sigma : (\exists k \geq n)[p_\sigma(k) \downarrow > f(k)]\}$ to show that there is a function g recursive in $\mathbf{a}$ with $g(k) \geq f(k)$ for infinitely many k. Here p_σ

enumerates $\{n : \sigma(n) = 1\}$ in increasing order.) It follows from [8, Lemma 1] that $\mathbf{a} \in \overline{GL_2}$. Thus we have an example of a result for $\overline{GL_2}$ degrees which does not extend to a.n.r. degrees. In fact, there is an a.n.r. degree $\mathbf{a}$ and a sequence of dense sets $\{S_n\}$ uniformly recursive in $\mathbf{0}'$ such that no function recursive in $\mathbf{a}$ meets each S_n. (Let $\mathbf{a}$ be a low a.n.r. degree, which exists by [5, Theorem 2.1] and Proposition 1.4.) Nonetheless, in the following result, we prove an appropriate analogue of [8, Lemma 6]. Call a sequence $\{S_n\}$ of sets of strings *uniformly wtt-dense* if there is a function $d \leq_{wtt} K$ such that for all strings $\sigma \in 2^{<\omega}$, $d(n, \sigma) \supseteq \sigma$ and $d(n, \sigma) \in S_n$.

Theorem 2.4. *Let $\mathbf{a}$ be an a.n.r. degree, and let $\{S_n\}$ be uniformly wtt-dense. Then there is a set A recursive in $\mathbf{a}$ which meets each S_n.*

Proof. Let $\mathbf{a}$ and S_n be as in the hypothesis of the theorem, and let $d(n, \sigma)$ witness the wtt-density of $\{S_n\}$. Let $\hat{d}(n, \sigma, s)$ and $b(n)$ be recursive functions such that $\lim_s \hat{d}(n, \sigma, s) = d(n, \sigma)$ and $|\{s : \hat{d}(n, \sigma, s) \neq \hat{d}(n, \sigma, s + 1)\}| \leq b(n)$. Let $h(n, \sigma) = (\mu s)(\forall t \geq s)[\hat{d}(n, \sigma, t) = \hat{d}(n, \sigma, s)]$. Let $f(n) = \max\{h(e, \sigma) : e \leq n, |\sigma| \leq n\}$. We have $f \leq_{wtt} K$ since $d \leq_{wtt} K$. Fix a function g recursive in $\mathbf{a}$ such that $g(n) \geq f(n)$ for infinitely many n.

We obtain A as $\cup_s \sigma_s$, where $|\sigma_s| = s$. The method of proof is familiar from [8], and we give only an informal sketch. The idea for making A meet S_e is to wait for a stage $n \geq e$ where $\hat{d}(e, \sigma_n, g(n)) \supseteq \sigma_n$ and then to set $\sigma_t = \hat{d}(e, \sigma_n, g(n)) \upharpoonright t$ for $n < t \leq |\hat{d}(e, \sigma_n, g(n))|$. However, this process may be interrupted if there is action for some S_m with $m < e$ before it is completed, and in this case we start over on S_e. Also, we say that the attack is *discredited* at stage u ($u > n$), if for some t, $g(n) < t \leq g(u)$ we have $\hat{d}(e, \sigma_n, t) \neq \hat{d}(e, \sigma_n, g(n))$. If this happens, we also start over on S_e.

To see that this strategy works and only acts finitely often, assume this for all $e' < e$, and consider $n > e$ with $g(n) \geq f(n)$ such that we do not act for any $e' < e$ at stage n or any subsequent stage. If we act as above for e at stage n, the process is clearly never interrupted or discredited, and it is clear that A meets S_e and that we never act again for S_e after the process is completed. If we did not act as above for e at stage n, there must have existed a stage $m < n$ such that we attacked e at stage m and the attack was not interrupted or discredited at any stage k, $m < k \leq n$. Since $g(n) \geq f(n) \geq f(m)$, it follows that the attack starting at stage m will never be discredited or interrupted. Thus also in this case, A meets S_e and we act only finitely often to achieve this. $\square$

The following application of the method of proof of Theorem 2.4 implies Theorem 2.1 and a special case of Theorem 2.2, as well as the fact that no degree is minimal among the a.n.r. degrees. The method of proof is basically that used by Cooper [1, Theorem 1] to show that there is a minimal pair of degrees with sup $\mathbf{0}'$.

Theorem 2.5. *Every a.n.r. degree is the supremum of two 1-generic a.n.r. degrees which form a minimal pair.*

Proof. Let $\mathbf{a}$ be a given a.n.r. degree and let A be a set of degree $\mathbf{a}$ such that A is recursive in every infinite subset of A. (The last condition entails no loss of generality by [3, P2 and T2].) We will construct a set $B = B_0 \oplus B_1$ so that the degrees of B_0 and B_1 have the desired properties. To make $A \leq_T B_0 \oplus B_1$ we require that $B_0 \cap B_1 \subseteq A$, and that $B_0 \cap B_1$ be infinite. Thus we must satisfy the following requirements for all $e \in \omega$ and all $i \leq 1$:

$G_e^i : (\exists \sigma \subseteq B_i)(\sigma \in W_e \vee (\forall \tau \supseteq \sigma)[\tau \notin W_e])$

$N_e : \{e\}^{B_0} = \{e\}^{B_1} = f, f \text{ total } \to f \text{ recursive}$

$A_e^i : (\exists k > e)[p_{B_i}(k) > m_K(k)]$

$P_e : |B_0 \cap B_1| \geq e$

Effectively list the requirements other than the P_e's as $R_0, R_1, \ldots$. We will define a function d such that $d(n, \sigma) \supseteq \sigma$ for all n and σ, $d \leq_{wtt} K$, and R_n is satisfied whenever $B \supseteq d(n, \sigma)$ for any string σ. It would then be a direct application of Theorem 2.4 to conclude that $\mathbf{a}$ bounds a minimal pair of a.n.r., 1-generic degrees. The requirements P_e are easily combined with these since there is an $\mathbf{a}$-recursive function p such that, for all σ, $p(e, \sigma) \supseteq \sigma$, and any set B which extends $p(e, \sigma)$ satisfies P_e. (The P_e's should be effectively meshed into the list of requirements, and values of p should be used in place of values of $\hat{d}$ when attacking P_e. Such attacks are never discredited but may be interrupted for requirements of higher priority.) However, we must also ensure that $B_0 \cap B_1 \subseteq A$. Call a string σ *admissible* if $\{n : \sigma(2n) = \sigma(2n + 1) = 1\} \subseteq A$. To make $B_0 \cap B_1 \subseteq A$, we also require that every string extended by B be admissible. The proof of Theorem 2.4 has the property that every string σ_t chosen during an attack for S_e is extended by $\hat{d}(e, \sigma_s, u)$ for some $s < t$, and some u. We will choose d and $\hat{d}$ so that if σ is admissible, then $\hat{d}(e, \sigma, s)$ is admissible for all e and s. Furthermore, the function p mentioned above can be chosen so that if σ is admissible, then $p(e, \sigma)$ is admissible for all e. Also, in the construction we let σ_0 be the empty string and, if no requirement requires attention when we define σ_{s+1}, we let $\sigma_{s+1} = \sigma_s^* < 0 >$. It is then clear by induction on s that σ_s is admissible.

Thus it suffices to define $p(e, \sigma), d(n, \sigma)$, and $\hat{d}(n, \sigma, s)$ as described above. The definition of p is left to the reader. We define d below and in all cases let $\hat{d}$ be the "natural" recursive approximation to d.

We now define $d(n, \sigma)$. Let $\sigma = \sigma_0 \oplus \sigma_1$, so $\sigma_j(n) = \sigma(2n + j)$ for $j \leq 1$. Suppose first that R_n is a genericity requirement G_e^0. If there is a string $\tau_0 \in W_e$ such that $\tau_0 \supseteq \sigma_0$ choose the first such τ_0 in an effective enumeration of W_e and let $d(n, \sigma) = \tau_0 \oplus \tau_1$, where τ_1 is obtained by concatenating σ_1 with an appropriate number of 0's. If no such τ_0 exists, let $d(n, \sigma) = \sigma$. Now suppose that R_n is a minimal pair requirement N_e. If there is no e-split pair (τ_0, μ_0) of extensions of σ_0, let $d(n, \sigma) = \sigma$. Otherwise, consider the first one (τ_0, μ_0) in an effective enumeration of all such e-splittings, and let k be the least argument on which τ_0, μ_0 e-split. We may assume $|\tau_0| = |\mu_0| = q$, and let τ_1 be an extension of σ_1 of length q obtained by concatenating σ_1 with an appropriate number of 0's. If there is no δ_1 extending τ_1 with $\{e\}^{\delta_1}(k) \downarrow$, let $d(n, \sigma) = \tau_0 \oplus \tau_1$. Otherwise, choose the first such δ_1. If τ_0 and δ_1 are e-split, let γ_0 be the extension of τ_0 of the same length as δ_1 obtained by concatenating 0's to τ_0, and set $d(n, \sigma) = \gamma_0 \oplus \delta_1$. Otherwise, μ_0 and δ_1 are e-split, and $d(n, \sigma)$ is defined similarly with μ_0 in place of τ_0. We leave the case where R_n is A_e^i to the reader.

Note that $d(n, \sigma)$ is always defined so that it does no new coding, i.e. if $d(n, \sigma)(2k) = d(n, \sigma)(2k+1) = 1$, then $2k+1 < |\sigma|$. We can easily arrange that this is also true of the approximations $\hat{d}(n, \sigma, s)$, so we are able to make all σ_s admissible as described above.

$\square$

3 GENERICITY AND CUPPING

We saw in the previous section that every a.n.r. degree bounds a 1-generic degree. On the other hand, 1-genericity is not a sufficiently strong form of genericity for all of our purposes. In this section we define a notion of genericity intermediate between 1-genericity and 2-genericity which corresponds more precisely to the a.n.r. property. Using this notion for convenience we prove that every a.n.r. degree can be nontrivially cupped to all larger degrees.

First, let $f \leq_{pb} C$ mean that f can be computed from oracle C by a reduction procedure with a primitive recursive bound on the use function. Note that $f \leq_{pb} K$ iff there is a recursive function $\hat{f}(n, s)$ and a primitive recursive function p such that $\lim_s \hat{f}(n, s) = f(n)$ and $|\{s : \hat{f}(n, s) \neq \hat{f}(n, s+1)\}| \leq p(n)$. It is then easy to see that the various functions $f \leq_{wtt} K$ mentioned in §2 actually satisfy $f \leq_{pb} K$. If S is a set of strings, call S *pb-dense* if there is a function $f \leq_{pb} K$ such that, for all strings σ, $f(\sigma) \supseteq \sigma$ and $f(\sigma) \in S$.

Definition 3.1. *A set A is pb-generic if A meets every pb-dense set of strings.*

For each e, $\{\sigma : \sigma \in W_e \vee (\forall \tau \supseteq \sigma)[\tau \notin W_e]\}$ is pb-dense, so each pb-generic set is 1-generic. If $f \leq_{pb} K$, then the range of f is Σ_2, so every 2-generic set is pb-generic.

Theorem 3.2. *(i) If* **a** *is a.n.r., there is a pb-generic set* A *recursive in* **a**. *(ii) If* A *is pb-generic, then* $deg(A)$ *is a.n.r.*

Proof. To show (i), we first observe that the functions $\leq_{pb} K$ are uniformly $\leq_{wtt} K$. Let $p_0, p_1, \ldots$ be a uniformly recursive listing of the primitive recursive functions. Let $g(\langle a, b \rangle, n) = \{b\}^K(n)$ if, for each $i \leq n$, $\{b\}^K(i) \downarrow$ with use at most $p_a(i)$. Otherwise, let $g(\langle a, b \rangle, n) = 0$. It is clear that $g \leq_{wtt} K$ and that each function $f \leq_{pb} K$ has the form $n \mapsto g(e, n)$ for some e. Let $h(e, \sigma) = g(e, \sigma)$ if $g(e, \sigma) \supseteq \sigma$, and otherwise let $h(e, \sigma) = \sigma$. Let $S_n = \{h(n, \sigma) : \sigma \in 2^{<\omega}\}$. Then the sets $S_0, S_1, \ldots$ are uniformly wtt-dense. Also each pb-dense set contains some S_i. By Theorem 2.4, there is a set A recursive in **a** which meets each S_n, and clearly A is pb-generic.

To prove (ii), assume that A is pb-generic. We show that the principal function p_A of A satisfies $p_A(n) \geq m_K(n)$ for infinitely many n. This suffices by Theorem 1.3 to conclude that $deg(A)$ is a.n.r. Let k be given. We define a pb-dense set S of strings so that every set A which meets S is such that $(\exists n \geq k)[p_A(n) \geq m_K(n)]$. Define:

$$S = \{\sigma : (\exists n \geq k)[p_\sigma(n) \geq m_K(n)]\}$$

(Here p_σ enumerates $\{n : \sigma(n) = 1\}$ in increasing order.)

Then S is witnessed pb-dense by f, where $f(\sigma) = \sigma 1^k 0^{m_K(n)} 1$, and $n = |\sigma^{-1}(1)| + k$. Note that if $A \supseteq f(\sigma)$, then $p_A(n) \geq m_K(n)$, where n is as in the definition of $f(\sigma)$ and, in particular, $n \geq k$. $\square$

The following corollary and the remarks after Definition 3.1 show that the concept of pb-genericity is strictly intermediate between 1-genericity and 2-genericity.

Corollary 3.3. *There is a 1-generic degree* **b** *which is not pb-generic. There is also a pb-generic degree* **d** *which is not 2-generic.*

Proof. Let **a** be a nonzero r.e. degree which is not a.n.r. (Such a degree exists by [5, Theorem 2.10] and Proposition 1.4.) Let **b** be a 1-generic degree $\leq$ **a** (see [14, Exercise VI.3.9]). Then **b** is 1-generic, but it is not pb-generic by Theorem 3.2 and the upward closure of the a.n.r. degrees. For the other half, it is clear from Theorem 3.2 and the fact that $\mathbf{0}'$ is a.n.r. that there is a pb-generic degree $\mathbf{d} \leq \mathbf{0}'$. Clearly **d** is not 2-generic, as any 2-generic degree is 1-generic relative to $\mathbf{0}'$ and hence not below $\mathbf{0}'$. $\square$

Work of S. Kurtz [11] gives further examples of 1-generic degrees which are not pb-generic. Specifically, Kurtz [11, Theorem 4.1] showed that for almost every degree **b** (in the sense of the measure on degrees induced by the usual product measure on 2^ω) there is a 1-generic degree $\mathbf{a} \leq \mathbf{b}$. In the other direction, Kurtz [11, Theorem 4.3] showed that there is a fixed function f recursive in $\mathbf{0}'$ such that for almost every degree **b**, every function g

recursive in **b** is such that $f(n) \geq g(n)$ for all sufficiently large n. Kurtz's proof actually shows that $f \leq_{wtt} K$. It follows that almost no degree is a.n.r., and thus that almost every degree bounds a 1-generic degree but no pb-generic degree.

We close by investigating cupping properties of a.n.r. degrees.

Theorem 3.4. *Let* **a** *be an a.n.r. degree, and suppose that* **c** > **a**. *Then there is a degree* **b** < **c** *such that* **a** ∪ **b** = **c**.

Proof. If we had the additional assumption that **a** is n-r.e. for some n, the result would be virtually a direct application of [8, Lemma 7] and Theorem 3.2. (Theorem 3.2 would need to be modified in an entirely trivial way to handle strings in $3^{<\omega}$.) However, to avoid making this assumption, we construct a minimal pair $\mathbf{b_0, b_1}$ of degrees such that $\mathbf{a} \cup \mathbf{b_i} = \mathbf{c}$ for $i = 0, 1$, so that the theorem holds for $\mathbf{b} = \mathbf{b_0}$ or $\mathbf{b} = \mathbf{b_1}$. To obtain $\mathbf{b_0}$ and $\mathbf{b_1}$, we let $F : \omega \to \{0, 1, 2\}$ be a function which is recursive in **a** and pb-generic for sets of strings in $3^{<\omega}$. (The definition of pb-genericity and the proof of Theorem 3.2 are modified in the obvious way for such strings.) Then F has the following properties:

(1) There are infinitely many even n and infinitely many odd n such that $F(n) = 2$.

(2) If $\hat{F}$ is obtained from F by replacing all values of 2 by either 0 or 1 in any way whatever, then the degrees of $\{n : \hat{F}(2n) = 1\}$ and $\{n : \hat{F}(2n + 1) = 1\}$ form a minimal pair.

Suppose that (1) and (2) have been established. To complete the proof of the result, let $n_0, n_1, \ldots$ be the even values of n with $F(n) = 2$ in natural order, and let $m_0, m_1, \ldots$ be the odd values of n with $F(n) = 2$ in natural order. Let C be a set of degree **c**. Define $\hat{F}(n_k) = \hat{F}(m_k) = C(k)$ and elsewhere let $\hat{F}(n) = F(n)$. Let $B_i = \{n : \hat{F}(2n + i) = 1\}$ for $i \leq 1$. Then the result holds with $\mathbf{b_i} = deg(B_i)$.

The proof of (1) and (2) is similar to the proof of Theorem 2.5. However, in defining $d(n, \sigma)$ when R_n is a minimal pair requirement, one must successively consider all ways of replacing the 2's in σ by 0's and 1's. This can be done without introducing any 2's into $d(n, \sigma)$ not already present in σ. We omit the details, but see [8, Lemma 7] for a similar argument. In fact, it is easily seen that $B_0 \oplus B_1$ is 1-generic. Thus, if **a** is any a.n.r. degree and $\mathbf{c} \geq \mathbf{a}$, then there exist $\mathbf{b_0}$ and $\mathbf{b_1}$ such that $\mathbf{b_0, b_1}$ form a minimal pair, $\mathbf{a} \cup \mathbf{b_0} = \mathbf{a} \cup \mathbf{b_1} = \mathbf{c}$, and $\mathbf{b_0, b_1, b_0} \cup \mathbf{b_1}$ are all 1-generic. □

It follows from the proof of Theorem 3.4 (or from Theorems 3.2 and 3.4) that every pb-generic degree can be nontrivially cupped up to all higher degrees. On the other hand, it is not true that every 1-generic degree can be nontrivially cupped up to all higher degrees. This follows from the theorem

of Cooper [2] and Slaman-Steel [13, Theorem 3.1] that there are r.e. degrees
a, c with $0 < a < c$ such that no degree $b < c$ satisfies $a \cup b = c$ and the
fact that every non-zero r.e. degree bounds a 1-generic degree [14, Exercise
VI.3.9].

We do not know whether the set of a.n.r. degrees is definable in the partial ordering of the degrees. One conceivable definition (in view of Theorem
2.5) is that the a.n.r. degrees are exactly those degrees a such that every
degree $b \geq a$ is the sup of a minimal pair.

REFERENCES

[1] S. B. Cooper, *Degrees of unsolvability complementary between recursively enumerable degrees*, Annals of Math. Logic **4** (1972), 31–73.

[2] S. B. Cooper, *The strong anticupping property for recursively enumerable degrees*, J. Symbolic Logic **54** (1989), 527–539.

[3] J. C. E. Dekker and J. Myhill, *Retraceable sets*, Canad. J. Math. **10** (1958), 357–373.

[4] R. Downey, *Array nonrecursive degrees and lattice embeddings of the diamond*, Ill. J. Math. **37** (1993), 349–374.

[5] R. Downey, C. Jockusch, and M. Stob, *Array nonrecursive sets and multiple per-mitting arguments*, Recursion Theory Week (Proceedings, Oberwolfach 1989) (K. Ambos-Spies, G. H. Müller, and G. E. Sacks, eds.); Lecture Notes in Math., vol. **1432**, Springer–Verlag, 1990, pp. 141–173.

[6] P. Fejer, *Embedding lattices with top preserved below non-GL_2 degrees*, Zeitschr. f. math. Logik und Grundlagen d. Math. **35** (1989), 3–14.

[7] C. Jockusch, *Degrees of generic sets*, Recursion Theory: Its Generalisations and Applications (F. Drake and S. S. Wainer, eds.); Proceedings of Logic Colloquium '79; London Mathematical Society Lecture Notes Series No. 45, Cambridge University Press, 1980, pp. 110–139.

[8] C. Jockusch and D. Posner, *Double jumps of minimal degrees*, J. Symbolic Logic **43** (1978), 714–724.

[9] M. Kumabe, *Degrees of generic sets*, This Volume.

[10] M. Kummer, *Kolmogorov complexity and instance complexity of recursively enumer-able sets*, SIAM Journal of Computing (to appear).

[11] S. Kurtz, *Randomness and Genericity in the Degrees of Unsolvability*, Ph. D. thesis, University of Illinois at Urbana-Champaign, 1981.

[12] T. Slaman, *A recursively enumerable degree that is not the top of a diamond in the Turing degrees*, to appear.

[13] T. Slaman and J. Steel, *Complementation in the Turing degrees*, J. Symbolic Logic **54** (1989), 160–176.

[14] R. I. Soare, *Recursively Enumerable Sets and Degrees*, Springer–Verlag, New York, 1987.

Mathematics Department, Victoria University, P.O. Box 600, Welling-
ton, New Zealand. *E-mail address*: downey@math.vuw.ac.nz

Department of Mathematics, University of Illinois, 1409 West Green
Street, Urbana, Illinois 61801. *E-mail address*: jockusch@math.uiuc.edu

Department of Mathematics, Calvin College, Grand Rapids, MI. *E-mail address*: stob@calvin.edu

Dynamic Properties of Computably Enumerable Sets

Leo Harrington
and
Robert I. Soare*

Abstract

A set $A \subseteq \omega$ is *computably enumerable (c.e.)*, also called *recursively enumerable, (r.e.)*, or simply *enumerable*, if there is a computable algorithm to list its members. Let $\mathcal{E}$ denote the structure of the c.e. sets under inclusion. Starting with Post [1944] there has been much interest in relating the definable (especially $\mathcal{E}$-definable) properties of a c.e. set A to its "information content", namely its Turing degree, $\deg(A)$, under $\leq_T$, the usual Turing reducibility. [Turing 1939]. Recently, Harrington and Soare answered a question arising from Post's program by constructing a nonemptly $\mathcal{E}$-definable property $Q(A)$ which guarantees that A is incomplete $(A <_T K)$. The property $Q(A)$ is of the form $(\exists C)[A \subset_m C \ \& \ Q^-(A, C)]$, where $A \subset_m C$ abbreviates that "A is a major subset of C", and $Q^-(A, C)$ contains the main ingredient for incompleteness.

A *dynamic* property $P(A)$, such as prompt simplicity, is one which is defined by considering how fast elements elements enter A relative to some simultaneous enumeration of all c.e. sets. If some set in $\deg(A)$ is promptly simple then A is *prompt* and otherwise *tardy*. We introduce here two new tardiness notions, small-tardy(A, C) and Q-tardy(A, C). We begin by proving that small-tardy(A, C) holds iff A is small in C $(A \subset_s C)$ as defined by Lachlan [1968]. Our main result is that Q-tardy(A, C) holds iff $Q^-(A, C)$. Therefore, the dynamic property, Q-tardy(A, C), which is more intuitive and easier to work with than the $\mathcal{E}$-definable counterpart, $Q^-(A, C)$, is exactly equivalent and captures the same incompleteness phenomenon.

*The first author was supported by National Science Foundation Grant DMS 92-14048, and the second author by National Science Foundation Grant DMS 91-06714 and DMS 94-00825.

1   Introduction

Warning. From now on all sets and degrees will be c.e. unless specified otherwise. Post [16] initiated the study of the relationship between definable properties of a c.e. set A and its "information content" as measured by its Turing degree, $\deg(A)$, under the usual Turing reducibility $\leq_T$. By the 1950's Myhill noticed that the c.e. sets form a lattice $\mathcal{E}$ under inclusion and from then on most definable properties considered for c.e. sets were $\mathcal{E}$-definable. An exception is hyper-simplicity.

Friedberg and Muchnik solved Post's problem by constructing an incomplete and nonrecursive c.e. set, and invented the priority method to do it. The method was quickly developed into more sophisticated forms (infinite injury and the $0'''$-method) and used to prove a number of theorems on c.e. sets and degrees. Sacks used the second method to construct an incomplete maximal set, Yates constructed a complete maximal set, and Martin [15] brought these results together and extended them in his beautiful theorem that the degrees of maximal sets are exactly $\mathbf{H}_1$, the high degrees. Then Lachlan [8] and Shoenfield [17] proved that the degrees of the atomless sets (those with no maximal supersets) are $\overline{\mathbf{L}}_2$, the complement of the low$_2$ degrees. Both properties of being maximal or atomless are $\mathcal{E}$-definable properties.

Meanwhile Soare [18] developed a new method for generating automorphisms of $\mathcal{E}$, and used it to show that maximal sets form an orbit. (The *orbit* of $A \in \mathcal{E}$ is the set of all sets B which are automorphic to A, written $A \simeq B$.) The question stemming from Post's program remained open of whether there was an $\mathcal{E}$-definable property $P(A)$ which guarantees that A is incomplete and nonrecursive. It seemed that automorphisms could be used to give a negative answer by showing that every nonrecursive set A has a complete set in its orbit. However, Harrington and Soare gave a negative answer to this question by proving the following.

Theorem 1.1 (Harrington-Soare [3]) *There is a nonempty $\mathcal{E}$-definable property $Q(A)$ such that every c.e. set A satisfying $Q(A)$ is noncomputable and Turing incomplete.*

The property, which we shall describe fully in §4, is in two parts,

$$Q(A) \iff A \subset_m B \;\&\; Q^-(A, C),$$

where $A \subset_m C$ abbreviates that "A is a major subset of C", and $Q^-(A, C)$, an $\mathcal{E}$-definable property with several quantifiers which contains the main ingredient for incompleteness. The property $Q^-(A, C)$ succeeds but it is not very intuitive or easy to work with. The main achievement of the present paper is to produce a simpler and dynamic property, called Q-tardy(A, C), and to prove

(1) $$Q^-(A, C) \iff Q - tardy(A, C).$$

Hence, the dynamic property Q-tardy(A,C) is exactly equivalent to $Q^-(A, C)$ (in the presence of $A \subseteq_m C$) and therefore captures the incompleteness phenomenon.

In §2 we discuss dynamic properties and particularly *promptness* properties, such as prompt simplicity, and their opposite, *i.e.*, *tardiness* properties. This will motivate our present tardiness property, Q-tardy(A, C).

The above result led us to a curious discovery Theorem 3.2 about the $\mathcal{E}$-definable and new dynamic definitions of small subsets. Lachlan first defined the notion of A being a small subset of C, written $A \subseteq_s C$, in connection with his decision procedure for part of the elementary theory of $\mathcal{E}$ as described in §3. This notion proved useful and other facts about small sets were added by Stob [20] (see [19, pp. 193–195]), and others. The property $\widehat{Q}(A) = (\exists C)[A \subseteq_s C]$ comes tantalizingly close to being a property like $Q(A)$ which guarantees A incomplete, but not quite. We note that $\widehat{Q}(A)$ implies that A is not a promptly simple *set* by Corollary 3.3, but does not ensure that A is not of promptly simple *degree*.

The investigation of tardy properties with an eye toward incompleteness led naturally to a new tardiness property, *small-tardy*(A, C). Our other main result in the present paper is that,

(2) $$A \subseteq_s C \iff \text{small-tardy}(A, C).$$

This property small-tardy(A, C) gave new insight into the nature of small subsets, and led to a brand new and simpler $\mathcal{E}$-definable definition for the relation $A \subseteq_s C$ which had been overlooked researchers for 25 years. The general point is that dynamic notions frequently are more intuitive and easier to work with than $\mathcal{E}$-definable ones. Each sheds light on the other, particularly when one can show equivalence of the two such notions.

We use the terms "computably enumerable (c.e.)" and "recursively enumerable (r.e.)" interchangably, and likewise "computable" and "recursive."

2 Dynamic Properties

Most properties of an r.e. set A are *static* properties in that they refer to A as a completed object without mention of the enumeration of A. Such include Post's properties of being simple or hh-simple, and Myhill's property of being maximal, all of which are also $\mathcal{E}$-definable properties. Another static property which is not $\mathcal{E}$-definable or even invariant under automorphisms is hyper-simplicity. A *dynamic* property on the other hand is one which is defined using an computable enumeration $\{A_s\}_{s \in \omega}$ of A.

2.1   The Extension Theorem and Automorphisms

The first essential use of a dynamic property was probably the covering hypothesis in the Extension Theorem of Soare's maximal set automorphism theorem [18]. Here there were several simultaneous enumerations of arrays of r.e. sets, $\{U_n\}_{n\in\omega}$ and $\{\widehat{V}\}_{n\in\omega}$, and it was important to measure for an element x which U_n sets it entered before entering certain $\widehat{V}_m$ sets.

2.2 d-simple sets

In 1980 Lerman and Soare [11] attempted to capture part of the dynamic property of the Extension Theorem with an $\mathcal{E}$-definable property which is called *d-simple*, but they succeeded in capturing only a small part.

Definition 2.1 A coinfinite set A is *d-simple* if for all X there exists $Y \subseteq X$ such that

(3) $$X \cap \overline{A} = Y \cap \overline{A}, \text{ and}$$

(4) $$(\forall Z)[(Z - X) \text{ infinite} \implies (Z - Y) \cap A \neq \emptyset].$$

The tension in constructing Y is that to meet (4) we wish to make Y as small as possible, but to meet (3) we must eventually put every element of $X - A$ into Y. Every hh-simple is d-simple, and every d-simple set is simple. The degrees of d-simple sets include $\mathbf{H}_1$ and split $\mathbf{L}_1$. Also a d-simple set cannot be small [11, p. 141]. (This old result takes on new significance in view of the present paper because d-simple sets behave like prompt sets and by the result here Theorem 3.2 on small sets, small sets must be tardy.) The major open question left over from Post's program is the following.

Question 2.2 *Find a necessary and sufficient condition on A for A to be automorphic to a complete set. In particular, is every d-simple set automorphic to a complete set?*

The second question is not of great intrinsic interest itself, but it appears to be on the cutting edge of the symmetry between the methodologies for generating automorphisms and for producing invariant properties (such as $Q(A)$), and may therefore be useful in gaining insight into the completeness phenomenon and the first part of the question.

2.3   Promptly Simple Sets

The next significant advance came with the following definition of promptly simple sets by Maass [12].

Definition 2.3 (i) A coinfinite r.e. set A is *promptly simple* if there is a computable function p and a computable enumeration $\{A_s\}_{s\in\omega}$ of A such that for every e,

(5) $$W_e \text{ infinite} \implies (\exists s)\,(\exists x)\,[x \in W_{e,\text{ at } s} \cap A_{p(s)}].$$

(ii) An r.e. set A is *prompt* if A has promptly simple degree namely, $A \equiv_T B$ for some promptly simple set B, and an r.e. degree is *prompt* if it contains a prompt set.

(iii) An r.e. set or degree which is not prompt is *tardy*.

By the Promptly Simple Degree Theorem [19, Theorem XIII.1.7(iii)] a set A being prompt is equivalent to the following property which we may take as the definition. Let $\{A_s\}_{s\in\omega}$ be any recursive enumeration of A. Then there is a recursive function p such that for all s, $p(s) \geq s$, and for all e,

(6) $$W_e \text{ infinite} \implies (\exists^\infty x)\,(\exists s)\,[x \in W_{e,\text{ at } s} \And A_s{\upharpoonright}x \neq A_{p(s)}{\upharpoonright}x],$$

namely infinitely often A "promptly permits" on some element $x \in W_e$.

Promptly simply sets and degrees helped bring some dramatic advances in the subject. Maass [12] proved that any two promptly simple low sets are automorphic and discovered other properties of these sets [13]. Ambos-Spies, Jockusch, Shore, and Soare [1] used prompt degrees to unify and extend results about r.e. degrees, and promptness has been very influential ever since. (See [19, Chap. XIII].)

2.4 Almost Prompt Sets and Degrees

The material from the next two subsections §2.4 and 2.5 is not strictly necessary for this paper but is helpful to understand other notions of promptness and tardiness.

Harrington and Soare [4, Theorem 1.2] proved that every prompt set is automorphic to a complete set. They noticed that the same proof would work for a strictly larger dynamically defined class of sets called *almost prompt*, which are defined in terms of n-r.e. sets.

Definition 2.4 (i) A set $X \leq_T K$ is *n-r.e.* if $X = \lim_s X_s$ for some recursive sequence $\{X_s\}_{s\in\omega}$ such that for all x, $X_0(x) = 0$ and

$$\operatorname{card}\{s : X_s(x) \neq X_{s+1}(x)\} \leq n.$$

(For example, the only 0-r.e. set is $\emptyset$, the 1-r.e. sets are the usual r.e. sets, and the 2-r.e. sets are the d.r.e. sets.)

(ii) Such a sequence $\{X_s\}_{s\in\omega}$ is called an *n-r.e. presentation* of X.

It is well-known and easy to show [19, Exercise III.3.8., p. 38] that for $n > 0$, X is n-r.e. iff

(7) $X = (W_{e_1} - W_{e_2}) \bigcup (W_{e_3} - W_{e_4}) \bigcup \cdots \bigcup W_{e_{2k+1}}$, or

(8) $X = (W_{e_1} - W_{e_2}) \bigcup (W_{e_3} - W_{e_4}) \bigcup \cdots \bigcup (W_{e_{2k+1}} - W_{e_{2k+2}})$,

according as $n = 2k + 1$ is odd or $n = 2k + 2$ is even.

Definition 2.5 For $n = 0$ let $X_0^0 = \emptyset$. For $n > 0$ and $e = \langle e_1, e_2, \ldots e_n \rangle$ define

(9) $X_e^n = (W_{e_1} - W_{e_2}) \bigcup \cdots$,

as in (7) or (8) according as n is odd or even. We say that $\langle n, e \rangle$ is an n-r.e. *index* for X_e^n. Let

(10) $X_{e,s}^n = (W_{e_1,s} - W_{e_2,s}) \bigcup \cdots$.

Definition 2.6 Let A be an r.e. set and let $\{A_s\}_{s \in \omega}$ be a recursive enumeration of A. We say A is *almost prompt*, abbreviated a.p., if there is a nondecreasing recursive function $p(s)$ such that for all n and e,

(11) $X_e^n = \overline{A} \implies (\exists x)(\exists s)[x \in X_{e,s}^n \ \& \ x \in A_{p(s)}]$.

Note that, as in the case of promptly simple, this definition is independent of the enumeration of A; if $p(s)$ works for the enumeration $\{A_s\}_{s \in \omega}$, and if $\{A'_s\}_{s \in \omega}$ is another enumeration of A, define $p'(s) = (\mu t)[A'_t \supseteq A_p(s)]$. We may think of Definition 2.6 as asserting that A will p-promptly hit every approximation $\{X_{e,s}^n\}_{s \in \omega}$ for every n-r.e. set $X_e^n = \overline{A}$ where the recursive approximation $X_{e,s}^n$ is determined by the *standard enumeration* $\{W_{e,s}\}_{e,s \in \omega}$ of the r.e. sets. In [4, Conversion Lemma 11.4] we prove that if we specify another collection of n-r.e. sets $\{\widehat{X}_e^n\}_{n,e \in \omega}$, by some recursive approximation $\{\widehat{X}_{e,s}^n\}_{n,e,s \in \omega}$, then there is a recursive function q such that A will q-promptly hit $\{\widehat{X}_{e,s}^n\}_{n,e,s \in \omega}$ if $\widehat{X}_e^n = \overline{A}$.

2.5   Very Tardy Sets

The negation of the property of almost prompt is called *very tardy*. An important special case of this is known as 2-tardy and is closely related to the property $Q(A)$.

Definition 2.7 Let A be an r.e. set and let $\{A_s\}_{s \in \omega}$ be a recursive enumeration of A.

(i) We say A is *very tardy* if A is not almost prompt, namely if for every nondecreasing recursive function $p(s)$,

$$(12) \qquad (\exists n)(\exists e)[X_e^n = \overline{A} \ \& \ (\forall y)(\forall s)[y \in X_{e,s}^n \implies y \notin A_{p(s)}]].$$

(ii) We say A is *n-tardy* if in (i) the fixed n works uniformly for *all* such functions p, namely for every nondecreasing recursive function $p(s)$,

$$(13) \qquad (\exists e)[X_e^n = \overline{A} \ \& \ (\forall y)(\forall s)[y \in X_{e,s}^n \implies y \notin A_{p(s)}]].$$

The main idea about a very tardy set A is that if $x \in X_{e,s}^n$ then x can later enter A eventually, but x must first undergo a delay until at least stage $p(s)+1$ before doing so. Since class of almost prompt sets is a strict extension of the class of prompt sets it follows that the class of very tardy sets is a strict subclass of the class of tardy sets, hence the name "very tardy." Note that A is 0-tardy iff $A = \omega$, and A is 1-tardy iff A is recursive. The 2-tardy sets play a special role in our work and have additional characterizations as follows, as we prove in [5].

Proposition 2.8 (Harrington-Soare [5]) *For an r.e. set A the following are equivalent:*
  (i) A is 2-tardy;
  (ii) For every nondecreasing recursive function $p(s)$,

$$(14) \quad (\exists W_i \supseteq \overline{A})(\exists W_e = A)(\forall y)(\forall s)[y \in W_{i,s} - W_{e,s} \implies y \notin A_{p(s)}]].$$

3 Small Subsets

Lachlan [9] introduced small sets in his program to construct canonical examples of certain diagrams and then rule out possible extensions so as to give a decision procedure for the $\overrightarrow{\forall}\,\overrightarrow{\exists}$-theory of the lattice of r.e. sets. The following definition is clearly equivalent to the standard definition as in [19, Definition 4.10, p. 193].

Definition 3.1 A subset $A \subset C$ is a *small subset* of C (written $A \subset_s C$) if $A \subset_\infty C$ and for all X and Y, if
  (i) $X \cap (C - A) \subseteq Y$, then
  (ii) $(\exists Z)_{Z \subseteq X} [Z \supseteq (X - C) \ \& \ (Z \cap C) \subseteq Y]$.

If A is both a small subset and major subset of C we say it is a *small major subset* and write $A \subset_{\mathrm{sm}} C$.

Note that the consequent of the implication in (ii) is equivalent to the property

(15) $(\forall Y \supseteq C - A)[Y \cup \overline{C}$ is r.e.].

It is interesting now to see that this important notion of small subset, Theorem 3.2(i) below, just like the $Q(A)$ property, has a dynamic equivalent, Theorem 3.2(iii), below which we now prove. It is particularly that the equivalent dynamic definition (iii) led to the discovery of another $\mathcal{E}$-definable definition (ii) below which is simpler than the original $\mathcal{E}$-definable one, but lay undiscovered for over 25 years.

Theorem 3.2 (Harrington and Soare) *Suppose $A \subset_\infty C$. Then the following are equivalent:*

(i) $A \subset_s C$;

(ii) $(\forall Y)[(C - A) \subseteq Y] \implies (\exists Z)[\overline{C} \subseteq Z \ \& \ Z \cap C \subseteq Y]$;

(iii) *small-tardy(A, C), namely:*

(16) $(\forall f)(\exists T)[\overline{C} \subseteq T \ \& \ (\forall x)[x \in (T \cap C)_{at\ s} \implies x \notin A_{f(s)}]]$.

(In (iii) it is understood that f ranges over recursive functions which are nondecreasing.)

Note that (ii) is equivalent to the property,

$$(\forall Y \supseteq C - A)[Y \cup \overline{C} \text{ is r.e. }].$$

We refer to the property (iii) on $A \subset_\infty C$ as *small-tardy(A, C)* because it is a dynamic property.

Proof. (i) $\implies$ (ii). Trivial. Let $X = \omega$.

(ii) $\implies$ (iii). Fix a recursive function f as in (iii). We (BLUE) will build $Y \supseteq (C - A)$, so by (ii) the opponent (RED) must reply with $Z = W_j$ for some j, satisfying (ii). Define $W_{g(j)} = W_j \smallsetminus C$. If $x \in (W_j \smallsetminus C)_{at\ s}$, then by the Recursion Theorem and Slowdown Lemma [19, Lemma XIII.1.5] we can compute $t = (\mu v)[x \in W_{g(j),v}]$, and know that $t > s$.

Namely, if $x \in C_{s+1} - C_s$ take all j such that $x \in W_{j,s}$ (necessarily $j \leq s$). For each j compute $t_{x,j} = (\mu v)[x \in W_{g(j),v}]$. Let $t = \max\{t_{x,j} : \text{all such } j\}$. If $x \notin A_{f(t)}$ then enumerate x in Y at stage $f(t) + 1$. Since every $x \in C - A$ enters Y after some finite delay we have,

(17) $C - A \subseteq Y$.

However, no element once in A ever enters Y, so

(18) $Y \cap A \subseteq Y \smallsetminus A$.

By (17) and (ii), RED must play some Z satisfying (ii). In (iii) we let $T = Z \setminus C$. Let $W_j = T$. Now $\overline{C} \subseteq T$ because $\overline{C} \subseteq Z$. But $T \cap C \subseteq Y$ by (ii) implies by (18) that $T \cap C \subseteq Y \setminus A$. Now $A \setminus Y = \emptyset$. Hence, for all x, if $x \in (T \cap C)_{\text{at } s}$, then

$$x \in (T \setminus C)_{\text{at } s},$$

$$x \in W_{g(j),t} \text{ for some } t > s,$$

$$x \notin Y_{f(t)} \text{ by (18) and definition of } Y,$$

$$x \notin A_{f(s)} \text{ since } s < t \text{ and } f \text{ is nondecreasing.}$$

(iii) $\implies$ (i). Fix $A \subset_\infty C$ satisfying (iii). Given X and Y satisfying Definition 3.1 (i), we (RED) define Z satisfying Definition 3.1 (ii) as follows. Define

$$f(s) = (\mu t > s)(\forall x)[x \in X_s \cap C_s \implies x \in A_t \cup Y_t].$$

Such t exists by Definition 3.1 (i). Choose T satisfying (iii). Enumerate

(19) $$x \in Z_s \iff x \in Z_{s-1} \lor x \in (X_s \cap T_s) - C_s.$$

Now by (iii) for all x,

$$x \in (Z_s \cap C_s) \implies x \in (X_s \cap T_s), \text{ and}$$

$$x \in (T \cap C)_{\text{at } s} \implies x \notin A_{f(s)}, \text{ so}$$

$$x \in (Z \cap C)_{\text{at } s} \implies x \in Y_{f(s)}$$

by definition of f. ∎

Consider the property $\widehat{Q}(A) : (\exists C)[A \subset C]$. This resembles the property $Q(A)$ because $\widehat{Q}(A)$ implies that A is not a promptly simple *set*. However, it does not guarantee that A is not of promptly simple *degree*, and therefore, unlike $Q(A)$ it does not ensure that the orbit of A contains only incomplete sets.

Corollary 3.3 *If $A \subset_s C$ then A is not a promptly simple set.*

Proof. Let $A \subset_s C$. Let $p(s)$ be any nondecreasing total recursive function. By Theorem 3.2 (iii) there exists $T \supseteq \overline{C}$ such that

$$(\forall x)[x \in (T \cap C)_{\text{at } s} \implies x \notin A_{f(s)}]].$$

Hence, $W_j = T \cap C$ witnesses that A fails to satisfy (5).

4 $Q(A)$ And Tardy Properties

In the following definition we separate the first property of $Q(A)$ into two parts: the first part $Q^-(A, C)$ which is equivalent to a purely dynamic property and is the key to satisfying tardiness and hence incompleteness; and the second part asserting $A \subseteq_m C$, which is entirely standard.

Definition 4.1 (i) $Q^-(A, C) : A \subseteq_\infty C$ & $(\forall B \subseteq C)(\exists D \subseteq C)(\forall S)_{S \sqsubseteq C}$ [

$$(20) \qquad\qquad [B \cap (S - A) = D \cap (S - A)]$$

$$(21) \qquad\quad \implies (\exists T)[\overline{C} \subset T \ \& \ A \cap (S \cap T) = B \cap (S \cap T)]].$$

$$\quad (ii) \ \ Q(A) : (\exists C)[A \subseteq_m C \ \& \ Q^-(A, C)].$$

Definition 4.2 If $A \subset C$ then $Q\text{-}tardy(A, C)$ holds if

$$(22) \qquad A \subseteq_\infty C \ \& \ (\forall f)(\exists I \supseteq \overline{C})(\exists E = A)_{(E \setminus C) \cap I = \emptyset}$$
$$(\forall x)[x \in (I \setminus C \setminus E)_{\text{at } s} \implies x \notin A_{f(s)}].$$

The main result of the present paper is the following.

Theorem 4.3 (Main Theorem) $Q^-(A, C) \iff Q\text{-}tardy(A, C).$

We prove this in the next two theorems. The first resembles Lemma 1 of [3], and the second Lemma 2 there.

Theorem 4.4 $Q^-(A, C) \implies Q\text{-}tardy(A, C).$

Proof. Fix A and $C \in \mathcal{E}$ such that A satisfies $Q(A)$ via C, and fix indices $W_a = A$ and $W_c = C$ such that $W_a \subseteq W_c \setminus W_a$, which we write,

$$(23) \qquad\qquad\qquad A \subseteq C \setminus A,$$

because we define $A_s = W_{a,s}$ and $C_s = W_{c,s}$. To utilize the hypothesis $Q(A)$ BLUE will first split C into the disjoint union of uniformly r.e. sets $\{S_i\}_{i\in\omega}$, written $C = \sqcup_{i\in\omega} S_i$, and then on S_i BLUE will play B against $D = W_i$ to satisfy (20). Since $Q(A)$ holds, RED must reply with $T = $ some W_j to satisfy (21). Now BLUE will use a Π_2^0 guessing procedure (described in §4.2 below) to determine the correct values of i and j. We let $\alpha = \langle i, j \rangle$.

To better explain the basic α-module we will assume in §4.1 two simplifying hypotheses (discharged later in §4.2), the first of which asserts that BLUE has fixed the correct i and j so that BLUE is playing single sets B and S and has the indices i and j (respectively) of single r.e. sets D and T such that if BLUE satisfies (20) then RED satisfies (21). Also all sets below except A, B, and C have subscript α which we drop for this subsection.

4.1   The basic α-module under simplifying assumptions

Now BLUE begins to satisfy (20) by first arranging that on $S - A$,

$$(24) \qquad\qquad B \subseteq (D \setminus B).$$

Hence, RED must ensure that on $S \cap T$,

$$(25) \qquad\qquad A \subseteq (B \setminus A),$$

because if $x \in (S \cap T \cap A) \setminus B$ then BLUE can restrain $x \in \overline{B}$ forever thereby refuting (21) while still maintaining (20) by ensuring (24) and (27) on $S - A$.
    Now (24) and (25) together ensure that on $T \cap S$,

$$(26) \qquad\qquad A \subseteq (D \setminus B \setminus A).$$

To achieve the rest of (20) for every x currently in $(D - B) \cap (S - A)$, after a finite number of stages of "restraint on x" BLUE will enumerate x in B. Thus, on $S - A$ BLUE will play

$$(27) \qquad\qquad D - B = \emptyset.$$

This will force RED to ensure (21) by enumerating in A all x currently in $(B - A) \cap (S \cap T)$ so that on $S \cap T$,

$$(28) \qquad\qquad B - A = \emptyset.$$

As a second simplifying assumption BLUE assumes in §4.1 that if (21) holds for T then (21) also holds with T replaced by a certain set $U \subseteq T$ which will be played by BLUE and which also satisfies

$$(29) \qquad\qquad (U \cap C) \subseteq^* S.$$

(BLUE will discharge this assumption in §4.2.)
    But $A \subset_m C$ and $\overline{C} \subseteq U$ (from (21)) imply

$$(30) \qquad\qquad \overline{A} \subseteq^* U,$$

so from (29) and (30) we get

$$(31) \qquad\qquad (C - A) \subseteq^* S.$$

4.2　Describing the α-module

We (BLUE) will define r.e. sets U_α, S_α, E_α, and B, whose indices we know in advance by the Recursion Theorem. Let $\{(D_i, T_j)\}_{i,j \in \omega}$ be an effective listing of all pairs of r.e. sets. Below BLUE will define r.e. sets $\{S_{i,j}\}_{i,j \in \omega}$ such that $C = \sqcup_{i,j \in \omega} S_{i,j}$. Now BLUE begins by playing for every i and j the set B on $S_{i,j}$ against D_i to satisfy (24) and (27) and therefore (20). Hence, (20) is also satisfied by the sets B, D_i, and $S_i = \sqcup_{j \in \omega} S_{i,j}$. Thus, for some j, T_j must satisfy (21) and hence (25) and (28) for B, D_i, and S_i, and therefore also for B, D_i, and $S_{i,j}$. Let $\alpha = \langle i, j \rangle$, and let D_α, S_α, and T_α denote $D_i = W_i$, $S_{i,j}$, and $T_j = W_j$, respectively, and $D_{\alpha,s} = W_{i,s}$ and $T_{\alpha,s} = W_{j,s}$. For each α the conjunction of all the conditions in the matrices of (20) and (21) (with D, S, T replaced by D_α, S_α, and T_α respectively) is a Π_2^0 condition $F(\alpha)$. Hence, there is an r.e. sequence of r.e. sets $\{Z_\alpha\}_{\alpha \in \omega}$ such that for every α, $F(\alpha)$ holds iff $|Z_\alpha| = \infty$.

Defining U_α.　Define r.e. set U_α by

(32)　　$x \in U_{\alpha,s} \Longleftrightarrow x \in U_{\alpha,s-1} \vee [x \in T_{\alpha,s} - C_s \ \& \ x \leq |Z_{\alpha,s}|].$

By the Recursion Theorem with parameter α and the Slowdown Lemma [19, Lemma XIII.1.5] there is an index u_α (which we know in advance) such that

(33)　　　　　　$U_\alpha = W_{u_\alpha} \ \& \ W_{u_\alpha} \subseteq (U_\alpha \setminus W_{u_\alpha}).$

Defining S_α.　If $x \in C_{s+1} - C_s$ choose the least α such that $x \in U_{\alpha,s}$, and enumerate x in $S_{\alpha,s+1}$. (If no such α exists enumerate x in $S_{x,s+1}$.) This defines an r.e. set S_α.

Defining E_α.　Using the enumerations above for C, A, D_α, S_α, and $W_{e_\alpha} = U_\alpha$ we now define the r.e. set,

(34)　　　　$E_\alpha = ((W_{u_\alpha} \cap S_\alpha) \setminus D_\alpha) \ \cup \ ((C \setminus W_{u_\alpha}) \setminus A).$

This defines a recursive enumeration $\{E_{\alpha,s}\}_{s \in \omega}$ of the r.e. set E_α. Again by the Recursion Theorem with parameter α and the Slowdown Lemma there is an index e_α such that $W_{e_\alpha} = E_\alpha$ and $W_{e_\alpha} \subseteq (E_\alpha \setminus W_{e_\alpha})$.

Defining B.　Fix a nondecreasing recursive function $p(s)$.

　1. If $x \in (W_{u_\alpha,s} - W_{e_\alpha,s}) \cap W_{e_\alpha,s+1}$, then α-restrain x from B_t for all $t \leq p(s)$.

　2. If $x \in (W_{u_\alpha,s} \cap S_{\alpha,s} \cap W_{e_\alpha,s+1}) - B_s$ and x is not α-restrained from B_{s+1} then enumerate x in B_{s+1}.

This defines a recursive enumeration $\{B_s\}_{s \in \omega}$ of the r.e. set B. Note that x can be α-restrained for only finitely many stages, starting when 1. first holds, and then never again after the α-restraint is dropped. Hence, there is no permanent restraint on x entering B so (27) holds. (Note that x can be α-restrained only if $x \in S_\alpha$ so x can never be α-restrained and also β-restrained for $\beta \neq \alpha$ because the S_α sets are disjoint. Thus, unlike the predecessor [3, Lemma 1], there is no injury and no conflict between α-strategies.)

Let α be the least β such that Z_β is infinite. Hence, D_α, S_α, T_α, and U_α satisfy the first two simplifying assumptions in §4.1 including (29), because by (32) Z_β and hence U_β and S_β are finite for every $\beta < \alpha$. Hence, (29), (30), and (31) hold. Define the finite set $\widehat{S}_\alpha = \cup_{\beta < \alpha} S_\beta$.

Lemma 4.5 *Q-tardy(A,C) holds.*

Proof. Fix a nondecreasing recursive function $p(s)$. Apply the above construction to produce W_{u_α} and W_{e_α} for the least α satisfying $F(\alpha)$. Now

$$(35) \qquad (W_{e_\alpha} \setminus C) \cap W_{u_\alpha} = \emptyset,$$

by (34). Next $W_{e_\alpha} = A$ and $W_{u_\alpha} = U_\alpha \supseteq \overline{C}$ by (32) because $T_\alpha \supseteq \overline{C}$. Define $k = \max(\widehat{S}_\alpha)$, and $G = [0,k] \cap A$. Let $I_\alpha = W_{u_\alpha} - G$. We claim that I_α and $E_\alpha = W_{e_\alpha}$ satisfy (22) so Q-tardy(A,C) holds.

Suppose $x \notin \widehat{S}_\alpha$ and

$$(36) \qquad x \in (W_{u_\alpha,s_1} - W_{e_\alpha,s_1}) \ \& \ x \in A_t.$$

Then
$$(37) \qquad (\exists s_2 < t)[x \in C_{s_2}],$$

by (23). Then $x \in U_{\alpha,s}$. But $x \in U_\alpha \setminus C$ because $C \setminus U_\alpha = \emptyset$ by (32). Furthermore, when $x \in U_\alpha$ enters C, x enters S_α since $x \notin \widehat{S}_\alpha$. However, on $S_\alpha \cap U_\alpha$ we know $A \subseteq D_\alpha \setminus B \setminus A$ by (26). Hence, we may assume

$$(38) \qquad (\exists s_3)_{s_1 \leq s_3 \leq t}[x \in W_{u_\alpha,s_3} \cap S_{\alpha,s_3} \cap \overline{W}_{e_\alpha,s_3} \cap W_{e_\alpha,s_3+1}]$$

(Namely, while in $W_{u_\alpha} \cap S_\alpha \cap \overline{W}_{e_\alpha}$, at stage s_3 element x "announces its intention" to eventually enter A by first entering $W_{e_\alpha,s+1}$.) By the action of the α-module, $x \notin B_t$ for all t, $s_3 + 1 \leq t \leq p(s_3)$. But then by (26), $x \notin A_t$, $s_3 + 1 \leq t \leq p(s_3)$. In (36) we must have $v > p(s_1)$ since p is nondecreasing. Hence, (22) so Q-tardy(A,C) holds. ∎

This completes the proof of Theorem 4.4. ∎

Theorem 4.6 *Q-tardy$(A,C) \implies Q^-(A,C)$.*

Proof. We let the opponent (BLUE) play one set B and we (RED) play one set D against B (rather than the infinitely many B_i and D_i as in Lemma 2 of [3]). Next we let $\{(S_j, \widehat{S}_j) : j \in \omega\}$ be an effective listing of all disjoint pairs of r.e. sets (*i.e.*, played by BLUE). RED must reply with a set $T_{\langle j,k \rangle}$ such that if B, D, and S_j satisfy (20) then $T_{\langle j,k \rangle}$ satisfies (21). Fix recursive enumerations A_s and C_s of A and C.

For each j define the nondecreasing partial recursive function $f_j(s)$ as follows. For each $x \leq s$ perform the following subroutine to obtain s'' depending on x:

1. If $x \in C_s$ define $s' = (\mu v \geq s)[x \in S_{j,v} \sqcup \widehat{S}_{j,v}]$.

2. If $x \in S_{j,s'} \cap D_{s'}$ let $s'' = (\mu v \geq s')[x \in B_v \cup A_v]$.

Define $f_j(s) = \max\{f_j(s-1),\ \max\{s''_x : x \leq s\}\}$.

If B, D, and S_j satisfy condition (20) of $Q(A)$, then $f_j(s)$ is total recursive. Now applying the hypothesis of Q-tardy(A, C) to C, A, and f_j, and letting $\alpha = \langle j, k \rangle$, we get a pair of sets I_α and E_α such that

$$(39) \qquad I_\alpha \supseteq \overline{A}\ \&\ E_\alpha = A\ \&\ I_\alpha \cap (E_\alpha \setminus C) = \emptyset$$
$$\&\ (\forall y)(\forall s)[y \in I_{\alpha,s} - E_{\alpha,s} \implies y \notin A_{f_j(s)}]].$$

For $\alpha = \langle j, k \rangle$ let S_α, $\widehat{S}_\alpha$, T_α, and f_α denote S_j, $\widehat{S}_j$, $T_{\langle j,k \rangle}$, and f_j, respectively. We now use I_α and E_α to build T_α which satisfies (21). For each $\alpha = \langle j, k \rangle$ the conjunction of: (20) for (B, D, S_α); $S_\alpha \sqcup \widehat{S}_\alpha = C$; $B \subseteq C$; and the conditions in (39) is a Π^0_2 condition $F(\alpha)$. Let $\{Z_\alpha\}_{\alpha \in \omega}$ be an r.e. array of r.e. sets such that $F(\alpha)$ holds iff $|Z_\alpha| = \infty$.

Define T_α by

$$(40) \qquad x \in T_{\alpha,s} \iff x \in T_{\alpha,s-1}\ \lor\ [x \in I_{\alpha,s} - C_s\ \&\ x \leq |Z_{\alpha,s}|\,].$$

Hence, $C \setminus T_\alpha = \emptyset$, $T_\alpha \subseteq I_\alpha$, and $T_\alpha \supseteq \overline{C}$ iff $|Z_\alpha| = \infty$.

Suppose x enters C at some stage t. (By hypothesis $x \notin E_{\alpha,t}$.) Choose the least α such that $x \in T_{\alpha,t}$. For all $s \geq t$ let $x \in D_s$ iff $x \in E_{\alpha,s}$. (Namely, for the least such α let α define D in the sense that we let D copy E_α on $T_\alpha \setminus C$.)

Lemma 4.7 $Q^-(A, C)$ *holds.*

Proof. Suppose (20) holds for (B, D, S_j). Let $\alpha = \langle j, k \rangle$ be the least β such that Z_β is infinite. We must show that (21) holds for (B, S_α, T_α). Now $S_\alpha \sqcup \widehat{S}_\alpha = C$, and f_α is total.

By the definition of $F(\alpha)$ the pair I_α and E_α witnesses that A is 2-tardy relative to f_α. Now $T_\alpha \subseteq I_\alpha$, and $T_\alpha \supset \overline{C}$ because $I_\alpha \supset \overline{C}$ and $|Z_\alpha| = \infty$. But the f_α delay ensures that on $S_\alpha \cap T_\alpha$ the sets obey the intended order of enumeration, namely $x \in A$ implies that $x \in D \setminus B \setminus A$, and hence $Q(A)$ holds. To verify this suppose $x \in T_\alpha \cap S_\alpha \cap A$. Then

$$x \in T_\alpha \setminus C \setminus S_\alpha \setminus A.$$

Hence,

$$x \in I_\alpha \setminus C \setminus S_\alpha \setminus A,$$

because $T \subseteq I_\alpha$, and $x \in T_\alpha$ implies $x \in I_\alpha \setminus C$. Hence,

$$x \in I_\alpha \setminus C \setminus E_\alpha \setminus A,$$

because $E_\alpha = A$ and $I_\alpha \cap (E_\alpha \setminus C) = \emptyset$ by (39). Hence,

$$x \in (I_\alpha \setminus C \setminus E_\alpha) \text{ at } s \implies x \notin A_{f_\alpha(s)},$$

by the 2-tardy assumption. Hence,

$$x \in (I_\alpha \setminus C \setminus E_\alpha) \text{ at } s \implies x \in B_{f_\alpha(s)},$$

by the definition of $f_\alpha(s)$. Therefore,

$$x \in (I_\alpha \setminus C \setminus E_\alpha) \text{ at } s \implies x \in B \setminus A.$$

∎

This completes the proof of Theorem 4.6. ∎

5 Relation of Q-tardy to 2-tardy

Note that Q-tardy(A, C) implies small-tardy(A, C). In [5, Theorems 3.3 and 3.8] we prove the following results.

Theorem 5.1 *(i)* $A \subset_{\mathrm{sm}} C \implies [Q(A) \iff A \text{ is 2-tardy}]$.
(ii) $Q(A) \iff (\exists C)[A \subset_{\mathrm{sm}} C \ \& \ A \text{ is 2-tardy}]$.

Thus, it is not true that $Q(A)$ holds iff A is 2-tardy, but this does hold if $A \subset_{\mathrm{sm}} C$ for some C.

What is the relation between Q-tardy(A, C) and and 2-tardy(A)? If Q-tardy(A, C) and $A \subset_x C$ where x denotes either major subset $\subset_{\mathrm{m}}$ or weak major subset $\subset_{\mathrm{wm}}$, a slightly weaker condition, then 2-tardy(A) holds. Also if 2-tardy(A) and $A \subset_s C$ then Q-tardy(A, C) holds. These are all fairly easy to prove, and they establish the relationship between 2-tardy(A) and Q-tardy(A, C).

In [5, Theorem 3.11] we prove that there is a maximal 2-tardy set and hence: (i) the property of A being 2-tardy does not guarantee that the orbit of A consists only of incomplete sets; and (ii) the property of A being 2-tardy is not $\mathcal{E}$-definable.

References

[1] K. Ambos-Spies, C. G. Jockusch, Jr., R. A. Shore and R. I. Soare, An algebraic decomposition of the recursively enumerable degrees and the coincidence of several degree classes with the promptly simple degrees, *Trans. Amer. Math. Soc.* **281** (1984), 109–128.

[2] P. A. Cholak, Automorphisms of the lattice of recursively enumerable sets, *Memoirs of the Amer. Math. Soc.* (1994), to appear.

[3] L. Harrington and R. I. Soare, Post's Program and incomplete recursively enumerable sets, *Proc. Natl. Acad. of Sci. USA*, **88** (1991), 10242–10246.

[4] L. Harrington and R. I. Soare, The Δ_3^0-Automorphism Method and Non-invariant Classes of Degrees, *Jour. Amer. Math. Soc.*, to appear.

[5] L. Harrington and R. I. Soare, Codable sets and orbits of computably enumerable sets, submitted for publication.

[6] L. Harrington and R. I. Soare, Definable properties of the computably enumerable sets, in preparation.

[7] L. Harrington and R. I. Soare, Noninvariance of the nonlow computably enumerable degrees, in preparation.

[8] A. H. Lachlan, Degrees of recursively enumerable sets which have no maximal superset, *J. Symbolic Logic* **33** (1968), 431–443.

[9] A. H. Lachlan, The elementary theory of recursively enumerable sets, *Duke Math. J.* **35** (1968), 123–146.

[10] A. H. Lachlan, On some games which are relevant to the theory of recursively enumerable sets, *Ann. of Math. (2)* **91** (1970), 291–310.

[11] M. Lerman, and R. I. Soare, d-Simple sets, small sets, and degree classes, *Pacific J. Math.* **87** (1980), 135–155.

[12] W. Maass, Recursively enumerable generic sets, *J. Symbolic Logic* **47** (1982), 809–823.

[13] W. Maass, Variations on promptly simple sets, *J. Symbolic Logic* **50** (1985), 138–148.

[14] W. Maass, R. A. Shore and M. Stob, Splitting properties and jump classes, *Israel J. Math.* **39** (1981), 210–224.

[15] D. A. Martin, Classes of recursively enumerable sets and degrees of unsolvability, *Z. Math. Logik Grundlag. Math.* **12** (1966), 295–310.

[16] E. L. Post, Recursively enumerable sets of positive integers and their decision problems, *Bull. Amer. Math. Soc.* **50** (1944), 284–316.

[17] J. R. Shoenfield, Degrees of classes of r.e. sets, *J. Symbolic Logic* **41** (1976), 695–696.

[18] R. I. Soare, Automorphisms of the recursively enumerable sets, Part I: Maximal sets, *Ann. of Math. (2)* **100** (1974), 80–120.

[19] R. I. Soare, Recursively Enumerable Sets and Degrees: A Study of Computable Functions and Computably Generated Sets, Springer-Verlag, Heidelberg, 1987.

[20] M. Stob, The Structure and Elementary Theory of the Recursively Enumerable Sets, Ph.D. Dissertation, University of Chicago, 1979.

[21] A. M. Turing, Systems of logic based on ordinals, *Proc. London Math. Soc.* **45** (1939), 161–228; reprinted in Davis [1965], 154–222.

AXIOMS FOR SUBRECURSION THEORIES

ANDREW J. HEATON AND STANLEY S. WAINER

1. INTRODUCTION

The subrecursive classes which motivate our interest here arise most naturally in a proof-theoretic context, as collections of total recursive functions whose computations can be proved to terminate in (subsystems of) first and second order arithmetic. These classes have a common form: given knowledge of the proof-theoretic ordinal τ of an arithmetical theory T, and given a "standard" presentation of it as a (primitive recursive) well ordering $\prec$ on $\mathbb{N}$, the class of functions provably recursive in T can be characterised hierarchically in two ways:

$$\mathcal{F}(\tau, \prec) = \{f : \exists \alpha \prec \tau (f \text{ elementary in } F_\alpha)\}$$
$$\mathcal{K}(\tau, \prec) = \{f : \exists \alpha \prec \tau (f \text{ elementary in } K_\alpha)\}$$

where F_α is the Fast Growing hierarchy and K_α is Kleene's enumeration hierarchy, both generated according to a fixed choice of fundamental sequences associated with the given well ordering $\prec$. For background see e.g. Buchholz-Wainer [BW87], Feferman [Fef62], Kleene [K58], Rose [R84], Schwichtenberg [S71], Weiermann [W95], and Zemke [Z77].

Although these characterisations deal only with classes of number-theoretic functions, further recent works of Feferman [Fef92] and Tucker-Zucker [TZ92] show how the notion of "provably recursive function" extends quite naturally to more abstract data-types.

This paper gives an overview of the first author's thesis-work, supervised by the second author, in which an attempt is made to elucidate the abstract recursion-theoretic structure of such subrecursive classes by *axiomatization*, following similar lines to those developed by Moschovakis [M69], [M71] and Fenstad [Fen80] in the case of "full" recursion theories. The crucial question is: when working over an abstract set A, how should the enumeration scheme be regulated in order to capture limited subclasses of total functions only? Our resulting scheme of "regulated enumeration" has its roots in the work of Schwichtenberg [S75] and Zemke [Z77].

2. SUBRECURSIVE JUMP OPERATOR

In this preliminary section we bring out the recursion-theoretic structure of the fast growing hierarchy by showing that it is elementarily equivalent to a hierarchy based on a natural subrecursive version of the ordinary jump operator.

As usual, $\{e\}^g(x)$ denotes computation of the e-th partial recursion with oracle g, on input x. Furthermore $[e]$ denotes the e-th *elementary recursive* function in some standard enumeration of them, defined so that

$$[e](x) = \begin{cases} f(x_1, ..., x_n) & \text{if } e \text{ is an elementary code for the } n\text{-ary} \\ & \text{function } f \text{ and } x = <x_1, ..., x_n> \\ 0 & \text{otherwise} \end{cases}$$

Similarly, for a given oracle g the e-th relativised elementary function is denoted by $[e]^g$.

Definition 1. We use a version O_E of Kleene's system of ordinal notations in which only elementary codes for fundamental sequences are used. O_E and $\prec_n$ for each $n \in \mathbb{N}$ are simultaneously inductively defined as follows:

(1) $0 \in O_E$ and $\neg a \prec_n 0$
(2) If $a \in O_E$ then $a +_0 1 =_{Def} 2(a+1) \in O_E$ and also
     $b \prec_n a +_0 1 \Leftrightarrow b \preceq_n a \Leftrightarrow_{Def} (b \prec_n a \text{ or } b = a)$
(3) If $a = 2e+1$ and $[e](0) \neq 0$ and $\forall m([e](m) \in O_E$ and
     $[e](m) +_0 1 \preceq_{m+1} [e](m+1))$ then we have $a \in O_E$ and also
     $b \prec_n a \Leftrightarrow b \preceq_n [e](n)$.

For $a = 2e+1 \in O_E$, we will often write a_n to denote $[e](n)$ and also write $a = \sup a_n$. We will refer to the condition in 3 that $[e](m) +_0 1 \preceq_{m+1} [e](m+1)$ for all $m \in N$ as the structuredness condition. It ensures that $b \preceq_n a \Rightarrow b \preceq_{n+1} a$, and we may define $b \prec a \Leftrightarrow \exists n(b \prec_n a)$.

Definition 2. We define the *subrecursive jump operator* sj as follows,

$$sj(f)(x) = \begin{cases} \{x_0\}^f(x_1) & \text{if } \{x_0\}^f(x_1) \text{ is defined in } x_2 \text{ steps} \\ 0 & \text{otherwise} \end{cases}$$

where $x \mapsto <x_0, x_1, x_2>$ is a standard bijection from $\mathbb{N}$ to $\mathbb{N}^3$.

Definition 3. The *subrecursive jump hierarchy* $\{J_a : a \in O_E\}$ is defined as

$$J_0(x) = 2^x, \quad J_{a+_01}(x) = sj(J_a)(x), \quad J_{\sup a_n}(x) = J_{a_{x_0}+_01}(x_1)$$

where $x \mapsto <x_0, x_1>$ is again a standard bijection from $\mathbb{N}$ to $\mathbb{N}^2$.

Definition 4. The *fast growing hierarchy* $\{F_a\}$ is defined so that for every $a \in O_E$ and $x \in \mathbb{N}$ we have

$$F_0(x) = 2^x, \quad F_{a+_01}(x) = It(F_a)(x), \quad F_{\sup a_n}(x) = F_{a_x}(x)$$

where $It(f)(x) = f^{x+1}(x)$, the $x+1$-th iterate of f applied to x.

Theorem 1. *The subrecursive jump and fast growing hierarchies are elementarily equivalent, i.e. there are elementary functions $P = [p]$ and $Q = [q]$ so that for every $a \in O_E$ and $x \in \mathbb{N}$ we have,*

$$F_a(x) = [P(a)]^{J_a}(x) \text{ and } J_a(x) = [Q(a)]^{F_a}(x).$$

Proof. We here give only an outline of the proof by showing that for every function f which has at least exponential rate of growth and is "honest" in the sense that its size reflects its computational complexity, the functions $It(f)$ and $sj(f)$ are elementarily equivalent. Using the uniformities that are apparent when working with subrecursive hierarchies, the elementary functions P, Q can then be defined as required by effective transfinite recursions based on the elementary recursion theorem.

Clearly, for increasing functions f of at least exponential size, $f^y(x)$ will bound the usage of f in the computation of $\{e\}^f(x)$ after y steps, i.e. the maximum value n such that $f(n)$ is an oracle call. Using a relativised version of Kleene's normal form theorem, we can then obtain $sj(f)$ as a function elementary in f and $It(f)$. But if f is "honest" then f will be elementary in $It(f)$. Hence we have uniformly obtained $sj(f)$ elementary in $It(f)$ as required.

To obtain $It(f)$ elementary in $sj(f)$, we can define $It(f)(x) = sj(f)(< e_0, x, g(x) >)$ where e_0 is a fixed code so that $\{e_0\}^f(x) = It(f)(x)$ and g is elementary. Hence we have a uniform elementary reduction of $It(f)$ from $sj(f)$. $\qquad\square$

3. Subrecursion Axioms

We present an axiomatization of subrecursive classes, motivated on the one hand by the Moschovakis-Fenstad axioms for "full" recursion theories, and on the other hand by the result above which, together with [S71] and [Z77], shows that naturally occurring subrecursive classes can alternatively be characterised in terms of iterated effective enumeration.

We study a relation over an abstract structure $\mathcal{A}$:

$$[e]_{\mathcal{A}}(a_1, ..., a_n) = b$$

which reads "the total function from A^n to A encoded by $e \in A$ gives output b on inputs $a_1, ..., a_n$". Note: the abstract notion $[e]_{\mathcal{A}}$ should not be confused with the enumeration of elementary functions, $[e]$.

The idea is to use a prewellordering on function codes in order to measure the complexity of functions in our theories, and to regulate the enumeration scheme of "full" recursion theories so as to keep all functions total.

The following terminology and notation comes from [M69].

For a given arbitrary set B, let 0 be some object not in B and let $B^0 = B \cup \{0\}$. We define the set B^* by the inductive clauses,

(1)
$$x \in B^0 \Rightarrow x \in B^*$$
$$x, y \in B^* \Rightarrow (x, y) \in B^*$$

where (x, y) is a set-theoretic operation representing the ordered pair of x and y, chosen so that no object in B^0 is an ordered pair. Thus if $z \in B^*$,

then either $z \in B^0$ or $z = (x, y)$ with uniquely determined $x, y \in B^*$.
We have a copy of the natural numbers $N = \{0, 1, 2, ...\}$ in B^* by defining

(2)
$$0 = 0$$
$$n + 1 = (0, n)$$

To each $z \in B^*$ we assign its components L, R by the induction

(3)
$$R(0) = L(0) = 0$$
$$R(z) = L(z) = 1 \quad \text{if } z \in B$$
$$R(z) = x, L(z) = y \quad \text{if } z = (x, y)$$

We also define the n-tuple operation $< b_1, ..., b_n >$ (for $n \geq 1$) on B^* as follows

(4)
$$< b_1 >= b_1$$
$$< b_1, ..., b_{n+1} >= (b_1, < b_2, ..., b_{n+1} >)$$

We assign inductively to each $z \in B^*$ the finite subset D_z of B^*

(5)
$$D_z = \{z\} \quad \text{if } z \in B^0$$
$$D_z = D_x \cup D_y \cup \{z\} \quad \text{if } z = (x, y)$$

So for every z, the "decomposition tree" D_z contains z plus all the elements of B^* that have been used to "build it up".
Also we define

(6)
$$ht(z) = 0 \quad \text{if } z \in B^0$$
$$ht(z) = \max(ht(x), ht(y)) + 1 \quad \text{if } z = (x, y)$$

Thus $ht(z)$ measures the height of the "decomposition tree" D_z.

Definition 5. A *computation domain* is a structure $< A, C, (,), L, R >$ with the following properties:

(1) $A = B^*$ for some set B
(2) $A \supseteq C \supseteq \emptyset^*$ and $\forall a, b \in A$ $(a, b \in C \Leftrightarrow (a, b) \in C)$
(3) $(,)$ satisfies (1) above and L, R satisfy (3)

Remark. C is called the code set, and by the above contains a copy of the natural numbers. Clearly, if $a \in C$ then either $a \in B^0$ or $a = (b, c)$ with uniquely determined $b, c \in C$. We note that $C = (B \cap C)^*$.

Definition 6. A *subrecursion structure* is a structure $\mathcal{A} =< A, C, (,), L, R, \tau, | \ |_A, [\ ]_A >$ with the following properties:

(1) $< A, C, (,), L, R >$ is a computation domain
(2) τ is an ordinal
(3) $| \ |_A$ maps C onto τ
(4) $[\ ]_A$ maps C into the set $\{f : A^k \to A \text{ for some } k \in \mathbb{N}\}$

We will denote arbitrary elements of A by $x, y, z, ...$, elements of C by $a, b, c, ...$, natural numbers by $i, j, k, l, m, n, ...$, and finite sequences (tuples) from A (including the empty tuple) by $\bar{x}, \bar{y}, \bar{z}, ...$. The concatenation of $\bar{x}$ and $\bar{y}$ is denoted by $\bar{x}, \bar{y}$. Lower case Greek letters shall denote ordinals.

We denote the set of n-ary functions in the range of $[\ ]_A$ by $Rec^n(A)$ and put $Rec(A)= \bigcup_{n \in N} Rec^n(A)$ and $C_A^{(n)} = \{a \in C : [a]_A \in Rec^n(A)\}$.

We say that f is A-computable iff $f \in Rec(A)$.

If $f \in Rec(A)$ we write $\hat{f}$ to denote any code for f, i.e. $\hat{f} \in C$ and $[\hat{f}]_A = f$. We say a function $f \in Rec(A)$ has rank α if there exists $\hat{f} \in C$ with $|\hat{f}|_A = \alpha$.

Definition 7. A *subrecursion theory* is a subrecursion structure A such that there exist total functions $p_1, ..., p_9 \in Rec(A)$, all with finite rank, so that the axioms 1 - 9 below are satisfied.

(1) *Constants*
$\forall n \geq 1 \in N \ \forall a \in C \ \forall \vec{x} \in A^n \ . \ p_1(n, a) \in C$ and $[p_1(n, a)]_A(\vec{x}) = a$ and $|p_1(n, a)|_A = 0$

(2) *Pairing Function*
$\forall n \in N \ \forall x, y \in A \ \forall \vec{z} \in A^n \ . \ p_2(n) \in C$ and $[p_2(n)]_A(x, y, \vec{z}) = (x, y)$ and $|p_2(n)|_A = 0$

(3) *Left Component*
$\forall n \in N \ \forall x \in A \ \forall \vec{y} \in A^n \ . \ p_3(n) \in C$ and $[p_3(n)]_A(x, \vec{y}) = L(x)$ and $|p_3(n)|_A = 0$

(4) *Right Component*
$\forall n \in N \ \forall x \in A \ \forall \vec{y} \in A^n \ . \ p_4(n) \in C$ and $[p_4(n)]_A(x, \vec{y}) = R(x)$ and $|p_4(n)|_A = 0$

(5) *Cases*
$\forall n \in N \ \forall w, x, y \in A \ \forall \vec{z} \in A^n \ . \ p_5(n) \in C$ and

$$[p_5(n)]_A(w, x, y, \vec{z}) = \begin{cases} x & \text{if } w = y \text{ and } x, y \in C \\ 0 & \text{otherwise} \end{cases}$$

and $|p_5(n)|_A = 0$

(6) *Projections*
$\forall n \geq 1 \in N \ \forall i < n \in N \ \forall x_0, ..., x_{n-1} \in A \ . \ p_6(n, i) \in C$ and $[p_6(n, i)]_A(x_0, ..., x_{n-1}) = x_i$ and $|p_6(n, i)|_A = 0$

(7) *Composition*
$\forall n \geq 1 \in N \ \ \forall m \leq n \in N \ \ \forall i \leq m \in N \ \ \forall g \in Rec^{m+1}(A) \ \forall h \in Rec^n(A) \ \forall \hat{g}, \hat{h} \in C \ \forall \vec{x} \in A^n \ . \ p_7(n, \hat{g}, \hat{h}, i, m) \in C$ and

$$[p_7(n, \hat{g}, \hat{h}, i, m)]_A(x_1, ..., x_n) = g(x_1, ..., x_i, h(x_1, ..., x_n), x_{i+1}, ..., x_m)$$

and $|p_7(n, \hat{g}, \hat{h}, i, m)|_A = \max(|\hat{g}|_A, |\hat{h}|_A) + 1$

(8) *Finite Recursion*
$\forall n \in N \ \forall g \in Rec^{n+1}(A) \ \forall h \in Rec^{n+4}(A) \ \forall \hat{g}, \hat{h} \in C \ \forall a \in C$

$\forall x \in A \; \forall \vec{y} \in A^n . p_8(n, \hat{g}, \hat{h}, a) \in C$ and $[p_8(n, \hat{g}, \hat{h}, a)]_A(\vec{x}, y) = f(\vec{x}, y)$
where

$$f(\vec{x}, y) = \begin{cases} g(\vec{x}, y) & \text{if } y \in B^0 \cap C \text{ or } y \notin D_a \\ h(f(\vec{x}, u), f(\vec{x}, v), \vec{x}, u, v) & \text{if } y = (u, v) \in D_a \end{cases}$$

and $|p_8(n, \hat{g}, \hat{h}, a)|_A = \max(|\hat{g}|_A, |\hat{h}|_A) + Card(D_a)$

(9) *Regulated Enumeration*
$\forall n \geq 1 \in \mathbb{N} \; \forall a \in C \; \forall x \in A \; \forall \vec{y} \in A^n$. $p_9(n, a) \in C$ and

$$[p_9(n, a)]_A(x, \vec{y}) = \begin{cases} [x]_A(\vec{y}) & \text{if } |a|_A \text{ is a limit ordinal and } x \in C_A^{(n)} \\ & \text{and } |x|_A \prec |a|_A \\ 0 & \text{otherwise} \end{cases}$$

and $|p_9(n, a)|_A = |a|_A + 1$

Remark. Our axioms are essentially those of Moschovakis-Fenstad but with the ordinary partial enumeration scheme replaced by regulated enumeration. Scheme (8) allows codes for finite recursions to be computed uniformly in their lengths. Together with the regulated enumeration scheme, this will allow definitions by primitive recursion. See lemma 4 below. The conditions placed on the prewellordering $c_1 \prec c_2 \Leftrightarrow |c_1|_A \prec |c_2|_A$ are designed so that the complexity of the functions $[c]_A$ in our theories are reflected by the sizes of their ordinals $|c|_A$.

4. BASIC PROPERTIES OF SUBRECURSION THEORIES

First we have a version of Kleene's S-m-n theorem.

Lemma 1. *Let A be a subrecursion theory. There is an A-computable function S with finite rank so that for all $n \geq 1 \in \mathbb{N}$, $m \leq n \in \mathbb{N}$, $a \in C$ and $\vec{x} \in A^n$, if f is an $n + 1$-ary A-computable function then we have*

$$[S(n, \hat{f}, m, a)]_A(\vec{x}) = f(x_1, ..., x_m, a, x_{m+1}, ..., x_n)$$

and $|S(n, \hat{f}, m, a)|_A = |\hat{f}|_A + 1$.

Proof. We put $S(n, \hat{f}, m, a) = p_7(n, \hat{f}, p_1(n, a), m, n)$. Hence we have
$S(n, \hat{f}, m, a)]_A(\vec{x}) = f(x_1, ..., x_m, [p_1(n, a)]_A(\vec{x}), x_{m+1}, ..., x_n)$
$= f(x_1, ..., x_m, a, x_{m+1}, ..., x_n)$.
The interested reader may check that we can put
$\hat{S} = p_7(4, p_7(5, \hat{p}_7, p_6(5, 0), 4, 4), p_7(4, \hat{p}_1, p_6(4, 3), 1, 1), 2, 4)$.   □

Another simple consequence of the axioms is an explicit definitions lemma.

Lemma 2. *Suppose f is defined so that*

$$f(x_1, ..., x_n) = Term(x_1, ..., x_n, c_1, ..., c_l, g_1, ..., g_k)$$

for given constants $c_1, ..., c_l \in C$ and A-computable functions $g_1, ... g_k$. Then f is A-computable with $|\hat{f}|_A \prec \max(|g_1|_A, ..., |g_k|_A) + \omega$.

Kleene's second recursion theorem for arbitrary recursion theories $\mathcal{A}$ now follows in the usual way:-

Theorem 2. *Let $\mathcal{A}$ be a subrecursion theory. For every $n+1$-ary $\mathcal{A}$-computable function f there exists an $a \in C$ so that for every $\vec{x} \in A^n$ we have*

$$[a]_{\mathcal{A}}(\vec{x}) = f(\vec{x}, a) \text{ and } |a|_{\mathcal{A}} \prec |\hat{f}|_{\mathcal{A}} + \omega.$$

Proof. Define S_1 so that $S_1(\vec{z}, y) = S(n, y, n, y)$ and let

$$a = S(n, p_7(n+1, \hat{f}, \hat{S}_1, n, n), n, p_7(n+1, \hat{f}, \hat{S}_1, n, n)).$$

Then using lemma 1 we get

$$
\begin{aligned}
[a]_{\mathcal{A}}(\vec{x}) &= [p_7(n+1, \hat{f}, \hat{S}_1, n, n)]_{\mathcal{A}}(\vec{x}, p_7(n+1, \hat{f}, \hat{S}_1, n, n)) \\
&= f(\vec{x}, S_1(\vec{x}, p_7(n+1, \hat{f}, \hat{S}_1, n, n))) \\
&= f(\vec{x}, a)
\end{aligned}
$$

$\square$

Note that the finite recursion and regulated enumeration axioms are not needed for the proof of the above theorem. We do, however, need them to obtain closure under primitive recursion from our axioms. This requires the following technical lemma.

Lemma 3. *Let $\mathcal{A}$ be a subrecursion theory. For all $a, b, c \in C$ and $\vec{x} \in A^n$ we have*

(1) *If $a \in D_b$ and $b \in D_c$ then $[p_8(n, \hat{g}, \hat{h}, b)]_{\mathcal{A}}(\vec{x}, a) = [p_8(n, \hat{g}, \hat{h}, c)]_{\mathcal{A}}(\vec{x}, a)$*
(2) *If $b = (b_0, b_1)$ then $[p_8(n, \hat{g}, \hat{h}, b_i)]_{\mathcal{A}}(\vec{x}, b_i) = [p_8(n, \hat{g}, \hat{h}, b)]_{\mathcal{A}}(\vec{x}, b_i)$, for $i = 0, 1$.*

Lemma 4. *(Primitive Recursion Lemma) Let $\mathcal{A}$ be a subrecursion theory. Suppose $g \in Rec^{n+1}(\mathcal{A})$, $h \in Rec^{n+4}(\mathcal{A})$ and $|\hat{g}|_{\mathcal{A}}, |\hat{h}|_{\mathcal{A}} \prec \lambda$ for some limit ordinal $\lambda \prec \tau$. Then the function f defined so that*

$$
f(\vec{x}, a) = \begin{cases} g(\vec{x}, a) & \text{if } a \in B^0 \cap C \text{ or } a \notin C \\ h(f(\vec{x}, a_0), f(\vec{x}, a_1), \vec{x}, a_0, a_1) & \text{if } a = (a_0, a_1) \in C \end{cases}
$$

is $\mathcal{A}$-computable and $|\hat{f}|_{\mathcal{A}} \prec \lambda + \omega$.

Proof. We put $f_a = [p_8(n, \hat{g}, \hat{h}, a)]_{\mathcal{A}}$ for $a \in C$ and define f' so that

$$
f'(\vec{x}, a) = \begin{cases} g(\vec{x}, a) & \text{if } a \notin C \\ [p_9(n+1, c_\lambda)]_{\mathcal{A}}(\hat{f}_a, \vec{x}, a) & \text{if } a \in C \end{cases}
$$

where c_λ denotes any element $c_\lambda \in C$ with $|c_\lambda|_{\mathcal{A}} = \lambda$. Clearly f' is $\mathcal{A}$-computable. We claim that $f' = f$.

For $a \notin C$, the result clearly holds.

For $a \in C$, $f'(\vec{x}, a) = [p_9(n + 1, c_\lambda)]_\mathcal{A}(\hat{f}_a, \vec{x}, a)$. Now since $|\hat{g}|_\mathcal{A}, |\hat{h}|_\mathcal{A} \prec \lambda$ and $Card(D_a)$ is finite, $|\hat{f}_a|_\mathcal{A} =$
$\max(|\hat{g}|_\mathcal{A}, |\hat{h}|_\mathcal{A}) + Card(D_a) \prec \lambda$, and so $f'(\vec{x}, a) = f_a(\vec{x}, a)$.

We now show $f_a(\vec{x}, a) = f(\vec{x}, a)$ for all $a \in C$ by induction on $ht(a) \in \mathbb{N}$.

Basis: $ht(a) = 0$, and so we have $a \in B^0 \cap C$ and therefore $f_a(\vec{x}, a) = g(\vec{x}, a) = f(\vec{x}, a)$.

Induction step: Here $ht(a) = j + 1$ for some $j \in \mathbb{N}$, and so we have $a = (a_0, a_1)$ for some $a_0, a_1 \in C$ with $ht(a_0), ht(a_1) \leq j$. Hence we get

$$
\begin{aligned}
f_a(\vec{x}, a) &= h(f_a(\vec{x}, a_0), f_a(\vec{x}, a_1), \vec{x}, a_0, a_1) \\
&= h(f_{a_0}(\vec{x}, a_0), f_{a_1}(\vec{x}, a_1), \vec{x}, a_0, a_1) \text{ by lemma 3} \\
&= h(f(\vec{x}, a_0), f(\vec{x}, a_1), \vec{x}, a_0, a_1) \text{ by the induction hypothesis} \\
&= f(\vec{x}, a)
\end{aligned}
$$

It is clear that we can construct a code $\hat{f}$ so that $|\hat{f}|_\mathcal{A} = \lambda + \alpha$ for some finite ordinal α, and hence this completes the proof. $\square$

Finally, note that every subrecursion theory satisfies a prewellordering property:-

Lemma 5. *Let $\mathcal{A}$ be a subrecursion theory. There is an $\mathcal{A}$-computable function w with finite rank and for every $a \in C$ an $\mathcal{A}$-computable function W_a so that $[w(a)]_\mathcal{A} = W_a$ and for all $b \in C$,*

$$
W_a(b) = \begin{cases} 1 & \text{if } |a|_\mathcal{A} \text{ is a limit ordinal and } |b|_\mathcal{A} \prec |a|_\mathcal{A} \\ 0 & \text{otherwise.} \end{cases}
$$

Furthermore, for all $a \in C$ we have $|w(a)|_\mathcal{A} \prec |a|_\mathcal{A} + \omega$.

Proof. We define the $\mathcal{A}$-computable function w as follows,

$$
\begin{aligned}
w(a) = \ &p_7(1, p_9(1, a), p_7(1, S(1, S(2, S(3, S(4, \hat{p}_7, 0, 1), 0, p_1(1, 1)), 1, 0), 1, 0), \\
&p_7(1, S(1, S(2, S(3, \hat{S}, 0, 1), 1, 0), 1, 0), S(1, \hat{p}_9, 0, 1), 0, 0), 0, 0), 0, 1)
\end{aligned}
$$

It is clear that $|w(a)|_\mathcal{A} \prec |a|_\mathcal{A} + \omega$ and that we can construct w so that $\hat{w}$ has finite rank. Putting $W_a = [w(a)]_\mathcal{A}$, from the above we get

$$
W_a(b) = [p_9(1, a)]_\mathcal{A}(p_7(1, p_1(1, 1), S(1, p_9(1, b), 0, 0), 0, 0), b)
$$

If $|a|_\mathcal{A}$ is a limit ordinal then we have

$$
|p_7(1, p_1(1, 1), S(1, p_9(1, b), 0, 0), 0, 0)|_\mathcal{A} \prec |a|_\mathcal{A} \Leftrightarrow |b|_\mathcal{A} \prec |a|_\mathcal{A}
$$

since $|p_7(1, p_1(1, 1), S(1, p_9(1, b), 0, 0), 0, 0)|_\mathcal{A} = |b|_\mathcal{A} + 3$. It is also clear that $p_7(1, p_1(1, 1), S(1, p_9(1, b), 0, 0), 0, 0) \in C_\mathcal{A}^{(1)}$ for all $b \in C$. Hence we have

$$
W_a(b) = \begin{cases} [p_7(1, p_1(1, 1), S(1, p_9(1, b), 0, 0), 0, 0)]_\mathcal{A}(b) & \begin{array}{l} \text{if } |a|_\mathcal{A} \text{ is a limit} \\ \text{and } |b|_\mathcal{A} \prec |a|_\mathcal{A} \end{array} \\ 0 & \text{otherwise} \end{cases}
$$

But $[p_7(1, p_1(1,1), S(1, p_9(1,x), 0, 0), 0, 0)]_\mathcal{A}(b) =$
$[p_1(1,1)]_\mathcal{A}([S(1, p_9(1,b), 0, 0)]_\mathcal{A}(b)) = 1$. Substituting this into the above equation we see that we have the required function. $\square$

Lemma 6. *Let $\mathcal{A}$ be a subrecursion theory. For every limit ordinal $\lambda \prec \tau$ there is an $\mathcal{A}$-computable function V_λ such that for all $a, b \in C$*

$$V_\lambda(a,b) = \begin{cases} 1 & \text{if } |b|_\mathcal{A} \text{ is a limit ordinal and } |a|_\mathcal{A} \prec |b|_\mathcal{A} \prec \lambda \\ 0 & \text{otherwise} \end{cases}$$

and $|\hat{V}_\lambda|_\mathcal{A} = \lambda + 2$.

Proof. Suppose λ is a limit ordinal $\prec \tau$. We put $V_\lambda(a,b) = [p_9(1, c_\lambda)]_\mathcal{A}(w(b), a)$ where w is defined in the first prewellordering lemma. Now $|\hat{w}|_\mathcal{A}$ is finite and it is easy to see that $\hat{V}_\lambda$ can be constructed so that $|\hat{V}_\lambda|_\mathcal{A} = \lambda + 2$. Hence, since V_λ is defined by the regulated enumeration scheme, we have

$$V_\lambda(a,b) = \begin{cases} [w(b)]_\mathcal{A}(a) & \text{if } w(b) \in C_\mathcal{A}^{(1)} \text{ and } |w(b)|_\mathcal{A} \prec \lambda \\ 0 & \text{otherwise} \end{cases}$$

Since $b \in C$, by the first prewellordering lemma we have $w(b) \in C_\mathcal{A}^{(1)}$. Also $|w(b)|_\mathcal{A} = |b|_\mathcal{A} + \alpha$ for some finite ordinal α and so $|w(b)|_\mathcal{A} \prec \lambda$ iff $|b|_\mathcal{A} \prec \lambda$. Hence

$$V_\lambda(a,b) = \begin{cases} W_b(a) & \text{if } |b|_\mathcal{A} \prec \lambda \\ 0 & \text{otherwise} \end{cases}$$

The result now easily follows from lemma 5. $\square$

5. THE FIRST RECURSION THEOREM

In this section we define a notion of "effective operator" in subrecursion theories, and then give what seems to be the appropriate version of Kleene's First Recursion Theorem. It amounts to the principle of transfinite recursion on rank.

Definition 8. Suppose $\mathcal{A}$ is a subrecursion theory and $f \mapsto \lambda\vec{x}.F(f,\vec{x})$ a functional of type $(A^n \to A) \to (A^m \to A)$. Then F is said to be $\mathcal{A}$-computable (or an $\mathcal{A}$-effective operator) if there exists an $\mathcal{A}$-computable function e_F such that for every $f \in Rec^n(\mathcal{A})$, $\hat{f} \in C$ and $\vec{x} \in A^m$ we have

$$F(f,\vec{x}) = [e_F(\hat{f})]_\mathcal{A}(\vec{x}).$$

The idea of our definition here, is that a functional F will be $\mathcal{A}$-computable in a subrecursion theory $\mathcal{A}$ if the function $\lambda\vec{x}.F(f,\vec{x})$ is $\mathcal{A}$-computable for every $\mathcal{A}$-computable function f, uniformly in $\hat{f}$.

Definition 9. Suppose $\mathcal{A}$ is a subrecursion theory with ordinal τ. We say that an *ordinal operation* $\kappa : \tau \to \tau$ is $\mathcal{A}$-computable with rank α if there is an $\mathcal{A}$-computable function $m_\mathcal{A}$ with rank α satisfying

$$|m_\mathcal{A}(a)|_\mathcal{A} = \kappa(|a|_\mathcal{A})$$

for every $a \in C$.

We now prove the "First Recursion Theorem" for subrecursion theories.

Theorem 3. *Let $\mathcal{A}$ be a subrecursion theory with ordinal τ closed under multiplication. Suppose F is an $\mathcal{A}$-computable functional such that e_F has rank $\prec \gamma$ and $|e_F(\hat{f})|_{\mathcal{A}} = \max(|\hat{f}|_{\mathcal{A}}, \gamma) + \alpha$, where α, γ are fixed ordinals $\prec \tau$. Also suppose that the ordinal operation $\zeta \mapsto (2 + \alpha + \gamma + \omega) \cdot (\zeta + 1)$ is $\mathcal{A}$-computable with rank $\prec \gamma$. Now consider the function f defined by recursion as follows:*

$$f(a, \vec{x}) = F(f_{\prec a}, a, \vec{x})$$

where for all $a, b \in C$, $\vec{y} \in A^n$ we have

$$f_{\prec a}(b, \vec{y}) = \begin{cases} f(b, \vec{y}) & \text{if } |b|_{\mathcal{A}} \prec |a|_{\mathcal{A}} \\ 0 & \text{otherwise} \end{cases}$$

Then for each fixed $a \in C$, the function $f_{\prec a}$ is $\mathcal{A}$-computable.

Proof. Using the fixed point theorem, we can define an $\mathcal{A}$-computable function φ so that

$$[\varphi(a)]_{\mathcal{A}}(b, \vec{x}) = \begin{cases} [e_F(\varphi(b))]_{\mathcal{A}}(b, \vec{x}) & \text{if } |b|_{\mathcal{A}} \prec |a|_{\mathcal{A}} \\ 0 & \text{otherwise} \end{cases}$$

We show by induction over $|a|_{\mathcal{A}} \prec \tau$ that $[\varphi(a)]_{\mathcal{A}}(b, \vec{x}) = f_{\prec a}(b, \vec{x})$ for every $b \in C$, $\vec{x} \in A^n$.
Basis: $|a|_{\mathcal{A}} = 0$. Clearly $[\varphi(a)]_{\mathcal{A}}(b, \vec{x}) = f_{\prec a}(b, \vec{x}) = 0$ for all $b \in C$, $\vec{x} \in A^n$.
Induction Step: $|a|_{\mathcal{A}} \succ 0$. By the induction hypothesis, if $|b|_{\mathcal{A}} \prec |a|_{\mathcal{A}}$ then $[\varphi(b)]_{\mathcal{A}}(b, \vec{x}) = f_{\prec b}(b, \vec{x})$ for all $\vec{x} \in A^n$. Hence for all $|b|_{\mathcal{A}} \prec |a|_{\mathcal{A}}$ it follows that $[e_F(\varphi(b))]_{\mathcal{A}}(b, \vec{x}) = F(f_{\prec b}, b, \vec{x})$. Therefore

$$[\varphi(a)]_{\mathcal{A}}(b, \vec{x}) = \begin{cases} F(f_{\prec b}, b, \vec{x}) & \text{if } |b|_{\mathcal{A}} \prec |a|_{\mathcal{A}} \\ 0 & \text{otherwise} \end{cases}$$

But $F(f_{\prec b}, b, \vec{x}) = f(b, \vec{x})$ and hence

$$[\varphi(a)]_{\mathcal{A}}(b, \vec{x}) = \begin{cases} f(b, \vec{x}) & \text{if } |b|_{\mathcal{A}} \prec |a|_{\mathcal{A}} \\ 0 & \text{otherwise} \end{cases}$$

and so we have $[\varphi(a)]_{\mathcal{A}} = f_{\prec a}$, as required. $\square$

6. MINIMAL MODELS AND THEIR PROPERTIES

In this section, we define inductive models for our axioms and then give an imbedding theorem showing them to be (in a sense) minimal among all others.

Definition 10. Let $\mathcal{D} =< A, C, (\ ,\ ), L, R >$ be a computation domain. Suppose τ is a limit ordinal, o a mapping from C onto τ and $\vec{f} = f_1, ..., f_l$ a list of functions on A so that each f_i is n_i-ary. We define the *inductive subrecursion theory* on $\mathcal{D}$ to be

$$Ind(\tau, o, \vec{f}\,) =< A, C, (\ ,\ ), L, R, \tau, |\ |_I, [\ ]_I >$$

where $|\ |_I$ and $[\ ]_I$ are defined separately for $a \in C$ in two steps according to the following inductive clauses. First $|a|_I$ is defined by induction on $ht(a) \in C$ and then $[a]_I$ is defined by induction on $|a|_I$.

(1) If $a =< 1, n, b >$ for $n \neq 0 \in \mathbb{N}$ and $b \in C$, then $[a]_I(\vec{x}) = b$ for all $\vec{x} \in A^n$ and $|a|_I = 0$.

(2) If $a =< 2, n >$ for $n \in \mathbb{N}$, then $[a]_I(x, y, \vec{z}) = (x, y)$ for all $x, y \in A$ and $\vec{z} \in A^n$ and $|a|_I = 0$.

(3) If $a =< 3, n >$ for $n \in \mathbb{N}$, then $[a]_I(x, \vec{y}) = L(x)$ for all $x \in A$ and $\vec{y} \in A^n$ and $|a|_I = 0$.

(4) If $a =< 4, n >$ for $n \in \mathbb{N}$, then $[a]_I(x, \vec{y}) = R(x)$ for all $x \in A$ and $\vec{y} \in A^n$ and $|a|_I = 0$.

(5) If $a =< 5, n >$ for $n \in \mathbb{N}$, then

$$[a]_I(w, x, y, \vec{z}) = \begin{cases} x & \text{if } w = y \text{ and } x, y \in C \\ 0 & \text{otherwise} \end{cases}$$

for all $w, x, y \in A$ and $\vec{z} \in A^n$ and $|a|_I = 0$.

(6) If $a =< 6, n, i >$ for $n \neq 0 \in \mathbb{N}, i < n \in \mathbb{N}$ then $[a]_I(x_0, ..., x_{n-1}) = x_i$ for all $x_0, ..., x_{n-1} \in A$ and $|a|_I = 0$.

(7) If $a =< 7, n, b, c, i, m >$ for $n \neq 0 \in \mathbb{N}$ and $0 \leq i \leq m \leq n \in \mathbb{N}$ where $[b]_I, [c]_I$ are of arities $m + 1, n$ respectively then

$$[a]_I(x_1, ..., x_n) = [b]_I(x_1, ..., x_i, [c]_I(x_1, ..., x_n), x_{i+1}, ..., x_m)$$

for all $x_1, ..., x_n \in A$ and $|a|_I = \max(|b|_I, |c|_I) + 1$.

(8) If $a =< 8, n, b, c, d >$ for $n \in \mathbb{N}$ and $d \in C$ where $[b]_I, [c]_I$ are of arities $n + 1, n + 4$ respectively then

$$[a]_I(\vec{y}, x) = \begin{cases} [b]_I(\vec{y}, x) & \text{if } x \in B^0 \cap C \text{ or } x \notin D_d \\ [c]_I([a]_I(\vec{y}, u), [a]_I(\vec{y}, v), \vec{y}, u, v) & \text{if } x = (u, v) \in D_d \end{cases}$$

for all $x \in A$ and $\vec{y} \in A^n$ and $|a|_I = \max(|b|_I, |c|_I) + Card(D_d)$.

(9) If $a =< 9, n, b >$ for $n \neq 0 \in \mathbb{N}$ and $b \in C$ then we have

$$[a]_I(x, \vec{y}) = \begin{cases} [x]_I(\vec{y}) & \text{if } |b|_I \text{ is a limit ordinal and } x \in C_I^{(n)} \\ & \text{and } |x|_I \prec |b|_I \\ 0 & \text{otherwise} \end{cases}$$

for all $x \in A$ and $\vec{y} \in A^n$ and $|a|_I = |b|_I + 1$.

(10) If $a =< 10, n_i, i >$ and the function f_i in the list $\vec{f} = f_1, ..., f_l$ is n_i-ary then $[a]_I(\vec{x}) = f_i(\vec{x})$ for all $\vec{x} \in A^{n_i}$ and $|a|_I = 0$.

(11) If $a =< 11, b >$ for $b \in C$ then $[a]_I(x) = 0$ for all $x \in A$ and $|a|_I = o(b)$.

(12) For all other $a \in C$, $[a]_I(x) = 0$ for all $x \in A$ and $|a|_I = 0$.

Remark. We note that the inductive clauses 1-9 above correspond to axioms 1-9 of definition 6. Clause 10 introduces the list $\vec{f}$ and clause 11 gives us the property that $|\ |_I$ maps C onto τ. Finally, clause 12 ensures that $|\ |_I$ and $[\ ]_I$ are defined everywhere on C.

Proposition. *Let $\mathcal{D} = <A, C, (\ ,\ ), L, R>$ be a computation domain. Suppose τ is a limit ordinal, o a mapping from C onto τ and $\vec{f} = f_1, ..., f_l$ a list of functions on A with f_i n_i-ary. Then the inductive subrecursion theory $Ind(\tau, o, \vec{f}) = <A, C, (\ ,\ ), L, R, \tau, |\ |_I, [\ ]_I>$ is a subrecursion theory with every function f_i in the list $\vec{f} = f_1, ..., f_l$ I-computable with rank 0.*

Definition 11. Let $\mathcal{A}$ and $\mathcal{B}$ be subrecursion theories on the same computation domain $\mathcal{D}$. We say that $\mathcal{A}$ is *imbeddable* in $\mathcal{B}$,

$$\mathcal{A} \leq \mathcal{B}$$

if there is a $\mathcal{B}$-computable function q such that for all $a \in C$,

$$[a]_\mathcal{A} = [q(a)]_\mathcal{B}$$

Hence every $\mathcal{A}$-computable function f is $\mathcal{B}$-computable and a code for f in $\mathcal{B}$ can be uniformly found in $\mathcal{B}$ from the code in $\mathcal{A}$. We now state an imbedding theorem for the inductive models.

Theorem 4. *Suppose $\mathcal{A}$ is a subrecursion theory on the computation domain $\mathcal{D}$ with ordinal τ and $\omega \prec \tau$. Let υ be a limit ordinal, o a mapping from C onto υ and $Ind(\upsilon, o)$ the inductive subrecursion theory on $\mathcal{D}$ for υ and o. Suppose there exists an $\mathcal{A}$-computable function π from C to C with finite rank satisfying*

(1) $\forall a, b \in C.\ o(a) \prec o(b) \Leftrightarrow |\pi(a)|_\mathcal{A} \prec |\pi(b)|_\mathcal{A}$
(2) $\forall a \in C.\ o(a)$ *a limit ordinal* $\prec \upsilon \Leftrightarrow |\pi(a)|_\mathcal{A}$ *a limit ordinal* $\prec \tau$

Then $Ind(\upsilon, o) \leq \mathcal{A}$.

Remark. (1) If π above exists then it follows that $\upsilon \preceq \tau$. (2) If we have $\upsilon = \tau$ and $o = |\ |_\mathcal{A}$ then putting $\pi(x) = x$ we would have π $\mathcal{A}$-computable with finite rank since we could put $\hat{\pi} = p_6(1, 0)$. Hence every subrecursion theory $\mathcal{A}$ has an inductive model imbeddable in it.

7. A CHARACTERISATION OF SUBRECURSIVE CLASSES ON ℕ

Here we show that for every limit ordinal τ closed under multiplication and represented as a well ordering $\prec$ on ℕ satisfying certain natural "regulatedness" conditions (1-6 below), there is an inductive subrecursion theory whose functions coincide with the subrecursive class $\mathcal{F}(\tau, \prec)$.

Definition 12. We use (elementary-) regulated well orderings as our system of ordinal notations. Here a well ordering $\prec$ on ℕ is defined to be regulated if $0 \preceq x$ for every $x \in \mathbb{N}$, and there exist elementary functions $ord_\prec, k_\prec, succ_\prec, pd_\prec, seq, reg$ so that for every $x, y \in \mathbb{N}$ we have

(1) $ord_\prec(x,y) = \begin{cases} 1 & \text{if } x \prec y \\ 0 & \text{otherwise} \end{cases}$

(2) $k_\prec(x) = \begin{cases} 0 & \text{if } x = 0 \\ 1 & \text{if } |x| \text{ is a successor ordinal} \\ 2 & \text{if } |x| \text{ is a limit ordinal} \end{cases}$

(3) $|succ_\prec(x)| = |x| + 1$

(4) If $|x| = |y| + 1$ then $pd_\prec(x) = y$

(5) $|x|$ a limit ordinal $\Rightarrow \forall n \in \mathbb{N}([seq(x)](n) \prec [seq(x)](n+1))$ and $|x| = \lim_n (|[seq(x)](n)|)$

(6) $x \prec y$ and $|y|$ is a limit $\Rightarrow x \prec [seq(y)](reg(x,y))$

We refer to condition 6 above as the regulatedness condition, similar to that introduced in [Z77].

Definition 13. For any given well ordering $\prec$ on $\mathbb{N}$ with order type τ we define $o_\prec$ from $\emptyset^*$ onto $\omega \cdot \tau$ as follows

$$o_\prec(a) = \begin{cases} \omega \cdot |x| & \text{if } a = <0, x> \text{ for } x \in N \\ o_\prec(b) + 1 & \text{if } a = <1, b> \text{ for } b \in \emptyset^* \\ 0 & \text{otherwise} \end{cases}$$

Recall that N denotes the copy of the natural numbers in $\emptyset^*$.

We use the mapping $o_\prec$ to define the inductive subrecursion theory $I = Ind(\omega \cdot \tau, o_\prec)$ over a computation domain $< \emptyset^*, \emptyset^*, (\ ,\ ), L, R >$ for suitable $(\ ,\ ), L$ and R.

Given a standard bijection M from $\mathbb{N}^2$ to $\mathbb{N}\backslash\{0\}$, let δ be the corresponding bijection from $\emptyset^*$ to $\mathbb{N}$ defined as follows,

$$\begin{aligned} \delta(0) &= 0 \\ \delta((a,b)) &= M(\delta(a), \delta(b)) \end{aligned}$$

We then have the following theorem giving a characterisation of the subrecursive classes $\mathcal{F}(\tau, \prec)$ in terms of $Ind(\omega \cdot \tau, o_\prec)$.

Theorem 5. *Suppose that $\prec$ is a regulated well ordering with limit order type τ closed under multiplication. Then for every function $f \in \mathcal{F}(\tau, \prec)$ the function $f_I = \delta^{-1} \circ f \circ \delta$ is I-computable, and for every I-computable function g the function $g_\mathbb{N} = \delta \circ f \circ \delta^{-1} \in \mathcal{F}(\tau, \prec)$.*

Proof. We merely indicate the main ideas, technical details can be found in [H95]. For the first part, we use the First Recursion Theorem to represent the fast growing functions in the inductive models. Since we can "compute" the bijections δ and δ^{-1}, and since the characteristic function of $N \subset \emptyset^*$ is I-computable, it turns out that we only need show the fast growing functions f to be I-computable on the copy of the integers, $N \subset \emptyset^*$ (because then f_I

will also be I-computable). Now define an I-computable functional F so that for all $x, y \in N$,

$$F(f, x, y) = \begin{cases} 2^y & \text{if } x = 0 \\ f_z^{y+1}(y) & \text{if } |x| = |z| + 1 \\ f([seq(x)](y), y) & \text{if } |x| \text{ is a limit} \end{cases}$$

where f_z^{y+1} denotes the $y + 1$-th iterate of $f_z = \lambda y. f(z, y)$. F is the recursion defining the fast growing hierarchy. We can obtain $|e_F(\hat{f})|_I = \max(|\hat{f}|_I, \omega \cdot \beta_0 + \beta_1) + \omega \cdot \beta_2 + \beta_3$ for finite ordinals $\beta_0, \beta_1, \beta_2, \beta_3$ and from the definition of $o_\prec$ it can be shown that the ordinal operation $\zeta \mapsto (\omega \cdot (\beta_0 + \beta_2 + 1)) \cdot (\zeta + 1)$ is I-computable with rank $\prec \omega \cdot \beta_2$. So we can apply the First Recursion Theorem to F, and it follows that f_I is I-computable for every fast growing function f. Hence because I is closed under elementary operations, f_I is I-computable for every $f \in \mathcal{F}(\tau, \prec)$.

For the second part, we use the results of Schwichtenberg [S71] and Zemke [Z77] that $\mathcal{F}(\tau, \prec) = \mathcal{K}(\tau, \prec)$. By the recursion theorem for primitive recursive functions, we can define primitive recursive D, E so that for every $Ind(\omega \cdot \tau, o_\prec)$-computable function g we have for all $\vec{x} \in \mathbb{N}^n$,

$$g_{\mathbb{N}}(\vec{x}) = \delta(g(\delta^{-1}(\vec{x}))) = [D(\delta(\hat{g}))]^{K_{E(\delta(\hat{g}))}}(\vec{x})$$

Hence for every I-computable function g, $g_{\mathbb{N}} \in \mathcal{K}(\tau, \prec)$.   □

8. Imbedding Theorem

Theorem 6. *Let $\prec_1, \prec_2$ be regulated well orderings with limit order types τ_1, τ_2 closed under multiplication, and suppose there is an order preserving map h from $\prec_1$ to $\prec_2$. Let $\mathcal{K}(\tau_2, \prec_2, h)$ denote the Kleene enumeration hierarchy over $\prec_2$, relativised to h as initial function.*

Then $\mathcal{K}(\tau_1, \prec_1)$ is imbeddable in $\mathcal{K}(\tau_2, \prec_2, h)$ in the sense that there are functions $r, s \in \mathcal{K}(\tau_2, \prec_2, h)$ such that:

$$\text{if } f = [e]^{K_a} \in \mathcal{K}(\tau_1, \prec_1) \text{ then } f = [r(e, a)]^{K_{s(e,a)}} \in \mathcal{K}(\tau_2, \prec_2, h).$$

Proof. Suppose $f = [e]^{K_a} \in \mathcal{K}(\tau_1, \prec_1)$. Then by Theorem 5, $f_I = \delta^{-1} \circ f \circ \delta$ is $Ind(\omega \cdot \tau_1, o_{\prec_1})$-computable with code $\hat{f}_I = p(e, a)$ computable from e, a. Now suppose we can construct a map $\pi : \emptyset^* \to \emptyset^*$ satisfying the conditions 1 and 2 of Theorem 4. Then f_I will be $Ind(\omega \cdot \tau_2, o_{\prec_2}, \pi)$-computable with code $q(\hat{f}_I)$, where q is the imbedding function. Hence by a relativised version of Theorem 5, $f = \delta \circ f_I \circ \delta^{-1}$ will be computable in $\mathcal{K}(\tau_2, \prec_2, \pi_{\mathbb{N}})$ as $f = [i]^{K_j}$ where $i = D(q(\hat{f}_I))$ and $j = E(q(\hat{f}_I))$ are also $\mathcal{K}(\tau_2, \prec_2, \pi_{\mathbb{N}})$-computable. If we can construct π such that $\pi_{\mathbb{N}}$ is primitive recursive in h then we will have $f = [r(e, a)]^{K_{s(e,a)}} \in \mathcal{K}(\tau_2, \prec_2, h)$ with r and s also computable in this theory.

We show below how π can be defined "primitive recursively from h" on $\emptyset^*$. This essentially uses the same method as defining $\pi_{\mathbb{N}}$ primitive recursively in h, but we avoid notational problems doing it this way and hence can give a clearer idea of how π is actually defined.

The conditions we need π to satisfy are as follows:

(1) $\forall a, b \in \emptyset^*.\ o_{\prec_1}(a) \prec_1 o_{\prec_1}(b) \Leftrightarrow |\pi(a)|_I \prec |\pi(b)|_I$

(2) $\forall a \in \emptyset^*.\ o_{\prec_1}(a)$ a limit ordinal $\Leftrightarrow |\pi(a)|_I$ a limit ordinal

where $|\ |_I$ denotes $|\ |_{Ind(\omega \cdot \tau_2, o_{\prec_2}, \pi)}$.

From the definition of $o_{\prec_1}$ we have $o_{\prec_1}(a)$ a limit ordinal $\Leftrightarrow a = <0, x>$ for some $x \neq 0 \in N$. Hence we can satisfy 2 by defining $\pi(a) = <11, 0, h(x)>$ since, by definition of the inductive models,

$|<11, 0, h(x)>|_I = o_{\prec_2}(<0, h(x)>) = \omega \cdot |h(x)|_{\prec_2}$.

When $o_{\prec_1}(a)$ is a successor ordinal, we have $a = <1, b>$. We preserve the successor operation here by defining $\pi(a) = <11, 1, \pi(b)>$ in this case, since

$|<11, 1, \pi(b)>|_I = o_{\prec_2}(<1, \pi(b)>) = o_{\prec_2}(\pi(b)) + 1$.

The remaining case is when $o_{\prec_1}(a) = 0$. Here we define $\pi(a) = <11, 0>$ and get $|<11, 0>|_I = o_{\prec_2}(0) = 0$.

Clearly the 3 cases above define π as a function "primitive recursive in h", and since h is order preserving the condition 1 above is also satisfied. $\square$

Corollary. *If the function h in the above is primitive recursive, then $\mathcal{K}(\tau_1, \prec_1)$ is imbeddable in $\mathcal{K}(\tau_2, \prec_2)$, and equivalently $\mathcal{F}(\tau_1, \prec_1)$ is imbeddable in $\mathcal{F}(\tau_2, \prec_2)$.*

REFERENCES

[BCW94] W. Buchholz, A. Cichon and A. Weiermann, *A uniform approach to fundamental sequences and subrecursive hierarchies*, Math. Logic Quarterly 40, 1994, pp273-286.

[BW87] W. Buchholz and S.S. Wainer, *Provably computable functions and the fast growing hierarchy*, Am. Math. Soc. Contemporary Mathematics Vol. 65, ed S.G. Simpson, 1987, pp179-198.

[FW95] M.V. Fairtlough and S.S. Wainer, *Hierarchies of provably recursive functions*, Handbook of Proof Theory, ed S.Buss, to appear.

[Fef62] S. Feferman, *Classifications of recursive functions by means of hierarchies*, Trans. Am. Math. Soc. 104, 1962, pp101-122.

[Fef92] S. Feferman, *Logics for termination and correctness of functional programs*, in Proof Theory, eds P. Aczel-H. Simmons-S. Wainer, Cambridge, 1992, pp195-225.

[Fen80] J.E. Fenstad, *General recursion theory - An axiomatic approach*, Springer-Verlag, Berlin Heidelberg New York, 1980.

[H95] A.J. Heaton, *Subrecursion Theory*, PhD Thesis, Leeds University, 1995.

[K58] S.C. Kleene, *Extension of an effectively generated class of functions by enumeration*, Colloq. Math. 6, 1958, pp67-78.

[M69] Y.N. Moschovakis, *Abstract first order computability I*, Trans. Am. Math. Soc. 138, 1969, pp427-464.

[M71] Y.N. Moschovakis, *Axioms for computation theories - First draft*, Logic Colloquium 69, eds R.O. Gandy-C.E.M. Yates, North Holland, Amsterdam, 1971, pp 199-255.

[R84] H.E. Rose, *Subrecursion: Functions and hierarchies*, Oxford Logic Guides 9 (1984), Clarendon Press Oxford.

[S71] H. Schwichtenberg, *Eine Klassifikation der ϵ_0-rekursiven Funktionen*, Zeit. Math. Logik. 17, 1971, pp61-74.

[S75]   H. Schwichtenberg, *Elimination of higher type levels in definitions of primitive recursive functionals by means of transfinite recursion*, Logic Colloquium 73, eds H. Rose-J. Shepherdson, N. Holland, 1975, pp279-303.

[TZ92]   J.V. Tucker and J.I. Zucker, *Provable computable selection functions on abstract structures*, in Proof Theory, eds P. Aczel-H. Simmons-S. Wainer, Cambridge, 1992, pp277-306.

[W95]   A. Weiermann, *How to characterise provably computable functions by local predicativity*, J. Sym. Logic, 1995, to appear.

[Z77]   F. Zemke, *P.R.-regulated systems of notation and the subrecursive equivalence property*, Trans. Am. Math. Soc. 234, 1977, pp89-118.

PURE MATHEMATICS DEPARTMENT, UNIVERSITY OF LEEDS, LEEDS, LS2 9JT

On the ∀∃ - Theory of the Factor Lattice by the Major Subset Relation

Eberhard Herrmann

Abstract

We consider a special subclass of ∀∃ - sentences and give a decision procedure for the truth of these sentences in the lattice of r.e. sets factorized by the major subset relation.

Introduction

In [La68b] Lachlan gave a decision procedure for the ∀∃ - theory of the lattice of r.e. sets under inclusion $\mathcal{E}$ and thus showed it to be decidable. This result was improved by Lerman and Soare in [LeSo80] by expanding the language of $\mathcal{E}$. Stob showed in [St79] that the ∀∃ - theories of the major subset intervals are all equal and decidable too. In a weaker language the same was proven for the factor lattice of $\mathcal{E}$ by the immune sets by Degtev in [De78]. But there are still many other sublattices and factor lattices of $\mathcal{E}$ for which this problem has not yet been considered. In this paper we will start to analyse the ∀∃ - theory of another factor lattice, which we get by factorizing $\mathcal{E}$ by the major subset relation. We shall also consider the ∀∃ - theory in the language used in [La68b] and in the other papers, since it seems to be very appropriate to do so for technical reasons and in order to avoid less interesting cases (see footnote 3). In the following we give a short consideration to it.

Let $\mathcal{L}$ be a distributive lattice. $I\!B A(\mathcal{L})$ will denote the Boolean algebra which is the (smallest) closure of $\mathcal{L}$ with respect to the operations $\wedge, \vee, ^-$. The structure $(I\!B A(\mathcal{L}), \mathcal{L})$ which denotes the Boolean algebra $I\!B A(\mathcal{L})$ expanded by a unary relation for the elements of $\mathcal{L}$ is called the *d-lattice* (of $\mathcal{L}$).

By the ∀∃ - theory of the lattice $\mathcal{L}$ we understand the family of sentences of the form:

$$(\forall x_0 \in \mathcal{L}) \cdots (\forall x_{n-1} \in \mathcal{L})(\exists y_0 \in \mathcal{L}) \cdots (\exists y_{m-1} \in \mathcal{L})$$
$$P(x_0, \ldots, x_{n-1}, y_0, \ldots, y_{m-1}),$$

where P is a quantifierless (q-less) formula in the language $\{\wedge, \vee, {}^-, 0, 1, \leq, L(\,)\}$, which are true in $(BA(\mathcal{L}), \mathcal{L})$.[1]

In this paper we only consider $\forall\exists$ - sentences of the form

$$(\forall x_0)(\forall x_1)(D(x_0, x_1) \rightarrow (\exists y_0)\ldots(\exists y_{m-1})P(x_0, x_1, y_0, \ldots, y_{m-1})), \quad (1)$$

where D is the formula

$$(0 < x_0 < x_1 < 1) \wedge (x_0 \vee \bar{x}_1 \notin \mathcal{L})^{2,3}$$

and P is any q-less formula.

Even though this is a very small subclass we shall see that in showing the decidability of these sentences we must prove some new structural properties of the lattice $\mathcal{E}$, resulting in a quite substantial paper. At the end of the paper we compare the truth of the sentences (1) in our factor lattice with those in other lattices and shall see that for each structure considered we can actually decide truth within this small class.

Let $(BA(\mathcal{L}), \mathcal{L})$ be a finite d-lattice. With it there is connected a partial ordering $\preceq$ between the atoms of $BA(\mathcal{L})$ defined by: If a, b are atoms of $BA(\mathcal{L})$ then $a \preceq b$ if

$$(\forall x \in \mathcal{L})(b \subseteq x \rightarrow a \subseteq x).$$

An atom a of $BA(\mathcal{L})$ is called *outermost* if a is maximal with respect to $\preceq$ and *innermost* if a is minimal with respect to $\preceq$. (It follows that the innermost atoms are always elements of $\mathcal{L}$.) $a \prec b$ means $a \preceq b \wedge a \neq b$, and $a \npreceq b$ means that not $a \preceq b$.

A *path* is a sequence of atoms $(a_0, a_1, \ldots, a_{n-1})$ with $a_0 \succeq a_1 \succeq a_{n-1}$, a_0 outermost and for every atom a of $BA(\mathcal{L})$

$$a_i \succeq a \succeq a_{i+1} \rightarrow (a_i = a \vee a = a_{i+1}).$$

A path is *maximal* if a_{n-1} is innermost.

---

[1] In the following we shall write $\forall x_i$ and $\exists y_j$ for $\forall x_i \in \mathcal{L}$ and $\exists y_j \in \mathcal{L}$ respectively, and also say "is true in $\mathcal{L}$" instead of saying "is true in $(BA(\mathcal{L}), \mathcal{L})$".

[2] If $\mathcal{L}$ is a distributive lattice with 1, and $x, y \in \mathcal{L}$ with $x \leq y$, $x \vee \bar{y} \in \mathcal{L}$ means ($\exists z \in \mathcal{L}$) ($z \vee y = 1 \wedge z \wedge y = x$). Moreover if such a z exists it is unique, since in distributive lattices we have: $(z_0 \vee y = z_1 \vee y) \wedge (z_0 \wedge y = z_1 \wedge Y) \Leftrightarrow z_0 = z_1$. This unique element is denoted by $x \vee \bar{y}$.

[3] Since in the language $\{\wedge, \vee, \leq, 0, 1\}$ "$x_0 \vee \bar{x}_1 \in \mathcal{L}$" is not q-less, D can only be "$0 < x_0 < x_1 < 1$". But then (1) is true iff only inside x_0 is taken place a consistent refinement and not in $\bar{x}_1$ and in $x_1 - x_0$. This is so, since for x_1 there can be chosen a maximal set and for x_0 a set maximal in this maximal set. So in this case the decidability of sentences of the form (1) is uninteresting.

A *component* of a d-lattice is a maximal connected part with respect to $\preceq$ of a d-lattice. Let $(B\!A(\mathcal{L}), \mathcal{L})$ and $(B\!A(\mathcal{L}'), \mathcal{L}')$ be d-lattices. We say that the first is a *sub-d-lattice* of the second one and the second is an *extension-d-lattice* of the first one if

$$B\!A(\mathcal{L}) \subseteq B\!A(\mathcal{L}') \quad \text{and}$$
$$\mathcal{L} = B\!A(\mathcal{L}) \cap \mathcal{L}'.$$

We see that if $(B\!A(\mathcal{L}), \mathcal{L})$ is a sub-d-lattice of $(B\!A(\mathcal{L}'), \mathcal{L}')$ then every atom of $B\!A(\mathcal{L}')$ is included in an atom of $B\!A(\mathcal{L})$, and for atoms a_0, a_1 of $B\!A(\mathcal{L})$ and atoms b_0, b_1 of $B\!A(\mathcal{L}')$ with $a_0 \subseteq b_0$, $a_1 \subseteq b_1$ we have $b_0 \prec b_1 \rightarrow a_1 \npreceq' a_0$ (where $\preceq'$ is the partial ordering defined in $(B\!A(\mathcal{L}'), \mathcal{L}')$).

A q-less formula $Q(x_0, \ldots, x_{n-1})$ is called a *diagram* if Q is consistent, i.e. there is a d-lattice $(B\!A(\mathcal{L}), \mathcal{L})$ and elements $a_0, a_1, \ldots, a_{n-1}$ from $\mathcal{L}$ such that $Q(a_0, a_1, \ldots, a_{n-1})$ holds in $(B\!A(\mathcal{L}), \mathcal{L})$ and for every other q-less formula $P(x_0, \ldots, x_{n-1})$ either $Q \wedge \neg P$ or $Q \wedge P$ is consistent. We see that every q-less and consistent formula is equivalent to a disjunction of diagrams in the same variables and this is effective. From the definition of a diagram it follows that with every diagram there is related in a unique way a d-lattice. (It has exactly the same number of atoms as there are terms of the form "$x_0^\pm \wedge x_1^\pm \wedge \ldots \wedge x_{n_1}^\pm \neq 0$", $x_i^+ = x_i$, $x_i^- = \bar{x}_i$, in the diagram). In the following we identify the diagram and its d-lattice and consequently say "a is an atom of the diagram Q".

From the above it follows that if for two diagrams $E(x_0, \ldots, x_{n-1})$ and $Q(x_0, \ldots, x_{n-1}, y_1, \ldots, y_{m-1})$ $Q \rightarrow E$ holds, then it follows that the d-lattice of E is a sub-d-lattice of that of Q.

$I\!N$ will denote the set of natural numbers. Notations such as $\subseteq^*, =^*, \mathcal{L}^*$ are used with their usual meanings, namely $\subseteq, =, \mathcal{L}$ respectively modulo finite differences. If X is an infinite subset of $I\!N$ then $\mathcal{P}_X$ denotes the *principal function of X*, i.e. the range of $\mathcal{P}_X$ is X and $\mathcal{P}_X(0) < \mathcal{P}_X(1) < \ldots$. For a partial function f we write $f(x) \downarrow$ if f is defined for this x and $f(x) \uparrow$ otherwise.

$(W_e)_{e \geq 0}$ denotes the standard sequence of all r.e. sets. If A is an infinite r.e. set then $(a_s)_{s \geq 0}$ means an effective enumeration of A (without repetition), $A_0 = \emptyset$ and $A_s = \{a_0, \ldots, a_{s-1}\}$, $s \geq 1$.

Let A be a coinfinite r.e. set. $(V_e)_{e \geq 0}$ will denote the sequence

$$(\{x : (\exists s)(\forall y \leq x(y \in A_s \cup W_{e,s}))\})_{e \geq 0},$$

where $(W_{e,s})_{e,s \geq 0}$ is some recursive simultanous enumeration of $(W_e)_{e \geq 0}$.

A *state* is a finite sequence of 0's and 1's. $<>$ denotes the empty state and $2^{<\omega}$ the set of all states. If σ is a state then $|\sigma|$ denotes the length of σ. σ is an *i-state* means that $|\sigma| = i$. For a state σ and a number $i < |\sigma|$ $\sigma(i)$ is

the ith member of σ. For the concatenation of the states σ and ν we write
$\sigma * \nu$. Examples of this are $\sigma * 0$ and $\sigma * 1$, $\sigma \in 2^{<\omega}$. $\sigma \preceq \nu$ means $\nu = \sigma * \nu'$
for some state ν'. For $\sigma \in 2^{<\omega}$, $\sigma \neq < >$ let $P(\sigma)$ be the state with $P(\sigma) \preceq \sigma$
and $|P(\sigma)| + 1 = |\sigma|$. For $n < |\sigma|$ let $\sigma \upharpoonright n$ be the state with $\sigma \upharpoonright n \preceq \sigma$ and
$|\sigma \upharpoonright n| = n$. $\sigma < \nu$ means $|\sigma| = |\nu|$ and

$$(\exists i < |\sigma|)(\sigma(i) = 0, \nu(i) = 1 \wedge (\forall j < i)(\sigma(j) = \nu(j))),$$

and $(\mu\sigma)(\ldots\sigma\ldots)$ denotes the smallest state satisfying $(\ldots\ldots)$. (σ is less
than ν if $|\sigma| < |\nu|$ or $\sigma < \nu$.)

A *state function* is a function from $I\!\!N$ into $\{0,1\}$, where for almost all n
$f(n) = 0$. For a state σ and a state function f we write $\sigma \preceq f$ if σ is equal to

$$(f(0), f(1), \ldots, f(|\sigma| - 1)).$$

Given $(V_e)_{e \geq 0}$ as above we get a state function st_V by

$$\mathrm{st}_V(e) = 1 \quad \leftrightarrow \quad \bar{A} \subseteq V_e$$
$$\leftrightarrow \quad V_e \text{ is infinite.}$$

We shall need st_V in almost all later theorems.

Given $(a_s)_{s \geq 0}$, $\mathrm{st}_V(a_s)$ denotes the state function for a_s defined by

$$\mathrm{st}_V(a_s)(n) = 1 \leftrightarrow a_s \in V_{n,s}.$$

$\mathrm{st}_V(a_s) \upharpoonright n$ denotes the state σ with $\sigma \preceq \mathrm{st}_V(a_s)$ and $|\sigma| = n$.

1 The major subset relation and the factor lattice $\mathcal{E}/\approx_{ms}$

We present here the notion which is fundamental to the paper, and give some
basic properties of it. These allow us to define a factor lattice which we want
to analyse.

Definition 1.1 *Let A be a nonrecursive r.e. set. An r.e. subset B of A is
called a* major subset *of A if $B \subset_\infty A$ (i.e. $A - B$ is infinite) and*

$$(\forall C \mathrm{r.e.})(A \cup C = I\!\!N \rightarrow B \cup C =^* I\!\!N). \tag{1.1}$$

This definition and the proof that every nonrecursive r.e. set has such a
special subset was given by Lachlan in [La68a].

We see that if we write "$A \cup C =^* I\!\!N$" instead of "$A \cup C = I\!\!N$" in (1.1)
this makes no difference. We use this in the Lemma below.

If B is a major subset of A we write $B \subset_m A$. $B \subset_m^0 A$ means $B \subset_m A$ or
$B \subseteq A$ and $B =^* A$.

Definition 1.2 *For r.e. sets A and B we write $A \approx_{ms} B$ if $A \cap B \subset_m^0 A \cup B$.*

Lemma 1.3 *Let A and B be r.e. sets. Then*

1) $A \cap B \subset_m^0 A \cup B$ *iff* $A \cap B \subset_m^0 A$ *and* $A \subset_m^0 A \cup B$.

2) $A \approx_{ms} B$ *iff* $(\forall C \text{ r.e.})$ $(A \cup C =^* \mathbb{N} \leftrightarrow B \cup C =^* \mathbb{N})$.

3) *If* $A \approx_{ms} B$ *and* C *r.e. then* $A \cup C \approx_{ms} B \cup C$ *and* $A \cap C \approx_{ms} B \cap C$.

Proof: 1) $\Longleftarrow$ Let C be r.e. with $C \cup A \cup B = \mathbb{N}$. Then $C \cup A =^* \mathbb{N}$ and thus $C \cup (A \cap B) =^* \mathbb{N}$.
$\Longrightarrow$ Obvious.

2) $\Longrightarrow$ Suppose $A \cup C =^* \mathbb{N}$ for some r.e. set C. Then by 1) $C \cup (A \cap B) =^*$ $\mathbb{N}$. But then, moreover, $C \cup B =^* \mathbb{N}$. The converse case is similar.
$\Longleftarrow$ Let C be r.e. with $C \cup (A \cup B) = \mathbb{N}$. Then $(C \cup B) \cup A = \mathbb{N}$. Hence $(C \cup B) \cup B =^* \mathbb{N}$. Thus $C \cup B =^* \mathbb{N}$ and this gives $C \cup A =^* \mathbb{N}$. From the two we get $C \cup (A \cap B) =^* \mathbb{N}$.

3) Assume given A, B, and let D be r.e. with $D \cup (A \cup C) =^* \mathbb{N}$. Then $(D \cup C) \cup A =^* \mathbb{N}$, and hence by 2) $(D \cup C) \cup B =^* \mathbb{N}$ and conversely. Thus $A \cup C \approx_{ms} B \cup C$.

Now suppose for A, B, C and D as above $D \cup (A \cap C) =^* \mathbb{N}$. Then $D \cup C =^* \mathbb{N}$ and $D \cup A =^* \mathbb{N}$. Hence $D \cup B =^* \mathbb{N}$ by 2) and thus $D \cup (B \cap C) =^* \mathbb{N}$. This and the converse implication gives $A \cap C \approx_{ms} B \cap C$.

Corollary 1.4 *The relation $\approx_{ms}$ is a congruence relation on $\mathcal{E}$.*

Proof: The properties $A \approx_{ms} A$, and $A \approx_{ms} B$ implies $B \approx_{ms} A$, are obvious.

Suppose $A \approx_{ms} B$ and $B \approx_{ms} C$. Then by using 2) we get at once $A \approx_{ms} C$.

Suppose $A_1 \approx_{ms} A_2$ and $B_1 \approx_{ms} B_2$. Using 3) we have

$$A_1 \cup B_1 \approx_{ms} A_1 \cup B_2 \approx_{ms} A_2 \cup B_2.$$

Thus $A_1 \cup B_1 \approx_{ms} A_2 \cup B_2$. Moreover, using 3),

$$A_1 \cap B_1 \approx_{ms} A_1 \cap B_2 \approx_{ms} A_2 \cap B_2,$$

which gives $A_1 \cap B_1 \approx_{ms} A_2 \cap B_2$.

Definition 1.5 *Let $\mathcal{E}/\approx_{ms}$ be the factor lattice of $\mathcal{E}$ with respect to $\approx_{ms}$.*

For $X \in \mathcal{E}$ we denote the equivalence class of X with respect to $\approx_{ms}$ by $[X] \approx_{ms}$. (In the following we just write $[X]$ for $[X] \approx_{ms}$.)

The relation $\approx_{ms}$ can be defined in an equivalent and straightforward way, which is more convenient to use as the definition.

Let $\mathcal{L}$ be a lattice of subsets of $I\!N$ (with the inclusion relation) and suppose $\emptyset \in \mathcal{L}$, $I\!N \in \mathcal{L}$ and $\mathcal{L}$ is $=^*$-closed (i.e. $X \in \mathcal{L}$, $Y =^* X \to Y \in \mathcal{L}$). Let $\mathcal{L}_r$ be the sublattice (of $\mathcal{L}$) consisting of the complemented elements of $\mathcal{L}$. For $X \in \mathcal{L}$ and $Y \in \mathcal{L}$ we write $X \approx_r Y$ if

$$(\forall R \in \mathcal{L}_r)(R \subseteq^* X \leftrightarrow R \subseteq^* Y).$$

Lemma 1.6 *Let $\mathcal{L}$ be a lattice as described above. Then*

1) *$\approx_r$ is an equivalence relation in $\mathcal{L}$*

2) *For all X and Y from $\mathcal{L}$ we have $X \approx_{ms} Y \to X \approx_r Y$. ($\approx_{ms}$ is defined in $\mathcal{L}$ by "$\forall C \in \mathcal{L}$" instead of "$\forall C$ r.e." in (1.1)).*

3) *If the reduction principle (see e.g. [So87], p. 30) holds in $\mathcal{L}$ then $\approx_{ms}$ and $\approx_r$ coincide.*

Proof: 1) $X \approx_r X$ and $X \approx_r Y \to Y \approx_r X$ are obvious. Suppose $X \approx_r Y$ and $Y \approx_r Z$. Then for $R \in \mathcal{L}_r$ we have

$$R \subseteq^* X \leftrightarrow R \subseteq^* Y \leftrightarrow R \subseteq^* Z.$$

Hence $X \approx_r Z$.

2) Let $R \in \mathcal{L}_r$ with $R \subseteq^* X$. It follows that $\bar{R} \cup X =^* I\!N$ and thus (since $\mathcal{L}$ is $=^*$ - closed) by part 2) of Lemma 1.3, $\bar{R} \cup Y =^* I\!N$. This gives $R \subseteq^* Y$. Similarly for the converse.

3) Suppose Z is from $\mathcal{L}$ with $Z \cup X = I\!N$. Apply to Z and X the reduction principle and let Z' and X' be these sets. Then both are from $\mathcal{L}_r$ (and $X' \subseteq X$). Thus $X' \subseteq^* Y$ and this gives $Z' \cup Y =^* I\!N$. But then, moreover, $Z \cup Y =^* I\!N$. Similarly for the converse. From both implications we get $X \approx_{ms} Y$.

Thus in $\mathcal{E}$ we have for r.e. sets A, B with $A \subset_\infty B$ the equivalence

$$A \subset_m B \text{ iff } (\forall R \text{ recursive})(R \subseteq B \to R \subseteq^* A). \tag{1.2}$$

2 Real r.e. subsets and the lattice $\mathcal{E}/\approx_{ms}$

In order to show the decidability of the sentences (1) in $\mathcal{E}/\approx_{ms}$ we need two new notions in $\mathcal{E}$, where we consider one of them here. This is motivated by the following sentence (which is of the form (1)):

$$(\forall x_0)(\forall x_1)(D(x_0, x_1) \to (\exists y)P(x_0, x_1, y)), \tag{2.1}$$

with P the formula:

$$(x_0 < y < x_1) \land (x_0 \lor \bar{y} \notin \mathcal{L}) \land (y \lor \bar{x}_1 \notin \mathcal{L}).$$

Even though it is quite easy to show that in $\mathcal{E}$ (2.1) is true (see e.g. [So87], p. 184 or 185, 2.13) in $\mathcal{E}/\approx_{ms}$ it is false. In order to show this we must first give the following definition:

Definition 2.1 *Let A be a nonrecursive r.e. set. An r.e. subset B of A is called a* real subset *of A if*

$$\neg(\exists R \ \ recursive)\,(R \subseteq A, B \cup R \subset_m^0 A). \tag{2.2}$$

If B is a real subset of A we write $B \subset_{\mathrm{real}} A$.

The real subset relation can be considered as the opposite relation to $\subset_m$. The most obvious real subset of A is $\emptyset$. But it is not difficult to construct a real subset of A which is also simple in A.

From Definition 2.1 we get the following Lemma:

Lemma 2.2 *Let A and B be r.e. sets with $B \subseteq A$. Then*

$$B \subset_{\mathrm{real}} A \quad iff \quad [B] \lor \overline{[A]} \notin \mathcal{E}/\approx_{ms}.$$

Proof: $\Longleftarrow$ Suppose B is not real in A. Let R be a recursive set such that $B \cup R \subset_m^0 A$. Then $[B \cup R] = [A]$. But $[B \cup R] = [B] \lor [R]$.

We show that $[B] \lor [\bar{R}] = [B] \lor \overline{[A]}$.[4] We have $[B] \lor [\bar{R}] \lor [A] = [B] \lor [\bar{R}] \lor [B] \lor [R] = [I\!N]$ and $([B] \lor [\bar{R}]) \land [A] = (\underline{[B]} \lor [\bar{R}]) \land ([B] \lor [R]) = [B] \land [I\!N] = [B]$. Since $[B] \lor [\bar{R}] \in \mathcal{E}/\approx_{ms}$, so is $[B] \lor \overline{[A]}$.

$\Longrightarrow$ Suppose $[B] \lor \overline{[A]} \in \mathcal{E}/\approx_{ms}$. Then there is an r.e. set C such that $[C] = [B] \lor \overline{[A]}$. This means that $[A] \lor [C] = [I\!N]$ and $[A] \land [C] = [B]$ [4]. $A \cup C =^* I\!N$ gives recursive reduction sets A' and C'. We show that $B \cup A' \subset_m^0 A$. Let W be r.e. with $\bar{A} \subseteq W$. Since $C' \subseteq C$, $\bar{C} \subseteq^* A'$ and thus we have $\bar{A} \cup \bar{C} \subseteq W \cup A'$. Thus $(A \cap C) \cup A' \cup W = I\!N$. But $A \cap C \approx_{ms} B$. Hence $B \cup (A' \cup W) =^* I\!N$ (see Lemma 1.3, 2)). This means $(B \cup A') \cup W =^* I\!N$. Thus from (1.1) we get $B \cup A' \subset_m^* A$.

Corollary 2.3 *Let A and B be r.e. sets. Then $D([B],[A])$ holds in $\mathcal{E}/\approx_{ms}$ iff $B \approx_{ms} A \cap B$ and $A \cap B$ is a real subset of A.*

Proof: $\Longrightarrow$ Suppose $D([B],[A])$. Then from $[B] < [A]$ we get $[B] = [B \cap A]$, hence $B \approx_{ms} A \cap B$. Further $[B] \lor \overline{[A]} \notin \mathcal{E}/\approx_{ms}$. Thus by using Lemma 2.2 $B \cap A$ is a real subset of A.

---

[4]See footnote 2.

$\Longleftarrow A \cap B$ is a real subset of A gives $[A \cap B] \vee \overline{[A]} \notin \mathcal{E}/ \approx_{ms}$ by Lemma 2.2. Further, since $[A \cap B] \leq [A]$, from above we get $[A \cap B] < [A]$. Thus $D([A \cap B], [A])$. But $B \approx_{ms} A$ gives $[B] = [A \cap B]$. Hence $D([B], [A])$.

Now we will consider the $\forall \exists$-sentence (2.1). In terms of real subsets it says that for all r.e. sets A, B with B a real subset of A we can find an r.e. set C with $B \leq C \leq A$ such that B is real in C and C is real in A. The following Theorem shows that in general this is not so. (So in particular Owings' Splitting Theorem (see e.g. [So87], p. 183) is also false in $\mathcal{E}/ \approx_{ms}$).

Theorem 2.4 *Let A be a nonrecursive r.e. set. Then there is a real subset B of A such that for every r.e. set C with $B \subseteq C \subseteq A$ either B is not real in C or C is not real in A.*

Proof: The proof is divided into two parts.
(Part A). We shall construct an r.e. sequence $(R_\sigma)_{\sigma \in 2^{<\omega}}$ with the following properties:

(i) $R_\sigma \subseteq A$, $\bigcup\limits_{\sigma \in 2^{<\omega}} R_\sigma = A$, $\sigma \neq \sigma' \to R_\sigma \cap R_{\sigma'} = \emptyset$.

(ii) $\sigma \neq < >$, then R_σ is recursive.

(iii) $a_s \in R_\sigma$ implies $\sigma \preceq \mathrm{st}_V(a_s)$.

(iv) For all σ with $\sigma \preceq \mathrm{st}_V \ R_\sigma$ is infinite.

Construction. Let $R_{\sigma,0} = \emptyset$ for all $\sigma \in 2^{<\omega}$.

At step $s+1$ the element a_s $((a_s)_{s \geq 0}$ an enumeration of A) is put in exactly one of the R_σ's. To do this we consider the set

$$\{\sigma : \sigma \preceq \mathrm{st}_V(a_s), \sigma \neq < >, \quad \max|R_{\sigma,s}| < a_s, \\ \mathrm{card}(R_{\sigma,s}) \leq \mathrm{card}(R_{P(\sigma),s}) - 2\}. \tag{2.3}$$

If the set (2.3) is empty put a_s into $R_{< >,s+1}$. Otherwise take the maximal element of (2.3) with respect to $\preceq$. Let this be σ_0 and put a_s into $R_{\sigma_0,s+1}$.

Result. Let $R_\sigma = \bigcup\limits_{s \geq 0} R_{\sigma,s}$.

First we have to show that the construction is effective, i.e. we can find the maximal element in (2.3). But, since $\bigcup\limits_{\sigma \in 2^{<\omega}} R_{\sigma,s} = \{a_0, \ldots, a_{s-1}\}$ and all are pairwise disjoint, almost all the sets $R_{\sigma,s}$ are empty. Further $R_{\sigma,s} = \emptyset$ implies $R_{\sigma',s} = \emptyset$ for all σ' with $\sigma \preceq \sigma'$. Thus the maximal element exists (if (2.3) is not empty) and can be found effectively.
(i) holds, since every a_s is put in exactly one $R_{\sigma,s}$.
(ii) follows from the requirement "$\max|R_{\sigma,s}| < a_s$" in (2.3) (and that $(R_\sigma)_{\sigma \in 2^{<\omega}}$ is r.e.).

(iii) follows from the requirement "$\sigma \preceq \mathrm{st}_V(a_s)$" in (2.3).

(iv) Since A is not recursive, for every $\sigma \preceq \mathrm{st}_V$ there are infinitely may a's with $\sigma \preceq \mathrm{st}_V(a_s)$. Otherwise the set

$$\{x \in I\!N : \sigma(e) = 1 \to x \in V_e, \text{ for all } e < |\sigma|\}$$

is r.e. and would be equal (mod $=^*$) to $\bar{A}$, hence A would be recursive. We first see that $R_{()}$ is infinite by the last requirement in (2.3). Assuming that for $\sigma \preceq \mathrm{st}_V$ $R_{P(\sigma)}$ is infinite we also show this for R_σ. Since the set $\{a_s : \sigma \preceq \mathrm{st}_V(a_s)\}$ is infinite (see (iii)), and hence includes sufficiently big elements, infinitely often σ belongs to (2.3). But infinitely often σ must also be the maximal element of (2.3) by the last requirement in (2.3).

In the following we shall use the following property of the sets R_σ for $\sigma \preceq \mathrm{st}_V$:

$$(\forall R \text{ recursive})(R \subseteq A \to \ (\exists n)(\forall \sigma)(n \geq |\sigma|, \atop \sigma \preceq \mathrm{st}_V \to R \cap R_\sigma = \emptyset)) \ . \tag{2.4}$$

(Part B) We will now construct the set B. Given $(R_\sigma)_{\sigma \in 2^{<\omega}}$ from above we see that the set

$$\{\sigma : R_\sigma \text{ is infinite} \wedge (\forall \sigma')(\sigma < \sigma' \to R_{\sigma'} \text{ is finite}\} \tag{2.5}$$

is just $\{\sigma : \sigma \preceq \mathrm{st}_V\}$, hence infinite and is in Δ_3^0. (The quantifier in (2.5) is bounded, since $\sigma < \sigma'$ implies $|\sigma| = |\sigma'| -$ see the Introduction).

Let $M^{(3)}$ be a Σ_3^0-maximal set[5] with $\overline{M^{(3)}}$ included in the set (2.5). Let $\lambda_\sigma(x, s)$ be a recursive approximation function for $M^{(3)}$, i.e.

$$\sigma \in M^{(3)} \to (\exists x)(\lambda_\sigma(x) = \omega)$$
$$\sigma \notin M^{(3)} \to (\forall x)(\lambda_\sigma(x) < \omega).$$

The basic idea for the following construction is the same as for the proof $(\Sigma_3^0, \Pi_3^0) \leq_m (\{e : W_e \text{ is cofinite}\}, \{e : W_e \text{ is maximal}\})$ (which is not sufficient for the estimation of the index set of the maximal sets, but one of the strongest for the index set of the cofinite sets and good for our purposes).

Let $B_0 = \emptyset$. The construction of B_{s+1} is divided into three cases.

In the following we use for $\mathcal{P}_{R_{\sigma,s} \cap \bar{B}_s}(x)$ the abbreviation $p(\sigma, s, x)$.

$(s+1) = 3 \cdot t)$ Let B_{s+1} be the union of B_s with

$$\{p(\sigma, s, x) : \lambda_\sigma(x, t+1) \neq \lambda_\sigma(x, t), |\sigma| \leq s, x \leq s, p(\sigma, s, x) \downarrow\}$$

---

[5] "Σ_3^0-maximal" means the analogy of "maximal" in $\mathcal{E}$ for the lattice of Σ_3^0-sets. We use an effective coding of the elements of $2^{<\omega}$ by numbers and can thus assume that $M^{(3)} \subseteq 2^{<\omega}$.

(This step ensures that $\sigma \in M^{(3)} \leftrightarrow B \cap R_\sigma =^* R_\sigma$).

$(s+1 = 3 \cdot t + 1)$ Let B_{s+1} be the union of B_s with

$$\{p(\sigma, s, 0) : (\exists \sigma')(\sigma < \sigma' \wedge \operatorname{card}(R_{\sigma', t+1}) \neq \operatorname{card}(R_{\sigma', t}), |\sigma| \leq s, p(\sigma, s, 0) \downarrow\}.$$

(This step ensures that for σ with $\sigma < \sigma'$, $\sigma' \preceq \operatorname{st}_V$ we not only have $R_\sigma \cap B =^* R_\sigma$ but further $R_\sigma \subseteq B$ – see (2.5)).

$(s+1 = 3 \cdot t + 2)$ Provide for all sets $R_{\sigma, s} \cap \bar{B}_s$, $|\sigma| \leq s$, one step of Friedberg's maximal set construction, i.e. let $\operatorname{st}_W(x, s)$ be the state function

$$\operatorname{st}_W(x, s)(i) \quad = \quad 1 : x \in W_{i,s}$$
$$= \quad 0 : \text{otherwise}$$

and ask whether

$$(\exists i)(\exists j)(i < j \leq s \quad , p(\sigma, s, j) \downarrow, \qquad \qquad (2.6)$$
$$\operatorname{st}_W(p(\sigma, s, i)) \upharpoonright i < \operatorname{st}_W(p(\sigma, s, j)) \upharpoonright i).$$

If yes, take the smallest such i, and then the smallest such j. Let i_0 and j_0 be these numbers and put all numbers $p(\sigma, s, k)$ with $i_0 \leq k < j_0$ into B_{s+1}.

If such numbers do not exist do nothing.

Let $B = \bigcup_{s \geq 0} B_s \cup R_{()}$.

We see that for all $\sigma \preceq \operatorname{st}_V$

$$\sigma \in M^{(3)} \quad \rightarrow \quad R_\sigma \cap B =^* R_\sigma \qquad \text{and}$$
$$\sigma \notin M^{(3)} \quad \rightarrow \quad R_\sigma \cap B \text{ is maximal in } B.$$

All $R_{\sigma'}$ with $\sigma' < \sigma \preceq \operatorname{st}_V$ are completely contained in B and all $R_{\sigma'}$ with $\sigma < \sigma'$ and $\sigma \preceq \operatorname{st}_V$ are finite.

B is a real subset of A, since for infinitely many $\sigma \preceq \operatorname{st}_V$ (namely all σ from $\overline{M^{(3)}}$) $R_\sigma \cap B \neq^* R_\sigma$, and by (2.4).

Let C be r.e. with $B \subseteq C \subseteq A$. We see that the set

$$\{\sigma : R_\sigma \subseteq^* C\}$$

belongs to Σ_3^0, since $(R_\sigma)_{\sigma \in 2^{<\omega}}$ and C are r.e.

Since $M^{(3)}$ is Σ_3^0-maximal, only two cases are possible.

Say for almost all $\sigma \in \overline{M^{(3)}}$ $R_\sigma \subseteq^* C$. Then C is not real in A, since the union of the remaining finitely many σ's is recursive and joined with C gives a major subset of A or is cofinite in A, by (2.4).

On the other hand, say for almost all $\sigma \in \overline{M^{(3)}}$ we do not have $R_\sigma \subseteq^* C$. Then for all these σ's $R_\sigma \cap C =^* R_\sigma \cap B$, since $R_\sigma - B$ is cohesive.

Again as above the union of the remaining R_σ's is recursive, and joined with B is a major subset of C, or is cofinite in C.

Remark. The author originally conjectured that the negation of Theorem 2.4 holds. However L. Harrington pointed out in conversation that Theorem 2.4 can be true, so stimulating the search for a proof.

Corollary 2.5 *In $\mathcal{E}/\approx_{ms}$ there are classes $[A]$ and $[B]$ with $D([B],[A])$ such that for every class $[C]$ with $[B] \leq [C] \leq [A]$*

$$[B] \vee \overline{[C]} \in \mathcal{E}/\approx_{ms} \quad \text{or} \quad [C] \vee \overline{[A]} \in \mathcal{E}/\approx_{ms}.$$

Remarks. 1) If B is a real subset of A then of course $B \cup \bar{A}$ is not r.e. Thus Owings' splitting holds for B and A. From Theorem 2.4 it then follows that there is a recursive set $R \subseteq A$ such that one splitting set joined with R is a major subset of A and $B \cup R$ a major subset of the second splitting set.

2) What can said about $\{C$ r.e. $: B \subseteq C \subseteq A\}$ for A and B as in Theorem 2.4? Do these properties already determine the above interval up to the isomorphism? If in addition A is maximal are all such B's automorphic?

3 Major subsets modulo r.e. sets

In order to analyse of the truth of the (1)'s in $\mathcal{E}/\approx_{ms}$ without the notion of real subset we will need another new notion which is a natural generalization of $\subset_m$.

Definition 3.1 *Let A and B be r.e. sets. We say that A is a major subset of B modulo r.e. sets if $A \subset_\infty B$ and*

$$(\forall W \quad r.e.)(\bar{B} \subseteq W \to (\exists C \text{ r.e.})(C \subseteq B - A \wedge A \cup C \cup W = I\!N)). \quad (3.1)$$

If in (3.1) we replace "$\exists C$ r.e." by "$\exists C$ finite" then we get the notion of major subset. Thus the notion of major subset modulo r.e. sets generalizes that of $\subset_m$. If we replace in (3.1) "$\exists C$ r.e." by "$\exists C$ recursive" we do not get a new notion. By using the reduction principle it is easy to see that both are equivalent. Thus "modulo recursive sets" gives the same as "modulo r.e. sets".

For "A is a major subset of B modulo r.e. sets" we write $A \subset_m B$ (mod r.e.).

Similarly to (1.2) we have: $A \subset_m B$ (mod r.e.) iff $A \subset_\infty B$ and

$$(\forall R \text{ recursive})(R \subseteq B \to R - A \text{ is recursive}). \quad (3.2)$$

$\Longrightarrow$ Let R be recursive with $R \subseteq B$. Then $\bar{B} \subseteq \bar{R}$ so that there is an r.e. set C with $C \subseteq B - A$ and $A \cup C \cup \bar{R} = I\!N$. From this we get $R - A \subseteq C$. Hence $R - A = C \cap R$ and so is r.e. $R - A$ as an r.e. half of a splitting of a recursive set is also recursive.

$\Longleftarrow$ Let W be r.e. with $\bar{B} \subseteq W$. Take reduction sets for B and W. Let B' and W' be these. Then B' is recursive and $B' \subseteq B$. Hence $B' - A$ is r.e. But $A \cup (B' - A) \cup W = I\!N$. Thus $A \subset_m B$ (mod r.e.).

Existence of major subsets modulo r.e. sets

Suppose $A \subset_m B$ (mod r.e.). If B is recursive then A also must be recursive. If B is not recursive then the following cases are possible:

(i) A is simple in B (then $A \subset_m B$)

(ii) A is a half of a splitting of B.

(iii) A is pseudosimple in B, i.e. there is an r.e. C with $A \cap C = \emptyset$ and $A \cup C \subset_m B$.

(iv) A is pseudocreative in B (i.e. $(\forall C$ r.e.$)$ $(C \subseteq B - A \to (\exists D$ r.e.$)$ $(D$ is infinite $\wedge D \subseteq B - A \wedge C \cap D = \emptyset)))$.

We give below a sketch of the easiest proof that case (iv) is possible. After this we give a theorem requiring a more complicated construction in order to obtain additional properties needed later.

Let T be the following subset of $2^{<\omega}$:

$$\{\sigma \in 2^{<\omega} : \sigma = \mathcal{B}_i * \mathcal{A}_j, i, j \geq\}$$

with $\mathcal{B}_i = (1 \cdots 1)$ (i-times) and $\mathcal{A}_j = (0 \cdots 0)$ (j-times). Further let T_i be the set

$$\{\sigma \in T : \sigma = \mathcal{B}_i * \mathcal{A}_j, \ j \geq 0\}, \quad i \geq 0.$$

Given an effective enumeration $(b_s)_{s \geq 0}$ of B and $\mathrm{st}_V(b_s)$ (defined similarly to $\mathrm{st}_V(a_s)$ with B instead of A – see the Introduction) we construct step by step A and a function T defined on T and with values in B.

Define $A_0 = T_0 = \emptyset$.

At step $s + 1$ assume we have b_s and ask if

$$(\exists \sigma \in T)(\sigma \in T_i \wedge T_s(\sigma) \downarrow \wedge \mathrm{st}_V(T_s(\sigma)) \restriction i < \mathrm{st}_V(b_s) \restriction i). \qquad (3.3)$$

If yes, take the smallest σ (in T) satisfying (3.3), $= \sigma_0$ say, and define $T_{s+1}(\sigma_0) = b_s$ and $A_{s+1} = A_s \cup \{T_s(\sigma_0)\}$. Otherwise define

$$T_{s+1}((\mu\sigma)(\sigma \in T \wedge T_s(\sigma) \uparrow)) = b_s$$

and $A_{s+1} = A_s$.

Let $A = \bigcup_{s \geq 0} A_s$ and $T = \lim_s T_s$. We see that if $\bar{B} \subseteq W_e$ then for almost all $\sigma \in T$ with $W_e \preceq \sigma$ we have $T(\sigma) \in W_e$. Moreover all $T(T_i)$ are r.e., since in (3.3) inside a T_i only the i-states are considered. Thus $T(T_0) \cup \ldots \cup T(T_{e-1})$ is r.e., is contained in $B - A$, and joined with A and W_e is equal to $I\!N$ (mod r.e.).

If C is an r.e. set with $C \subseteq B - A$ then $C \subseteq T(T_0) \cup \ldots \cup T(T_i)$ for some i. Otherwise, since the sets $\bigcup_{s \geq 0} T_s(T_i)$, $i \geq 0$, are pairwise disjoint, we can construct an infinite, recursive subset R of C (hence $R \subseteq B - A$) and $R \not\subseteq \bigcup_{j \leq i} T(T_j)$ for any i. But then $\bar{R}$ contradicts the above property, since $\bar{B} \subseteq \bar{R}$. Thus A is also pseudocreative in B.

Remark. A d-lattice $(I\!\!BA(\mathcal{L}), \mathcal{L})$ is called $\exists$-*complete* if every consistent q-less formula is satisfied in $(I\!\!BA(\mathcal{L}), \mathcal{L})$. From the $\exists$-completeness it follows in particular that the $\exists$-theory of this d-lattice is decidable.

By using the above construction we can show that as for $\mathcal{E}$ and others we also have that $\mathcal{E}/\approx_{ms}$ is $\exists$-complete.

By using classes with recursive sets for different components of a separated diagram, classes with maximal sets for the outermost, not innermost atoms, classes with Friedberg splitting sets and major subsets (mod r.e.) in an r.e. set[6] for all other, not innermost atoms we can realize every diagram in $\mathcal{E}/\approx_{ms}$. By the considerations in [La68b] this is sufficient.

We say that an r.e. subset B of A is r-*separable in* A if for every C r.e. with $C \subseteq A - B$ there is a recursive set R with $R \subseteq A - B$ and $C \subseteq R$.

Theorem 3.2 *Let A be a nonrecursive r.e. set. Then there is an r.e. subset B of A with the following properties:*

1) $B \subset_m A$ *(mod r.e.).*

2) B *is pseudocreative in* A.

3) B *is r-separable in* A.

4) $(\forall e)(B \subseteq W_e \subseteq A \rightarrow (\exists R \quad recursive) (R \subseteq A - B \wedge (B \cup R \subset_m^0 W_e R \vee W_e \cup R \subset_m^0 A)))$.

Proof. We shall construct step by step the set B along with a function T defined on $\mathcal{T}$ with values in A.

Let $B_0 = T_0 = \emptyset$ and suppose B_s and T_s are already constructed and T_s has the following properties:

- T_s is a bijection between $\mathrm{dom}(T_s)$ (the domain of T_s) and $A_s - B_s$.

- For every $\sigma \in \mathrm{dom}(T_s)$ (Suppose $\sigma \in T_i$)

$$(\forall \sigma' \in T_i)(\sigma' \preceq \sigma \rightarrow \sigma' \in \mathrm{dom}(T_s)).$$

---

[6]Let A, B, C r.e. sets with $A \subset_\infty B \subseteq C$ and $C - B$ not r.e. We say $A \subset_m B$ (mod r.e.) relative to C if in (3.1) instead of "$\bar{B} \subseteq W$" we take "$C - B \subseteq W$". The existence of such an A for given $B \subseteq C$, $C - B$ not r.e. follows easily from the proof before this remark.

Priority property:

Every element a_t (in $A_s - B_s$) has two state functions. These are $\mathrm{st}_V(a_t)$ (see the Introduction), which is constant for the whole construction, and $\mathrm{st}_W(a_t, s)$ (which is defined by using $(W_e)_{e \geq 0}$, see step: $s + 1 = 3n + 2$). For two pairs of states (σ_0, σ_1) and (ν_0, ν_1) all of the same length we write $(\sigma_0, \sigma_1) < (\nu_0, \nu_1)$ if $\sigma_0 < \nu_0$ or $\sigma_0 = \nu_0 \wedge \sigma_1 < \nu_1$ (and $(\sigma_0, \sigma_1) = (\nu_0, \nu_1)$ if $\sigma_0 = \nu_0 \wedge \sigma_1 = \nu_1$).

For every i and all $\sigma, \sigma' \in (T_i \cap \mathrm{dom}(T_s))$ with $\sigma \preceq \sigma'$:

$$(\mathrm{st}_V(T_s(\sigma')) \upharpoonright i, \mathrm{st}_W(T_s(\sigma'), s) \upharpoonright i) < (\mathrm{st}_V(T_s(\sigma)) \upharpoonright i, \mathrm{st}_W(T_s(\sigma), s) \upharpoonright i) \tag{3.4}$$

or both pairs of states are equal and $T_s(\sigma) < T_s(\sigma')$.

This means that $T_i \cap \mathrm{dom}(T_s)$ is divided into blocks with elements $T_s(\sigma)$ of the same $(\mathrm{st}_V, \mathrm{st}_W)$-states. The states of the blocks are decreasing and inside every block the elements $T_s(\sigma)$'s are increasing. Further the st_V-states have a higher priority than the st_W-states.

Construction. We describe step $s + 1$.

$(s + 1 = 3 \cdot n + 1)$ Take a_n, let $\mathrm{st}_W(a_n, s) = 0$ and consider the set

$$\{i : (\forall \sigma \in T_i \cap \mathrm{dom}(T_s))(\mathrm{st}_V(T_s \upharpoonright \sigma)) \upharpoonright i = \mathrm{st}_V(a_n) \upharpoonright i \wedge \\ \mathrm{st}_W(T_s(\sigma), s)i = 0 \rightarrow T_s(\sigma) < a_n)\}. \tag{3.5}$$

(Since all i with $\mathrm{dom}(T_s) \cap T_i = \emptyset$ satisfy (3.5), almost all i belong to (3.5)).

For every i from (3.5) let σ_i be

$$(\mu\sigma \in T_i)((\forall \sigma' \in T_i)(\mathrm{st}_V(T_s(\sigma')) \upharpoonright i \geq \mathrm{st}_V(a_n) \upharpoonright i \rightarrow \sigma' \prec \sigma).$$

From among all σ_i, $i \in (3.5)$, take the smallest one, say $\hat{\sigma}$, and define

$$T_{s+1}(\hat{\sigma}) = a,$$

and for all other σ's let T_{s+1} be equal to T_s.

$$\begin{aligned} B_{s+1} &= B_s \cup \{T_s(\hat{\sigma})\} &&: T_s(\hat{\sigma}) \downarrow \\ &= B_s &&: \text{otherwise.} \end{aligned}$$

$(s + 1 = 3 \cdot n + 2)$ We assume that (a, m) with $a \in W_m$ is enumerated at the n-th enumeration step.
If

$- a \notin A_n - B_s$ or

   $- a = T_s(\sigma)$, $\sigma \in T_i$, $i < m$ or

     $- \mathrm{st}_W(a, s)(m) = 1$ do nothing.

Otherwise we have $a = T_s(\sigma)$, $\sigma \in T_i$ and $m \leq i$.

Define

$$\text{st}_W(a, s+1)(k) = 1 \qquad : k = m$$
$$= \text{st}_W(a, s)(k) \quad : k \neq m$$

and see if there is a $\sigma' \in T_j$, $j \leq i$, $|\sigma'| = |\sigma|$ with

$(\forall \sigma'' \in T_j)$ (the pair of states (3.4) with $T_s(\sigma'')$ and $T_s(\sigma)$ (with $\text{st}_W(a, s+1)$) are equal $\rightarrow (\sigma'' \prec \sigma' \; T_s(\sigma'') < T_s(\sigma)))$

(We see that without the requirement "$T_s(\sigma'') < T_s(\sigma)$" above, such σ' always exists for $j = i$.)

If such a σ' exists choose the smallest one, say σ'_0, and define

$$T_{s+1}(\sigma'_0) = a$$
$$T_{s+1}(\sigma * \mathfrak{B}_j) = T_s(\sigma * \mathfrak{A}_j * 0) \quad : j \geq 0$$
$$T_{s+1}(\nu) = T_s(\nu) \qquad\qquad : \text{otherwise}$$
$$B_{s+1} = B_s \cup \{T_s(\sigma'_0)\}.$$

If such a σ' does not exist define

$$T_{s+1}(\sigma * \mathfrak{B}_j) = T_s(\sigma * \mathfrak{B}_j * 0) \quad : j \geq 0$$
$$T_{s+1}(\nu) = T_s(\nu) \qquad\qquad : \text{otherwise} \qquad\qquad (3.6)$$
$$B_{s+1} = B_s \cup \{a\}$$

$(s+1 = 3 \cdot n + 3)$ See if

$$(\exists e)(W_{e,s} \subseteq \bigcup_{\substack{\sigma \in T_j \\ j > e}} T_s(\sigma) \wedge W_{e,s} \neq \emptyset \wedge W_{e,s} \cap B_s = \emptyset). \qquad (3.7)$$

If not do nothing. Otherwise take the smallest such e and the smallest σ with $T_s(\sigma) \in W_{e,s}$, σ_0 say, and define

$$B_{s+1} = B_s \cup \{T_s(\sigma_0)\}$$
$$T_{s+1}(\sigma_0 * \mathfrak{B}_j) = T_s(\sigma_0 * \mathfrak{B}_j * 0) \quad : j \geq 0$$
$$T_{s+1}(\nu) \qquad\quad = T_s(\nu) \qquad : \text{otherwise}.$$

Result. Let $B = \bigcup_{s \geq 0} B_s$. We see that B is an r.e. subset of A. Then

1. $\text{Lim}_s T_s(\sigma) \downarrow$ for all $\sigma \in T$. (Proof by induction).
$\sigma = < >$. We see that $T_1(< >) = a_0$ and $T_s(< >) \downarrow \rightarrow T_{s+1}(< >) \downarrow$. If $T_{s+1}(< >) \neq T_s(< >)$ then the 0-state pair of $T_{s+1}(< >)$ is greater than that of $T_s(< >)$. But this can happen at most three times.
Take σ and suppose that for all σ' less than σ we have that $\lim_s T_s(\sigma')$ exists. Let $\sigma \in T_i$. We choose s_0 such that

(i) $T_{s_0}(\sigma') = \lim_s T_s(\sigma')$ $(= T(\sigma'))$ for all σ' less than σ.

(ii) For all $s \geq s_0$ and $s+1 = 3 \cdot n + 1$ $a_n > T(\sigma')$ for all such σ'.

(iii) For all $s \geq s_0$ and $s+1 = 3 \cdot n + 3$ no e with $e < i$ satisfies (3.7).

We see that such an s_0 exists, since the set of $T(\sigma')$, σ' less than σ, is finite, since step $3 \cdot n + 3$ of the construction can happen at most once for each e, and by the assumption of the induction.

If for some $s > s_0$ $T_s(\sigma) \uparrow$ then at the next step of the form $t+1 = 3 \cdot n + 1$ we have $T_{t+1}(\sigma) \downarrow$, by (ii) and (i) and the minimality of σ. Here we use the fact that in the case $s+1 = 3 \cdot n + 2$ σ_0' is less than σ. If $T_s(\sigma)$ infinitely often becomes undefined then for some sufficiently large s $T_s(\sigma) \downarrow$ and $T(\sigma') < T_s(\sigma)$ for all lesser σ'. But for the next $s' > s$ with $T_{s'}(\sigma) \uparrow$ this can only happen via the case $3 \cdot n + 2$ (since in the definition of σ_i in $s+1 = 3 \cdot n + 1$ we have $\sigma' \prec \sigma$). But this contradicts the choice of s_0. Thus for some $s_1 \geq s_0$ we have $T_s(\sigma) \downarrow$ for all $s \geq s_1$. But $T_{t+1}(\sigma) \neq T_t(\sigma)$ for $t \geq s_1$ only if the i-state pair of $T_{t+1}(\sigma)$ is greater than that of $T_t(\sigma)$. This is only possible finitely often, hence $\lim_s T_s(\sigma)$ exists.

2. For every infinite V_e, and for almost all $\sigma \in T_j$, $j \geq e$, we have $T(\sigma) \in V_e$.

This can be shown by induction over the V_e's. We assume that this holds for all V_i, $i < e$. Thus there is a $\sigma_e \in T_j$ $j \geq e$ such that for all σ which are greater than σ_e and from T_j, $j \geq e$ $\mathrm{st}_V(T(\sigma) \in V_e$ and for infinitely many not then there are two which are greater than σ_e and contradicting step $s+1 = 3 \cdot n + 1$ (also the requirement $T_t(\sigma) < /t_s(\sigma') = a_n$, since there are infintely many).

3. For every W_e for almost all j we either have $T(T_j) \cap W_e$ finite, or cofinite in $T(T_j)$.

This also is proven by induction over e. We assume this for all W_i, $i < e$. Let j_o be such that $(\forall i < e)((\forall j \geq j_o T(T_j) \cap W_i$ is finite) or $(\forall j \geq j_0)(T(T_j) \cap W_e =^* T(T_j)))$. If for some j_1, j_2 with $j_0 \leq j_1 \leq j_2$ $T(T_{j_1}) \cap W_e$ is finite and $T(T_{j_2}) \cap W_e$ is infinite, then this contradicts $s+1 = 3 \cdot n + 2$. (We cannot show for W_e the same as for V_e in 2., since for every $j \geq e$ only V_i $i \leq e$ has a higher priority than V_e. For W_e and $j \geq e$ not only V_i $i \leq e$ and W_i $i < e$, but also $V_{e+1}, \ldots, V_j$ are more important than W_e).

4. All sets $T(T_j)$, $j \geq 0$ are recursive sets.

Since for every j for almost all $T(\sigma)$, $\sigma \in T_j$ the st_V-state and the final st_W-states are equal, but inside this block the elements are increasing, these are recursive. We first show it for $T(T_0)$, then for $T(T_1)$, and so on, and use the fact that in $s+1 = 3 \cdot n + 2 \sigma'$ one has to satisfy "$\sigma' \in T_j$, $j \leq i$, $|\sigma'| \leq |\sigma|$".

The required properties of B follow.

1) 2 ensures that every recursive subset of A is contained in $B \cup T(T_0) \cup \cdots \cup T(T_i)$ for some i. Thus from 4 and (3.2) we get 1).

3) From step $s + 1 = 3 \cdot n + 2$ we further obtain

$$W_e \not\subseteq T(T_0) \cup \cdots \cup T(T_e) \rightarrow W_e \not\subseteq A - B.$$

Thus from 4 we get 3).

2) All sets $T(T_j)$ are infinite. If some $W_e \subseteq A - B$, then by 3) $T(T_{e+1}) \cap W_e = \emptyset$ and $T(T_{e+1}) \subseteq A - B$. Thus $B \cup W_e$ is not simple or cofinite in A.

4) From 2 it follows that

$$(\forall j)(T(T_j) \cap W_e =^* T(T_j)) \rightarrow B \cup W_e \subset^0_m A \text{ and}$$
$$(\forall j)(T(T_j) \cap W_e - \text{finite}) \rightarrow B \subset^0_m B \cup W_e.$$

Thus, from 3 and 4 we get 4).

The index set $\{e : W_e \subset_m A \text{ (mod r.e.)}\}$

That there are very many r.e. sets which are major subsets (mod r.e.) of a nonrecursive r.e. set follows from the following Theorem which improves a result of Lempp (see [Lem86]).

Theorem 3.3 *Let A be a nonrecursive r.e. set. Then $(\Pi^0_4, \Sigma^0_4) \leq_m (\{e : W_e \subset_m A\}, \{e : \text{not } W_e \subset_m A \text{ (mod r.e.)}\})$.*

Proof: That both sets $\{e : W_e \subset_m A\}$ and $\{e : W_e \subset_m A \text{ (mod r.e.)}\}$ belong to Π^0_4 is easy to see.

We have "$W_e \subset_m A$" if

$$W_e \subset_\infty A \wedge (\forall f)(A \cup W_f = \mathbb{N} \rightarrow (\exists k)(W_k \text{ is finite} \wedge$$
$$W_k \subseteq A - B \wedge W_k \cup W_e \cup W_f = \mathbb{N})). \quad (3.8)$$

Thus (3.8) has the quantifier form

$$\Pi^0_3 \wedge \forall(\Pi^0_2 \rightarrow \exists(\Sigma^0_2 \wedge \Pi^0_2 \wedge \Pi^0_2)) \equiv \forall \exists \Pi^0_2. \quad (3.9)$$

For "$W_e \subset_m A$ (mod r.e.)" replace "finite" by "r.e." in (3.8). Then the result will still be Π^0_4.

We now show the (Π^0_4, Σ^0_4)-completeness of the pair in the Theorem. Let $Q \in \Pi^0_4$. Then there is a recursive approximation function $\lambda_i(x, y, s)$ such that

$$Q(i) \rightarrow (\forall x)(\exists y)(\lambda_i(x, y) = \omega)$$
$$\neg Q(i) \rightarrow (\forall x)(\exists y)(\lambda_i(x, y) < \omega)^7$$

We shall construct an r.e. sequence $(U_i)_{i \geq 0}$ such that

$$Q(i) \quad \rightarrow \quad U_i \subset_m A$$
$$\neg Q(i) \quad \rightarrow \quad (\exists R \text{ recursive})(R \subseteq A \wedge R - U_i \text{ is not recursive}).$$

The latter implies that $\neg U_i \subset_m A \pmod{\text{r.e.}}$ – see (3.2).

We fix an i and omit it in the following. (Thus we write U and $\lambda(x, y, s)$ instead of U_i and $\lambda_i(x, y, s)$ respectively.)

Let $(a_s)_{s \geq 0}$ be an enumeration of A. We first construct step by step functions T_s with

- $T_s \subseteq T_{s+1}$

- $T_s : \mathbb{N}^2 \rightarrow A_s$, partial, bijective between $\text{dom}(T_s)$ and A_s.

- $T_s(n, k) \downarrow \wedge i \leq k. \rightarrow T_s(n, i) \downarrow$ and $T_s(n, 0) \downarrow \wedge i \leq n. \rightarrow T_s(i, 0) \downarrow$.

Then using $(T_s)_{s \geq 0}$ we construct U.

Let $T_0 = \emptyset$ and suppose T_s with the above properties has already been constructed.

Construction of T_{s+1}: We look for the place for a_s. For this we consider the set

$$\{i : (\forall z)(\text{st}_V(a_s) \restriction i \leq \text{st}_V(T_s(i, z)) \restriction i \rightarrow T_s(i, z) < a_s\}. \tag{3.10}$$

We see that at least all i with $T_s(i, 0) \uparrow$ belong to (3.10). Let φ_s be the function defined for the i's from (3.10) by

$$\varphi_s(i) = \max\{z : \text{st}_V(a_s) \restriction i \leq \text{st}_V(T_s(i, z)) \restriction i : i \in (3.10)\}.$$

From the set $\{i + \varphi_s(i) : i \in (3.10)\}$ choose the smallest number and for this the greatest i such that $i + \varphi_s(i)$ is this smallest one. Let i_0 be this number. Now we define

$$T_{s+1}(i_0, (\mu z)(T_s(i_0, z) \uparrow)) = a_s,$$

and for all other pairs equal to the value of T_s.

Result. 1. Let $T = \lim_s T_s$. Since $T_s \subseteq T_{s+1}$, T obviously exists.

Further $T : \mathbb{N}^2 \rightarrow A$ is partial recursive, bijective between $\text{dom}(T)$ and A and $T(n, k) \downarrow, i \leq k. \rightarrow T(n, i) \downarrow$.

2. Let R_n be the set $\{T(n, k) : T(n, k) \downarrow, k \geq 0\}$. We see that all R_n's are r.e., disjoint and $\bigcup_{n \geq 0} R_n = A$.

---

[7] We claim that $\lambda_i(x, 0, s) = 0$ for all $s \geq 0$.

We show that $T(n,k) \downarrow$ for all pairs $(n,k) \in \mathbb{N}^2$. We see that for every n there are infinitely many a_s's with $\mathrm{st}_V(a_s) \upharpoonright n = \mathrm{st}_V \upharpoonright n$. Thus the set of these a_s's is unbounded. This means that if $T(n,k) \downarrow$ for some k the number n would belong infinitely often to (3.10). There would hence be an i_0 at some step $s+1$.

3. All sets R_n are recursive. Since in (3.10) we claim $T_s(i,z) < a_s$, the a_s's in R_n with the same n-states are increasing and thus form a recursive subset of R_n. But there are only finitely many n-states. Thus R_n as a union of finitely many recursive sets is also recursive.

4. For every n R_n includes infinitely many a_s's with $\mathrm{st}_V(a_s) \upharpoonright n = \mathrm{st}_V \upharpoonright n$.

This follows from the fact that infinitely many a_s's with $\mathrm{st}_V(a_s) \upharpoonright n = \mathrm{st}_V \upharpoonright n$ are enumerated (see 2), with greater n-states only finitely many, and in (3.10) the only $T_s(i,z)$'s considered are those with i-states greater than or equal to $\mathrm{st}_V(a_s) \upharpoonright i$.

Construction of U: Inside every R_n we provide a construction very similar to that of the proof for $(\Sigma_3^0, \Pi_3^0) \leq_m (\{e : W_e =^* \mathbb{N}\}, \{e : W_e \text{ is not recursive}\})$, but with a small modification.

For one n we do the following:

$(s+1 = 2 \cdot t)$ If for a $y \leq s \lambda(n,y,t+1) \neq \lambda(n,y,t)$ choose the smallest such y, say y_0.

Let $p_{s,n}$ be the principal function of $\{k : T(n,k) \notin U_s\}$ and write: $\tilde{T}(n,k,s) = T(n, p_{s,n}(k))$. Define

$$z_0 =_{\mathrm{df}} \begin{cases} (\mu z)(z \leq y_0 + 1 \wedge \quad \mathrm{st}_V(\tilde{T}(n,z,s)) < \mathrm{st}_V(\tilde{T}(n,y_0+1,s)) \\ \qquad\qquad : \text{if such } z \text{ exists} \\ y_0 + 1 \end{cases}$$

(We take $y_0 + 1$ instead of y_0 to ensure that $R_n - U \neq \emptyset$).

Now define $U_{s+1} = U_s \cup \{\tilde{T}(n, z_0, s)\}$.

If such a y does not exist let $U_{s+1} = U_s$.

$(s+1 = 2 \cdot t + 1)$ Ask if

$$(\exists e \leq s)(W_{e,s} \cap U_s = \emptyset \wedge (\exists z \leq s)(\tilde{T}(n,z,s) \in W_{e,s} \wedge$$
$$(\exists^{e+1} z' < z)(\mathrm{st}_V(\tilde{T}(n,z',s)) \geq \mathrm{st}_V(\tilde{T}(n,z,s)))) \quad (3.11)$$

(where $\exists^{e+1} z' < z$ means that there are $e+1$ many numbers less than z such that $\ldots$).

If yes, look for the smallest such e and then for the smallest corresponding z, z_0 say. Define

$$U_{s+1} = U_s \cup \{\tilde{T}(n, z_0, s)\}.$$

If such an e does not exist let $U_{s+1} = U_s$.

Result. Let $U = \bigcup_{s \geq 0} U_s$. Then

1. If for some y $\lambda(n, y) = \omega$ then $U \cap R_n =^* R_n$, $R_n - U \neq \emptyset$ and if $a \in R_n - U$ then $\text{st}_V(a) \restriction n \geq \text{st}_V \restriction n$.

Let y_0 be the smallest y such that $\lambda(n, y) = \omega$. So infinitely often $\lambda(n, y_0, t+1) \neq \lambda(n, y_0, t)$. And so infinitely often $\tilde{T}(n, z_0, s)$ with $z_0 \leq y_0 + 1$ enters U. But this means that $U \cap R_n =^* R_n$.

We see that

$$T(n, \mu z(\text{st}_V(T(n, z) \restriction n \geq \text{st}_V(T(n, z') \restriction n \text{ for all } z')) \qquad (3.12)$$

never enters U (by $y_0 + 1$ and $\exists^{e+1}$). Thus $R_n - U \neq \emptyset$. Let $a \in R_n - U$. Then there is a k_0 and s_0 such that $a = \tilde{T}(n, k_0, s)$ for all $s \geq s_0$.

Let $s_1 \geq s_0$ be such that if in the case $s + 1 = 2 \cdot t + 1$ e exists then $e > k_0$. (Since (3.11) holds for each e only finitely often, that is until it is the smallest such one, such an s_1 exists.)

If $\text{st}_V(a) \restriction n < \text{st}_V \restriction n$ then $b = \tilde{T}(n, k, s')$, $k > k_0$, $s' > s_1$ with $\text{st}_V(b) \restriction n \geq \text{st}_V \restriction n$ cannot enter U via $s + 1 = 2 \cdot t$. This implies that infinitely many such b's remain in $R_n - U$. But $R - U$ is finite. Hence such a's are not in $R_n - U$.

2. Suppose for all y $\lambda(n, y) < \omega$.

First we show that $R_n - U$ is infinite.

Take the element (3.12) and let s_0 and k_0 be such that (3.12) is equal to $\tilde{T}(n, k_0, s)$ for all $s \geq s_0$, if for (3.11) e exists then $e > k_0$ and for $k \leq k_0$ $\lambda(n, k, s_0) = \lambda(n, k)$. Then similarly to (3.12) the number

$$\tilde{T}(n, (\mu z)(z > k_0 \wedge \text{st}_V(\tilde{T}(n, z, s_0) \restriction n \geq \text{st}_V(T(\tilde{n}, z', s_0) \restriction n), \text{ for all } z' \geq z, s_0)$$

never enters U. By iterating this we get that $R_n - U$ is infinite.

$R_n - U$ is not r.e. Suppose $R_n - U = W_e$. Since W_e is infinite, for sufficiently large s (3.11) must hold for e with e the smallest such number. Hence $W_e \cap U \neq \emptyset$, which contradicts the assumption. (We have only used $W_e \subseteq R_n - U$ and that W_e is infinite. Thus $R_n \cap U$ is even simple in R_n.)

Corollary 3.4 1) (Lempp) $\{e : W_e \sqsubseteq_m A\}$ *is* Π_4^0-*complete.* 2) $\{e : W_e \sqsubseteq_m A \pmod{r.e.}\}$ *is* Π_4^0-*complete.*

4 The refinement Theorem for $\mathcal{E}/\approx_{ms}$

We will now give a necessary and sufficient criterion for the truth of a sentence (1) in $\mathcal{E}/\approx_{ms}$.

Theorem 4.1 *The sentence (1) is true in $\mathcal{E}/\approx_{ms}$ iff*
1) *P is consistent with D, i.e. $P(x_0, x_1, y)$ is equivalent to*

$$Q_1(x_0, x_1, \bar{y}) \vee \cdots \vee Q_s(x_0, x_1, \bar{y})... \tag{4.1}$$

where the Q_i's, $i = 1, \ldots, s$ in (4.1) are all diagrams with $Q_i \rightarrow D$ and
2) *Among $Q_1, \ldots, Q_s$ there are two diagrams Q' and Q'' with the following properties:*

    2.) *Q' has a component C which has a maximal path p with $\bar{x}_1 \in p$, $p \cap (x_1 - x_0)$ consisting of exactly one atom and $p \cap x_0 \neq \emptyset$. All other atoms in $x_1 - x_0$ are outermost, but not innermost.*

    2.2) *Q'' consists of a component C with the property: $\bar{x}_1$ is an atom of C, C has an atom $a \subseteq x_1 - x_0$ such that if p is a path with $\bar{x}_1 \subseteq p$, $p \cap x_0 \neq \emptyset$ then $p \cap (x_1 - x_0) = \{a\}$. All other components of Q'' are completely included in either $x_1 - x_0$ or x_0.*

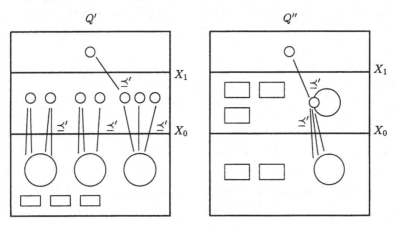

◯   denotes part of a component (maybe with outermost and innermost atoms)

▭   denotes a (complete) component

○   denotes a single atom

Proof: $\Longrightarrow$ Suppose the sentence (1) is true in $\mathcal{E}/\approx_{ms}$.

1. Take A to be a maximal set and B to be an r.e. subset of A as in Theorem 2.4. Then $D([B], [A])$ (see Corollary 2.5). Let $Y_0, \ldots, Y_{m-1}$ be r.e. sets such that

$$P([B], [A], [Y_0], \ldots, [Y_{m-1}]).$$

Then, since (1) is true, there is a Q_0 with $Q_0 \rightarrow D$, $Q_0 \rightarrow P$ and $Q_0([B], [A], [Y_0], \ldots, [Y_{m-1}])$. But by the properties of B in Theorem 2.4 all atoms in $A - B$ (of the d-lattice of Q_0), except one, are equivalence classes with a q-cohesive set. Thus they can be only outermost, not innermost atoms. Hence Q_0 must be as the Q' described in 2.1).

2. Take A maximal and B as in Theorem 3.2. Of course $D([B],[A])$. Again let $Y_0, \ldots, Y_{m-1}$ be r.e. sets such that $Q_1([B],[A],[Y_0],\ldots,[Y_{m-1}])$ for some diagram Q_1 with $Q_1 \rightarrow P$. Let R be a recursive set such that $[R]$ is the component with $[\bar{A}] \subseteq [R]$. Then $\bar{R} \subseteq A$, so that $R - B$ is recursive (see (3.2)). Thus all components contained in $[\bar{R}]$ are completely contained in either $[A] - [B]$ or $[B]$.

Let a be an atom in $[A] - [B]$ with $a \preceq [\bar{A}]$ and $C \subseteq A$ an r.e. set with $[C] = \cup\{b - \text{atom} : b \not\preceq a\}$. Then, by 4) in Theorem 3.2 there is a recursive set $S \subseteq A - B$ such that $C \cup S \subset_m^0 A$. Thus $[C]$ includes an r.e. set D such that $S \cup D = A$. This gives that all other atoms, except a, in $[R] \cap ([A] - [B])$ are contained in $[S]$, hence they are not in the relationship $\preceq$ with $[\bar{A}]$ *and with atoms of* $[B]$: But this is exactly how Q'' was defined in 2.2).

$\Longleftarrow$ Suppose that for P 1) and 2) are satisfied. We then show that the sentence (1) is true.

Let A and B be r.e. sets such that $D([B],[A])$ holds. We can assume that $B \subseteq A$, and know by Corollary 2.3 that for B: (Case 1)

$(\forall R \text{ recursive}, \subseteq A)(\exists R' \text{ recursive}, \subseteq A)(R \cap R' = \emptyset \wedge R' - B$
is not recursive (hence not r.e.)).

Then we easily find r.e. sets $Y_0, \ldots, Y_{m-1}$ such that Q' is satisfied. Suppose $[A] - [B]$ consists of $n + 1$ atoms (of Q'). Then take n many disjoint recusive sets $R_0, \ldots, R_{n-1}$ with $R_i - B$ not r.e., $i = 0, 1, \ldots, n - 1$. Then $[R_i] - [B]$, $i = 0, \ldots, n-1$ give all these atoms, except a, but is equal to $[A] - [\bigcup_{i=0}^{n-1} R_i \cup B]$. Inside $[B]$ every diagram can be easily realized – see the remark concerning the existence of major subsets (mod r.e.).

(Case 2) Suppose that case 1 does not hold. Then there is a recursive set $R \subset A$ such that $B \cup R \subset_m A$ (mod r.e.) and $B \cup R$ is pseudocreative in A, since $D([B],[A])$. Take two infinite, recursive sets R_0, R_1 both in $A - B$ and $R_0 \cap R_1 = \emptyset$, $R_i \cap R = \emptyset$, $i = 0, 1$. Inside R_0 we can realize all components of Q'' which are inside $x_1 - x_0$ and different from C. Inside R_1 we realize $C \cap (x_1 - x_0)$. (The set $A - (R_0 \cup R_1 \cup B)$ is taken to be the atom a.) Inside $[B]$ we can again realize every diagram, so that we also have $Q'' \wedge x_0$.

5   The lattice $\mathcal{E}$

We will now analyse the truth of sentences of the form (1) for the whole lattice $\mathcal{E}$ and compare this with that for $\mathcal{E}/\approx_{ms}$. To do this we essentially use the results of Lachlan in [La68b]. The truth in $\mathcal{E}/\approx_{ms}$ and in $\mathcal{E}^*$ of the

(1)'s, as we shall see, is quite different.

Theorem 5.1 *The sentence (1) is true in $\mathcal{E}^*$ iff*
1) P is consistent with D, i.e. P is equivalent to

$$Q_1(\vec{x}, \vec{y}) \vee \cdots \vee Q_s(\vec{x}, \vec{y}) \vee \cdots, \tag{5.1}$$

where the Q_i's, $i = 1, \ldots, s$ are all diagrams in (5.1) with $Q_i \to D$.
2) Among the $Q_1, \ldots, Q_s$ are diagrams Q' and Q'' such that
2.1) Q' is a "Lachlan diagram", i.e. $\bar{x}_1$ is an atom of Q' and every maximal path p either $p \subseteq x_0$ or $\bar{x}_1 \in p$, $p \cap (x_1 - x_0) \neq \emptyset$ and $p \cap x_0 \neq \emptyset$.
2.2) Q'' has x_1 as atom and there is an atom $a \subseteq x_1 - x_0$ such that

- *($\forall b$) (b-atom, $b \subseteq x_1 - x_0$, $b \not\preceq a$, b-minimal in $x_1 - x_0$, then b is innermost)*

- *($\forall b$)($\forall b'$)(b, b'-atoms, $b \subseteq x_1 - x_0, b' \subseteq x_0, b' \preceq b$, then $b' \preceq a \wedge a \preceq b$)*

All other components of Q'' are completely contained in either $x_1 - x_0$ or x_0.

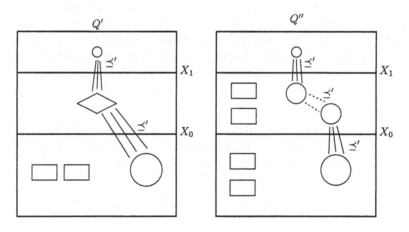

 denotes part of a component without innermost and outermost atoms

means that $\preceq'$ can hold between atoms of both parts

Proof: $\Longrightarrow$ Q' must exist (see [La68b]) since the truth of (1) implies the potential truth, which exactly says that a diagram such as Q' must exist.

Q'' must exist. Let A be maximal, R an infinite recursive subset of A and B be a set maximal in R.

Then $D(B^*, A^*)$, since $B \cup \bar{A}$ is not r.e.

- Since $\bar{A}$ is cohesive, $\bar{x}_1$ must be an atom.

- Let C^* be an atom with $R - B \subseteq C$ and D^* be an atom with $D \subseteq A - B$ and $D^* \preceq C^*$ and minimal inside $A - B$. Then $D \subseteq^* A - R$. Let X be the smallest (mod $=^*$) r.e. set with $D \subseteq X$. then $X \cap (A - B) = D$, by the minimality of D and X. Hence $D = X \cap \bar{R}$ and this is r.e.

- Take C as above and suppose D^*, D'^* are atoms with $D \subseteq A - B$, $D' \subseteq B$ and $D'^* \preceq D^*$.

  Let X be r.e. with $D \subseteq X$ (and X is a union of atoms). Then we also have $D' \subseteq X$ (since $D'^* \preceq D^*$). Suppose $X \cap C =^* \emptyset$. Then $X' = \bar{R} \cap X$ is r.e., but $X' \cap D' = \emptyset$. This contradicts $D'^* \preceq D^*$.

  Suppose $D' \not\preceq C$. Hence there is an r.e. set X with $C \subseteq X$ and $D' \cap X = \emptyset$. Let Y be r.e. with $D \subseteq Y$. Then $Z = X \cup (Y \cap \bar{R})$ is r.e. and $D, C \subseteq Z$, but $D' \cap Z = \emptyset$. This contradicts $D'^* \preceq D^*$.

$\Longleftarrow$ Since the existence of Q' is claimed, the sentence is potentially true ((1) is a separated sentence). So by Lachlan's refinement Theorem (1) is true iff all sentences of lower characteristics are true (i.e. all sentences $\forall \vec{x} \forall z (E(\vec{x}, z) \to \exists \vec{y} P(\vec{x}, \vec{y}, z))$) with $E \to D$ and E with a lower characteristic than D and with "$\bar{z} \in \mathcal{L}$" belonging to E). But there is only the diagram $E(x_0, x_1, z)$ of the form

$$D(x_0, x_1) \wedge x_0 < z < x_1 \wedge \bar{z} \in \mathcal{L}$$

with lower characteristic and implying D.

Among all refinements of Q'' by the variable z there is a $Q_0''(x_0, x_1, \vec{y}, z)$ having the same atoms as Q'', with only a becoming split nontrivially. Then Q_0'' is a Lachlan diagram (in its simplest version) namely

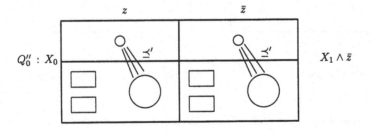

Hence this sentence is true in $\mathcal{E}^*$.

Remark. If $s = 1$, i.e. P is a diagram itself with $P \to D$, then the sentence (1) is true iff $\bar{x}_1$ is an atom and in $x_1 - x_0$ there is an atom a as in 2.2) which additionally is minimal inside $x_1 - x_0$. If additionally P is separated then $x_1 - x_0$ is refined by atoms which are linearly ordered by $\preceq'$.

P is a diagram P is a separated diagram

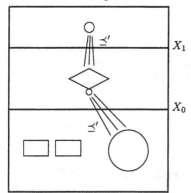
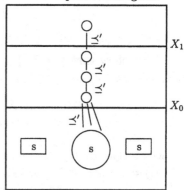

 denotes part of a separated component
(hence is a finite tree with finitely many roots)

denotes a separated component
(hence is a finite tree with one root)

Definition 5.2 *The $\forall\exists$-theory of a d-lattice $(\mathbb{B}A(\mathcal{L}), \mathcal{L})$ is n-bounded if for every $\forall\exists$-sentence of the form*

$$\forall x_0 \cdots \forall x_{n-1}(D(\vec{x}) \to \exists y_0 \cdots \exists y_{m-1} P(\vec{x}, \vec{y})) \tag{5.2}$$

with D a diagram and P a q-less formula if (5.2) is true in $\mathcal{L}$ then there are n diagrams $Q_1, \ldots, Q_n$ with $Q_n \to P$ such that

$$(\forall \vec{x})(D(\vec{x}) \to (\exists \vec{y})(\bigvee_{i=1}^{n} Q_i(\vec{x}, \vec{y}))$$

is true in $\mathcal{L}$.

Lemma 5.3 *The $\forall\exists$-theory of $(\mathbb{B}A(\mathcal{E}^*), \mathcal{E}^*)$ is not n-bounded for any n.*

Proof: Let D be the diagram

$$0 < x_0 < x_1 < \cdots < x_n < 1 \land (\forall i)(\forall y \in \mathcal{L})(x_{i+1} - x_i \leq y \to x_i \leq y)$$

and P be the following q-less formula:

Let $F : \{1, \ldots, n\} \to \{0, 1\}$ be a function and
$Q_f(x_0, \ldots, x_n, y_0, y_1, \ldots, y_{2 \cdot n - 1})$ be the diagram:

$$D(x_0, \ldots, x_n) \quad \land \bigwedge_{i=1}^{n} \{y_{2 \cdot i - 2} \land y_{2 \cdot i - 1} = x_{i-1} \land y_{2 \cdot i - 2} \lor y_{2 \cdot i - 1} = x_i \land$$

$$y_{2 \cdot i - 2} - x_{i-1} \notin \mathcal{L} \land y_{2 \cdot i - 1} - x_{i-1} \notin \mathcal{L} : f(i) = 1\} \land$$

$$\bigwedge_{i=1}^{n} \{y_{2 \cdot i - 2} \land y_{2 \cdot i - 1} = x_{i-1} \land y_{2 \cdot i - 2} \lor y_{2 \cdot i - 1} = x_i \land$$

$$y_{2 \cdot i - 1} - x_{i-1} \in \mathcal{L} : f(i) = 0\}.$$

Let $P(x_0, \ldots, x_n, y_0, \ldots, y_{2 \cdot n - 1})$ be

$$\bigvee_f Q_f(\vec{x}, \vec{y}).$$

1)   We show that $\forall \vec{x}(D(\vec{x}) \rightarrow \exists \vec{y} P(\vec{x}, \vec{y}))$ is true in $\mathcal{E}^*$. Take r.e. sets $X_0, X_1, \ldots, X_n$ such that $D(X_0^*, \ldots, X_n^*)$. We define a function f by

$$
\begin{aligned}
f(i) &= 1 \quad \text{if } X_i - X_{i-1} \text{ is immune} \\
&= 0 \quad \text{otherwise.}
\end{aligned}
\tag{5.3}
$$

We now see that $\exists \vec{y} Q_f(\vec{X}^*, \vec{y})$ holds.

If $f(i) = 1$ take for $Y_{2 \cdot i - 2}$, $Y_{2 \cdot i - 1}$ Owing's splitting sets for X_{i-1}, X_i and for $f(i) = 0$ let $Y_{2 \cdot i - 2}$ be an infinite r.e. set contained in $X_i - X_{i-1}$ and $Y_{2 \cdot i - 1} = X_i - (X_{i-1} \cup Y_{2 \cdot i - 2})$.

2) Let P' be P where one of the Q_f's, say Q_{f_0}, is omitted. We show that $\forall \vec{x}(D(\vec{x}) \rightarrow \exists \vec{y} P'(\vec{x}, \vec{y}))$ is false in $\mathcal{E}^*$.

Take $X_0, X_1, \ldots, X_n$ such that (5.3) holds for these sets and f_0 and additionally:

If $f_0(i) = 0$ then X_{i-1} is maximal in R and R is a recursive subset of X_i.

Then for every $f \neq f_0$ there is an i such that $f(i) \neq f_0(i)$. If $f(i) = 1$ but $f_0(i) = 0$

$$Y_{2 \cdot i - 2} - X_{i-1} \text{ is r.e. or } Y_{2 \cdot i - 1} - X_{i-1} \text{ is r.e.}$$

If $f(i) = 0$ but $f_0(i) = 1$

$$X_i - X_{i-1} \text{ is immune,}$$

and hence doesn't include an infinite r.e. set.

Final remarks. 1)   An interesting fact is that shown in [St79], namely that the $\forall \exists$-theory of the major subset intervals is even 1-bounded. The results there particularised for the sentences (1) imply that (1) is true iff there is a diagram Q with $Q \rightarrow D$ and $Q \rightarrow P$ such that for every maximal path (of atoms of Q) all sets

$$p \cap \bar{x}_1, p \cap (x_1 - x_0), p \cap x_0$$

are not empty (see the schema below).

The same feature of 1-boundedness of the $\forall \exists$-theory also pertains to the structure $(\mathcal{D}, \leq, \vee, 0)$, where $\mathcal{D}$ is the set of all T - degrees, $\leq$ - the T-ordering, $\vee$ -the join operation and 0 - a constant symbol for the smallest degree. This was shown in [JoSl93].

2) Let $\mathcal{S}^{0*}$ be the sublattice of $\mathcal{E}^*$ consisting of all simple and cofinite sets factorized by $=^*$. Using [La68b] it is easy to see that a sentence (1) is true in

S^{0*} iff there is a diagram Q with $Q \to D$ and $Q \to P$ such that Q consists of only one component,
- $\bar{x}_1$ is an atom of Q
- Q has only one innermost atom which is in x_0
- in $x_1 - x_0$ there are no outermost and no innermost atoms.

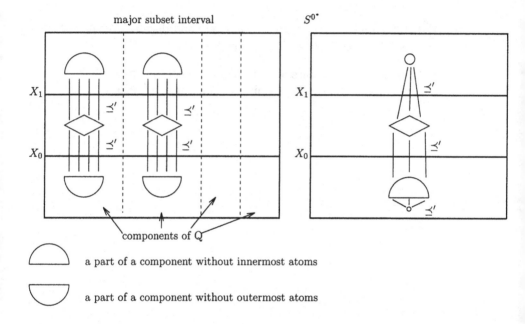

components of Q

a part of a component without innermost atoms

a part of a component without outermost atoms

References

[De78] A.N. Degtev, Decidability of the $\forall\exists$-theory some quotient lattice of recursively enumerable sets. Algebra i Logika 17 (1978), 134-143

[He81] E. Herrmann, Verbandseigenschaften der rekursiv aufzählbaren Mengen. Seminarbericht Nr. 36, Sektion Mathematik, Humboldt-Universität, Berlin (1981)

[JoSl93] C.G. Jockusch, T.A. Slaman, On the Σ_2-theory of the upper semi-lattice of Turing degrees. Journ. Symb. Logic 58 (1993), 193-204

[La68a] A.H. Lachlan, On the lattice of recursively enumerable sets. Trans. Amer. Math. Soc. 130 (1968), 1-37

[La68b] A.H. Lachlan, The elementary theory of recursively enumerable sets. Duke Math. Journ. 35 (1968), 123-146

[Lem86] S. Lempp, Topics in recursively enumerable sets and degrees. Dissertation, University of Chicago (1986)

[LeSo80] M. Lerman, R.I. Soare, A decidable fragment of the elementary theory of the lattice of recursively enumerable sets. Trans. Amer. Math. Soc. 257 (1980), 1-37

[So87] R.I. Soare, Recursively enumerable sets and degrees. Persp. in Mathematical Logic, Springer Verlag Berlin (1987)

[St79] M. Stob, The structure and elementary theory of the recursively enumerable sets. Dissertation, University of Chicago (1979)

Degrees of Generic Sets

Masahiro Kumabe
University of the air,
2-11, Wakaba, Mihama-ku,
Chiba 261, Japan

1 Introduction

We consider genericity in the context of arithmetic. A set $A \subseteq \omega$ is called n-generic if it is Cohen-generic for n-quantifier arithmetic. By *degree* we mean Turing degree (of unsolvability). We call a degree n-*generic* if it has an n-generic representative. For a degree a, let $D(\leq a)$ denote the set of degrees which are recursive in a. Since the set of n-generic sets is comeager, if some property is satisfied in $D(\leq a)$ with a any generic degree, then in the sense of Baire category, we can say that it is satisfied in $D(\leq a)$ for almost every degree a. So the structure of generic degrees plays an important role when we study the structure of D, the set of all degrees. For example, Slaman and Woodin [38] showed that there is a generic degree a such that if f is an automorphism of D and $f(a) = a$ then f is identity. In this paper we mainly survey $D(\leq a)$ when a is n-generic, as well as the properties of generic degrees in D. We assume the reader is familiar with the basic results of degree theory and arithmetical forcing. Feferman [4], Hinman [8], Hinman [9], and Lerman [25] are good references in this area. Odifreddi [29] is a good survey for basic notions and results for forcing and reducibilities. Jockusch [11] is a pioneering work in this area.

Our notations are almost standard. Let $A \oplus B = \{2n \mid n \in A\} \cup \{2n + 1 \mid n \in B\}$ for any sets A and B. Lower-case Greek letters other than ω denote strings. Fix a recursive enumeration of all strings. For strings σ and ν, $\sigma \geq \nu$ denotes that σ extends ν, and in this case we say that ν is a substring of σ. Further σ and ν are said to be *comparable (or compatible)* if either extends the other. If σ and ν are incomparable we denote this by $\sigma \mid \nu$. We identify a set $A \subseteq \omega$ with its characteristic function. So $\sigma \leq A$ means that the characteristic function of A extends the string σ and in this case we say that σ is a beginning of A or an initial segment of A. We write $\sigma * \nu$ for the usual concatenation of σ and ν. We identify $0, 1$ with the corresponding strings $0, 1$ of length 1. For $i = 0, 1$, let $[i] = 1 - i$. $\emptyset$ denotes the empty string. For each

n, $i^{(n)}$ denotes a string σ of length n such that $\sigma(m) = i$ for all $m < n$. For a string σ, $|\sigma|$ denotes the length of σ. For two strings σ and ν, $\sigma \cap \nu$ is the substring λ of σ such that $\sigma(m) = \nu(m)$ for all $m < |\lambda|$, and $\sigma(|\lambda|) \neq \nu(|\lambda|)$ or at least one of them is not defined. For a string σ and a natural number n such that $n < |\sigma|$, let $\sigma[n]$ be the substring of σ of length n. Let Φ_n be n-th partial (or, reduction) operator for some fixed recursive enumeration of all such operators. Let $\Phi_n(\sigma)(x) = y$ mean that the n-th reduction operator with oracle σ and input $x < |\sigma|$ yields output y in at most $|\sigma|$ steps and further that $\Phi_n(\sigma)(u)$ is defined for all $u < x$. Of course B is recursive in A iff for some e, $\Phi_e(A) = B$.

2   Forcing and Turing degrees

Let $\mathcal{L}$ be a language of first order number theory with equality such that $\mathcal{L}$ contains a constant $\tilde{n}$ for each natural number n, a set constant X, and the membership relation $\in$. Let ψ be a sentence of $\mathcal{L}$, and A be a subset of ω. Then let $A \models \psi$ mean that ψ is true in the standard model of number theory obtained by interpreting X as A. For a string σ, "σ forces ψ" (written as $\sigma \Vdash \psi$) is defined as usual by induction on sentences as follows.

If ψ is an atomic sentence and does not have X, then $\sigma \Vdash \psi$ if ψ is true in arithmetic.

If ψ is of the form $\tilde{n} \in X$, then $\sigma \Vdash \psi$ if $\sigma(n) = 1$.

If ψ is of the form $\neg\phi$, then $\sigma \Vdash \psi$ if for any extension ν of σ, $\nu \nVdash \psi$.

If ψ is of the form $\phi_0 \vee \phi_1$, then $\sigma \Vdash \psi$ if $\sigma \Vdash \phi_0$ or $\sigma \Vdash \phi_1$.

If ψ is of the form $\exists x \phi$, then $\sigma \Vdash \psi$ if for some n, $\sigma \Vdash \phi(\tilde{n})$.

Define $A \Vdash \psi$ by $\sigma \Vdash \psi$ for some $\sigma < A$. Then we have the following definition of generic sets for arithmetic.

Definition 2.1 *Let A be a set of natural numbers. A is called generic if for every sentence ψ of $\mathcal{L}$, either $A \Vdash \psi$ or $A \Vdash \neg\psi$.*

Jockusch [11] characterizes genericity as follows:

Lemma 2.1 *Jockusch [11]. Let A be a set of natural numbers. The following are equivalent.*

   i. A is generic.

   ii. For any arithmetical set S of strings, there is a string $\sigma < A$ such that either $\sigma \in S$ or no extension of σ is in S.

   iii. For every comeager arithmetical subset $\mathcal{A}$ of $P(\omega)$, $A \in \mathcal{A}$.

Given a set A, A', *the completion of A*, is defined as $\{e \mid \Phi_e(A)(e) \downarrow\}$. The completion operator can be iterated by defining $A^{(0)} = A$ and $A^{(n+1)} = (A^{(n)})'$. The completion operator is Turing degree invariant, so gives rise to the jump operator. Hence if A has a degree a, then $a^{(n)}$ will denote the degree of $A^{(n)}$. In particular, we start with degree 0 of the empty set $\emptyset$, and use the jump operator to generate a strictly increasing sequence of degrees $\{0^{(n)} \mid n \in w\}$. Post's theorem says that B is Δ_{n+1}^A iff B is Turing reducible to $A^{(n)}$. The arithmetical degrees are those below $0^{(n)}$ for some n. At level ω, we let $\emptyset^{(\omega)} = \{\langle n, e \rangle \mid e \in \emptyset^{(n)}\}$ and let $0^{(\omega)}$ be the degree of $\emptyset^{(\omega)}$.

Let $S \subseteq \omega^\omega$ be given. We say that a is an *upper bound for S* if every element of S has degree $\leq a$. We say a is a *minimal upper bound for S* if a is an upper bound for S, and no upper bound for S has degree $< a$. We say that a is a *uniform upper bound for S* if there is a function f of degree $\leq a$ such that $S = \{f^{[i]} \mid i \in \omega\}$, where we define $f^{[i]}$ by $f^{[i]}(x) = f(\langle i, x \rangle)$. Let AR be the class of arithmetical functions.

By an easy direct construction, there are generic degrees below $0^{(\omega)}$. We consider upper bounds for the arithmetical degrees. Jockusch noted that a is a uniform upper bound for AR iff there is a function of degree $\leq a$ which dominates every arithmetical function. In Kumabe [20] it is shown that if a function f dominates every arithmetical function then f computes a generic set. So if a is a uniform upper bound for AR then a bounds a generic degree.

Lemma 2.2   *i. there is a generic degree below $0^{(\omega)}$.*

  ii. Kumabe [20]. If a is a uniform upper bound for AR then there is a generic degree below a.

On the other hand, Kumabe [20] showed that given any countable set of increasing functions $\{g_n\}_{n \in \omega}$, there is a minimal upper bound $a \leq deg((\oplus_n g_n) \oplus \emptyset^{(\omega)})$ for the arithmetical degrees such that a does not bound a generic degree, and that there is a function f of degree $\leq a$ such that f is not dominated by any function g_n.

When we consider some property of A, if it is arithmetical and invariant under finite changes in A, then either all generic sets satisfy that property or all generic sets satisfy its negation. But we do not have to assume full genericity of A. So we consider a weaker notion of genericity.

Definition 2.2 *Let A be a set of natural numbers. A is called n-generic if for every Σ_n^0 sentence ψ of $\mathcal{L}$, either $A \Vdash \psi$ or $A \Vdash \neg\psi$.*

The next result is a characterization of n-genericity by Jockusch [11].

Lemma 2.3 *Jockusch [11]. Let A be a set of natural numbers. The following are equivalent.*

i. A is n-generic.

ii. For any Σ_n^0 set S of strings, there is a string $\sigma < A$ such that either $\sigma \in S$ or no extension of σ is in S.

For $n \geq 1$, let GL_n be the set of degrees a such that $a^{(n)} = (a \cup 0')^{(n-1)}$. Let GH_n be the set of degrees a such that $a^{(n)} = (a \cup 0')^{(n)}$. Clearly for all n, $GL_n \subseteq GL_{n+1}$, $GH_n \subseteq GH_{n+1}$, and $GL_i \cap GH_j = \emptyset$ for all i and j. By relativization, for $n \geq 1$, let $GL_n(a)$ be the set of degrees $b \geq a$ such that $b^{(n)} = (b \cup a')^{(n-1)}$. Let $GH_n(a)$ be the set of degrees $b \geq a$ such that $b^{(n)} = (b \cup a')^{(n)}$. Sacks [32] showed that for all n, $GL_{n+1} - GL_n \neq \emptyset$ and $GH_{n+1} - GH_n \neq \emptyset$.

Lemma 2.4   *i. For each $n \geq 1$, there is an n-generic degree below $0^{(n)}$.*

ii. If A is n-generic, $A^{(n)} \equiv_T A \oplus \emptyset^{(n)}$, so the degree of A is in GL_n.

The following are generalizations of Friedberg's Completeness Criterion.

Theorem 2.1   *i. Friedberg [5] and Selman [34]. For each n, if $a \geq 0^{(n)}$ then there is an n-generic b such that $b^{(n)} = b \cup 0^{(n)} = a$.*

ii. Macintyre [28]. If $a \geq 0^{(\omega)}$ then there is a generic b such that $b^{(\omega)} = b \cup 0^{(\omega)} = a$.

The next proposition (well known) and theorem are on degrees bounding 1-generic degrees.

Proposition 2.1 *Any non-zero r.e. degree bounds a 1-generic degree.*

Theorem 2.2 *Jockusch and Posner [12]. Every degree not in GL_2 bounds a 1-generic degree.*

3   The structure of generic degrees

We study the strucure of generic degrees. The next proposition shows that the theory of $D(\leq a)$ is independent of the choice of the generic degree a.

Proposition 3.1 *If a and b are generic degrees, then the structures of $D(\leq a)$ and $D(\leq b)$ are elementarily equivalent.*

Proof. Let ψ be a sentence of the language of partial orderings. Then there is a sentence ϕ of $\mathcal{L}$ such that ψ is true in $D(\leq a)$ iff $A \models \phi$. Let a generic degree a be given, and A be a generic set of degree a. Note that there is a σ such that $\sigma \Vdash \phi$ or $\sigma \Vdash \neg\phi$. If $\sigma \Vdash \phi$, then let A^* be such that σ is an initial segment of A^*, and for all $n \geq |\sigma|$, $A^*(n) = A(n)$. Then A^* is generic and has the same degree as A. If $\sigma \Vdash \phi$ then $A^* \Vdash \phi$, and so $A^* \models \phi$. So ψ is true in $D(\leq a)$. If $\sigma \Vdash \neg\phi$, $A^* \Vdash \neg\phi$, and so $A^* \models \neg\phi$. So $\neg\psi$ is true in $D(\leq a)$. As a was any generic degree, for every generic degree a, ψ is true in $D(\leq a)$, or for every generic degree a, ψ is false in $D(\leq a)$.

We do not know whether $D(\leq a)$ and $D(\leq b)$ are isomorphic whenever a and b are generic.

Definition 3.1 *A family $\{A_i\}_{i\in I}$ is independent if for all finite subsets F of I and all $i \in I - F$, $A_i \not\leq_T \oplus\{A_j \mid j \in F\}$.*

Given a set A, let $A_i = \{k \mid \langle i, k \rangle \in A\}$. Note that if A is 1-generic then $\{A_i\}_{i\in\omega}$ is independent.

We show that for each 1-generic degree a, $D(\leq a)$ is not a lattice.

Theorem 3.1 *Jockusch [11]. For each 1-generic degree a, $D(\leq a)$ is not a lattice.*

Proof. Let A be a 1-generic set of degree a. Let

$$F_i = \{j \mid \langle 3i+1, j \rangle \in A \ \& \ (\forall k \leq j)[\langle 3i+2, k \rangle \in A]\}.$$

As A is 1-generic, A has no infinite r.e. subset. So for each i, F_i is finite. Define $B = \Gamma(A)$ and $C = \Theta(A)$ by

$$(\Gamma(A))_i = (A)_{3i},$$
$$(\Theta(A))_i = (A)_{3i} \triangle F_i(A),$$

where $X \triangle Y$ is the symmetric difference of X and Y. For strings σ, define $\Gamma(\sigma)$ and $\Theta(\sigma)$ in the obvious way. We show that if $\Phi_b(B)$ and $\Phi_c(C)$ are total and equal, then it is recursive in some finite join of sets of the form $(A)_{3i}$. Note for each i, $(A)_{3i}$ is recursive in both B and C. As $\{(A)_{3i}\}_{i\in\omega}$ is independent, it follows that degrees of B and C have no greatest lower bound.

Now assume $\Phi_b(B)$ and $\Phi_c(C)$ are total and equal. Let S be the set of strings σ such that $\Phi_b(\Gamma(\sigma))$ and $\Phi_c(\Theta(\sigma))$ are incompatible. Note S is recursive. As $\Phi_b(\Gamma(A))$ and $\Phi_c(\Theta(A))$ are total and equal, by the 1-genericity of A, there is a $\sigma < A$ such that no extension of σ is in S. We show that $\Phi_b(\Gamma(A))$ is recursive in $\{(A)_{3i}\}_{i\leq|\sigma|}$. Given k, to compute $\Phi_b(\Gamma(A))$, find a string $\nu \geq \sigma$ such that

1. $\Phi_b(\Gamma(\nu))(k)$ is defined, and

2. ν is compatible with the restriction of the characteristic function of A to $\{\langle 3i, j \rangle \mid i \leq |\sigma| \ \& \ j \in \omega\}$.

Then $\Phi_b(\Gamma(\nu))(k) = \Phi_b(\Gamma(A))(k)$. (If not there is an initial segment $\mu \geq \sigma$ of A such that $\Phi_b(\Theta(\mu))(k) \neq \Phi_b(\Gamma(\nu))(k)$. By the definition of $\Gamma(A)$ and $\Theta(A)$, and by (2) above, it is straightforward to see that there is a $\delta \geq \sigma$ such that $\Gamma(\delta) = \Gamma(\nu)$ and $\Theta(\delta) = \Theta(\mu)$. So $\Phi_b(\Gamma(\delta))$ and $\Phi_c(\Gamma(\delta))$ are incompatible. This is a contradiction.)

In the construction of a minimal degree, given σ we extend σ to ν so that ν is in the (splitting or nonsplitting) subtree of a given tree. But in the construction of a generic set, given σ we extend σ to ν to meet the given dense set. So these two construction are quite different. So we would like to know given a generic (or n-generic) degree a, whether a bounds a minimal degree. The following theorem of Jockusch [11] is based on a result of Martin, and shows a downward homogenious property of $D(\leq a)$ when a is 2-generic.

Theorem 3.2 *Jockusch [11]. For each $n \geq 2$, each n-generic degree a, and any non-zero degree $b \leq a$, there is an n-generic degree $c \leq b$.*

A *chain* of $D(\leq a)$ is a set C of degrees $\leq a$ such that any two elements of C are comparable. A maximal chain C of $D(\leq a)$ is one which is not contained in a strictly larger chain of $D(\leq a)$. From the above theorem, every maximal chain of $D(\leq a)$ is infinite. As any 1-generic degree is not minimal, any 2-generic degree bounds no minimal degrees. For 1-generic degree below $0'$, Chong and Jockusch [2] showed the same result as Theorem 3.2. But Chong and Downey [1] and Kumabe [16] independently showed by different method that there is a 1-generic degree which bounds a minimal degree.

Theorem 3.3   i. *Chong and Jockusch [2]. For each 1-generic degree $a <$ $0'$, and any non-zero degree $b < a$, there is a 1-generic degree $c \leq b$.*

   ii. *Chong and Downey [1] and Kumabe [16]. There is a 1-generic degree below $0''$ which bounds a minimal degree. (In Chong and Downey [1] it is shown that there is a 1-generic degree $a < 0''$ and minimal degree $m < 0'$ such that $m < a$.)*

So we see that the initial segments below 1-generic degrees are not order isomorphic. Haught [7] sharpened Theorem 3.3-(i) as follows.

Theorem 3.4 *Haught [7]. If $0 < a < b < 0'$ and b is 1-generic then a is also 1-generic.*

Another downward homogenious property of $D(\leq a)$ when a is 2-generic is given by

Theorem 3.5 *Kumabe [22] For any degree a and any generic degree b, there is a generic degree c such that for any non-zero degrees $b_0 \leq b$ and $c_0 \leq c$, $b_0 \cup c_0 \geq a$.*

It is easy to show that 1-generic degrees are not only non-r.e., but also they do not bound r.e. degrees.

Proposition 3.2 *Any 1-generic degree does not bound any nonzero r.e. degree.*

Proof. Let A be a 1-generic set. Assume that for some r.e. set E, $E \leq_T A$. Let Φ be such that $\Phi(A) = E$. As E is r.e., we can assume that for each σ and k, if $\Phi(\sigma)(k) = 1$ then k is enumerated into E by the end of stage $|\sigma|$. Let S be the set of strings σ such that for some k, $\Phi(\sigma)(k) = 0$ but $E(k) = 1$. Then S is Σ_1^0. As $\Phi(A) = E$, there is a $\sigma < A$ such that no extension of σ is in S. We show E is recursive. To compute E, given k, find a string $\nu \geq \sigma$ such that $\Phi(\nu)(k)$ is defined. Then $\Phi(\nu)(k) = 1$ iff $E(k) = 1$. So E is recursive.

As 1-generic degrees are not r.e., we would like to know relative recursive enumerability of n-generic degrees.

Definition 3.2 *A set A is immune if A is infinite but has no infinite recursive subset.*

Note that if A is 1-generic then A and its complement are immune.

Theorem 3.6 *Jockusch [11]. If a is 1-generic, there is a c < a such that a is r.e. in c.*

Proof. Let A be 1-generic. Let $p(i,j) = 2^i 3^j$. For any string σ, let $\Phi(\sigma)$ be the string ν of the same length as σ such that

$$\nu^{-1}(1) = \{p(i,j) \mid \sigma(i) = 1 \ \& \ \sigma(p(i,j)) = 0\}.$$

Define $\Phi(A)$ by obvious way. As A is immune, for every i, there is j such that $p(i,j) \notin A$. So A is r.e. in $\Phi(A)$. We show A is not recursive in $\Phi(A)$.

Lemma 3.1 *Let σ and τ be strings such that $\tau \leq \sigma$, and for some $n \geq |\tau|$ not in the range of p, $\sigma(n) = 0$. Then there is a string $\nu \geq \tau$ such that $\nu(n) = 1$ and $\Phi(\nu) \geq \Phi(\sigma)$.*

Proof. Let T be the smallest set with respect to inclusion such that $n \in T$ and such that if $i \in T$, $\sigma(i) = 0$, and $p(i,j) < |\sigma|$, then $p(i,j) \in T$. Let ν be the string of the same length of σ such that $\nu^{-1}(1) = \sigma^{-1}(1) \cup T$. $\nu \geq \tau$, since every element of T is greater than or equal to $|\tau|$. As $n \in T$, $\nu(n) = 1$. It remains to show $\Phi(\nu) \geq \Phi(\sigma)$. Let $k < |\Phi(\sigma)| = |\sigma|$ be given. If k is not of the form $p(i,j)$ then $\Phi(\nu)(k) = \Phi(\sigma)(k) = 0$. Assume $k = p(i,j)$ for some i,j.

If $\Phi(\sigma)(k) = 0$, then either $\sigma(i) = 0$ or $\sigma(p(i,j)) = 1$. First assume $\sigma(i) = 0$. If $i \in T$ then $p(i,j) \in T$ by the definition of T. So $\nu(p(i,j)) = 1$. Hence $\Phi(\nu)(p(i,j)) = 0$. If $i \notin T$ then $\nu(i) = 0$. So $\Phi(\nu)(p(i,j)) = 0$. Next assume $\sigma(p(i,j)) = 1$. Then clearly $\nu(p(i,j)) = 1$. So $\Phi(\nu)(p(i,j)) = 0$. Hence if $\Phi(\sigma)(k) = 0$, then $\Phi(\nu)(p(i,j)) = 0$.

If $\Phi(\sigma)(k) = 1$, then $\sigma(i) = 1$ and $\sigma(p(i,j)) = 0$. As $\sigma(i) = 1$, $\nu(i) = 1$. Also by the definition of T, $p(i,j) \notin T$. Hence $\nu(p(i,j)) = 0$. As $\nu(i) = 1$ and $\nu(p(i,j)) = 0$, $\Phi(\nu)(k) = 1$. Hence if $\Phi(\sigma)(k) = 1$, then $\Phi(\nu)(p(i,j)) = 1$. This completes the proof of the lemma.

Now we complete the proof of the theorem. Assume for a contradiction that for some Ψ, $\Psi(\Phi(A)) = A$. Let S be the set of strings μ such that μ is incompatible with $\Psi(\Phi(\mu))$. Clearly S is recursive. As A is 1-generic, there is an initial segment α of A such that no extension of α is in S. Let $n \geq |\alpha|$ be such that $n \notin A$, and n is not of the form $p(i,j)$ for any i,j. As $\Psi(\Phi(A)) = A$, let β be such that $\Psi(\Phi(\beta))(n) = 0$. By the above lemma, there is a string $\gamma \geq \alpha$ such that $\gamma(n) = 1$ and $\Phi(\gamma) \geq \Phi(\beta)$. Then $\Psi(\Phi(\gamma))(n) = 0$ and $\gamma(n) = 1$. This is a contradiction.

The above theorem has many corollaries.

Corollary 3.1 *If a is 1-generic, then $D(\leq a)$ is not dense. In fact no non-trivial initial segment of $D(\leq a)$ is dense.*

Proof. Let a be 1-generic. Let $b < a$ be such that a is r.e. in b. By relativizing to b the proof that every non-zero r.e. degree bounds a minimal degree in Yates [42], there is a degree c such that c is a minimal cover of b. So $D(\leq a)$ is not dense. The second sentence of the corollary follows from Theorem 3.2.

We say a set A is n-generic over B if for every set S of strings which is Σ_n^0 in B, there is a string $\sigma < A$ such that either $\sigma \in S$ or no extension of σ is in S. By Post's Hierarchy Theorem, a set A is $n+1$-generic iff A is 1-generic over $\emptyset^{(n)}$. Note that if A is n-generic and B is n-generic over A then $A \oplus B$ is n-generic.

Corollary 3.2 *If a is 2-generic, there is a degree $b < a$ such that $b \in GL_2 - GL_1$.*

Proof. If a is 2-generic, then a is 1-generic over $0'$. By relativizing the proof of Theorem 3.6, it follows that there is a degree $b < a$ such that a is r.e. in b, and $a \not\leq b \cup 0'$. So $a \leq b'$. As a is 2-generic, $a'' = a \cup 0''$ by Lemma 2.4. So $b'' \leq a'' = a \cup 0'' \leq b' \cup 0'' \leq (b \cup 0')'$. Hence $b'' \leq (b \cup 0')'$, and so $b \in GL_2$. As $a \not\leq b \cup 0'$ and $a \leq b'$, $b' \not\leq b \cup 0'$. So $b \in GL_2 - GL_1$.

As any 1-generic degree is in GL_1, we have the following corollary.

Corollary 3.3 *If a is 2-generic, there is a nonzero degree $b < a$ which is not 1-generic.*

By Proposition 3.1, if a and b are generic degrees, the structure of $D(\leq a)$ and $D(\leq b)$ are elementarily equivalent. So the above corollary raises the following question by Jockusch:

Question: If a is generic, is $D(\leq a)$ elementarily equivalent to $D(\leq b)$ for every non-zero $b \leq a$?

Martin showed that if $\mathcal{A}$ is a meager set of degrees not containing 0, and $\mathcal{A} \cup \{0\}$ is an initial segment of the degrees, then the upward closure of $\mathcal{A}$ is also meager. The following corollary contrasts with it.

Corollary 3.4 *Martin. There is a meager set $\mathcal{A}$ of non-zero degrees such that the upward closure of $\mathcal{A}$ is not meager.*

Proof. Let $\mathcal{A}$ be the set of degrees not in GL_1. As $\mathcal{A}$ is disjoint from the set of 1-generic degres, $\mathcal{A}$ is meager. But the upward closure of $\mathcal{A}$ is not meager, since by Corollary 3.2 it contains the comeager set of 2-generic degrees .

Kumabe sharpened Theorem 3.6 as follows.

Theorem 3.7 *Kumabe[17]. For all $n \geq 1$ and for any n-generic degree a there is an n-generic degree $c < a$ such that a is r.e. in c.*

The following theorem raises Theorem 3.6 by one level of the arithmetical hierarchy.

Theorem 3.8 *Jockusch [11]. If A is 2-generic, there is a set $B \leq_T A$ such that A is Σ_2^0 in B but not Π_2^0 in B.*

Corollary 3.5 *Jockusch [11]. If a is 3-generic, there is a degree $< a$ which is in $GL_3 - GL_2$.*

Proof. If a is 3-generic, then a is 2-generic over $0'$. By relativizing the proof of Theorem 3.8 (as in the proof of Corollary 3.2), it follows that there is a degree $b < a$ such that a is Σ_2^0 in $b \cup 0'$ but not Π_2^0 in $b \cup 0'$. So $a \leq b^{(2)}$ but $a \not\leq (b \cup 0')'$. As a is 3-generic, $a^{(3)} = a \cup 0^{(3)}$ by Lemma 2.4. So $b^{(3)} \leq a^{(3)} = a \cup 0^{(3)} \leq b^{(2)} \cup 0^{(3)} \leq (b \cup 0')^{(2)}$. Hence $b^{(3)} \leq (b \cup 0')^{(2)}$, and so $b \in GL_3$. As $a \leq b^{(2)}$ and $a \not\leq (b \cup 0')'$, $b^{(2)} \not\leq (b \cup 0')'$. So $b \in GL_3 - GL_2$.

As any 3-generic degree bounds no minimal degrees, we have the following corollary.

Corollary 3.6 *Jockusch [11]. There is a degree in $GL_3 - GL_2$ which bounds no minimal degrees.*

The above corollary is in contrast to Theorem 2.2 which says that every degree not in GL_2 bounds a 1-generic degree. Further Lerman [26] showed that the class of degrees bounding no minimal degrees contains degrees in each classes $GL_{n+1} - GL_n$ and $GH_{n+1} - GH_n$ for $n \geq 1$ as well as degrees not in GH_n and GL_n for any n.

From Theorem 3.6 and Theorem 3.8, Jockusch [11] conjectured the following:

Conjecture. If A is n-generic, then there is a set $B \leq_T A$ such that A is Σ_n^0 in B but not Π_n^0 in B.

Conjecture. If a is $n + 1$-generic, there is a degree $< a$ which is in $GL_{n+1} - GL_n$. Also every generic degree bounds a degree which is not in GL_n for all n.

We say a degree a has the *cupping property* if for any $b > a$, there is a $c < b$ such that $b = a \cup c$. Friedberg's Complete Criterion says that for every $a > 0'$, there is a 1-generic b such that $a = b' = b \cup 0'$. So $0'$ has the cupping property. Jockusch and Posner [12] showed that every degree not in GL_2 has the cupping property. By Corollary 3.5, every 3-generic degree has the cupping property. Jockusch [11] showed that every 2-generic degree has the cupping property.

Theorem 3.9 *Jockusch [11]. Let A be a set which satisfies that*

    i. For any Π_1^0 set S of strings, there is a string $\sigma < A$ such that either $\sigma \in S$ or no extension of σ is in S.

Then A has the cupping property. Hence every 2-generic degree has the cupping property.

Corollary 3.7 *Jockusch [11]. If a is 2-generic, and b is any degree which fails to have the cupping property, then a and b form a minimal pair.*

Proof. If a and b do not form a minimal pair, then let $c \neq 0$ be such that $c \leq a, b$. By Theorem 3.2 let $d < c$ be 2-generic. Then d has the cupping property. So b has the cupping property, which is a contradiction.

Corollary 3.8 *Jockusch [11]. If a is 2-generic and $0 < c < b < a$, then there exists d such that $c \cup d = b$.*

Proof. By Corollary 3.7, c has the cupping property.

By Theorem 3.9, every 2-generic degree has the cupping property. For 1-generic degrees, there is a 1-generic degree without the cupping property.

Theorem 3.10 *Slaman and Steel [37] and Cooper [3]. There are recursively enumerable degrees a and b with $0 < b < a$ such that no degree $c < a$ joins b above a.*

Corollary 3.9 *There is a 1-generic degree $< 0'$ without the cupping property.*

Proof. Let a and b as Theorem 3.10. By Theorem 2.1, take a 1-generic degree g below b. Then there is no $c < a$ such that $g \cup c = a$. So g does not have the cupping property.

We say a is a *strong minimal cover of g* if $a > g$ and every degree strictly below a is less than or equal to g. Kumabe [21] showed that

Theorem 3.11 *Kumabe [21]. There is a degree $a < 0'$ and a 1-generic degree $g < a$ such that a is a strong minimal cover of g. So g does not have the cupping property.*

For a degree a, we say $D(\leq a)$ is *complemented* if for every $b < a$ there is a c such that $b \cap c = 0$ and $b \cup c = a$. Posner [30] showed $D(\leq 0')$ is complemented by a nonuniform method. Given $a < 0'$ he constructed a $b < 0'$ such that $a \cup b = 0'$ and $a \cap b = 0$ by different methods depending on whether a satisfies $a'' = 0''$ or not. Slaman and Steel [37] showed by a uniform method that for each $a < 0'$, there is a 1-generic $b < 0'$ such that $a \cup b = 0'$ and $a \cap b = 0$. Further Seetapun and Slaman [33] showed that for each $a < 0'$, there is a minimal $b < 0'$ such that $a \cup b = 0'$. For 2-generic degrees a, Kumabe [19] showed that $D(\leq a)$ is complemented.

Theorem 3.12 *Kumabe [19]. For each $n \geq 2$, any n-generic degree a, and any non-zero degree $b < a$, there are an n-generic degree c and an n-generic degree $d < b$ such that for any non-zero degree $e \leq c$ and any degree f such that $d \leq f < a, e \cup f = a$ and $e \cap f = 0$.*

Corollary 3.10 $D(\leq a)$ *is complemented for any 2-generic degree a.*

By the result of Haught [7], for 1-generic degrees $a, b < 0'$, two structures $D(\leq a)$ and $D(\leq b)$ look same. But Kumabe [23] showed that there are two 1-generic degree $a, b < 0'$ such that $D(\leq a)$ and $D(\leq b)$ are not isomorphic.

Theorem 3.13 *Kumabe [23].*

    i. *There is a 1-generic degree $a < 0'$ such that for any non-zero degree $b < a$, there is $c < a$ such that for any $d \geq b$ and non-zero $e \leq c$, $d \cap e = 0$ and $d \cup c = a$.*

ii. There are 1-generic degrees $b < a < 0'$ such that for any non-zero degree $c < a$, $b \cap c \neq 0$.

Corollary 3.11 *There are two 1-generic degrees $a, b < 0'$ such that $D(\leq a)$ and $D(\leq b)$ are not isomorphic.*

We say that a degree a is a *minimal cover* if there is a $b < a$ such that for no c, $b < c < a$. Kumabe [18] showed that any 2-generic degree is a minimal cover.

Theorem 3.14 *Kumabe [18]. For each $n \geq 2$, every n-generic (or generic) degree is a minimal cover of an n-generic (or generic, respectively) degree.*

This result is in contrast to some other results. One is the Sacks Density Theorem [31] that for any two r.e. degrees a and b such that $a < b$ there is a r.e. degree c such that $a < c < b$. By using this result, Jockusch and Soare [15] showed that for all $n \geq 1$, $0^{(n)}$ is not a minimal cover. We say a set A of degrees is a *cone* if there is a degree b such that $A = \{a \mid a \geq b\}$. Harrington and Kechris [6] found a Σ_1^0 game which implied the existence of a cone of minimal covers, and computed a vertex of this cone, the degrees of Kleen's O, a complete Π_1^1 set. Jockusch and Shore [14] then showed that the set of degrees $\geq 0^{(\omega)}$ is a cone of minimal covers.

Assume $A \in a$ is 1-generic. As $\{A_i\}_{i \in \omega}$ is independent, any finite lattice is embeddable into $D(\leq a)$. So the Σ_1^0 theory of $D(\leq a)$ is decidable. The construction of minimal degree by Spector [41] is extended to show many embedding theorems. For example, Lerman [24] showed that any finite lattice is embeddable as an initial segment in D. By usins this he and Shore [36] independently showed that the Σ_2-theory of D is decidable. It seems that the construction of minimal cover theorem could be extended to the embedding theorem that any finite lattice is embeddable as a filter in $(\leq a)$, where a is 2-generic. So we conjecture that the Σ_2-theory of $D(\leq a)$ for a 2-generic is decidable, and that this theory is independent of a for a 2-generic. Lerman [25] showed that for any r.e. $a > 0$, any finite distributive lattice is embeddable into $D(\leq a)$ as an initial segment. By using this, he showed the theory of $D(\leq a)$ is undecidable. By Theorem 2.1, for any 1-generic a, there is $b < a$ such that a is r.e. in b. By relativization, we can see that for any 1-geneic a, the theory of $D(\leq a)$ is undecidable. By the method of Slaman and Woodin [39], we can code a standard model of arithmetic in $D(\leq a)$. Hence we see that for any arithmetical 1-generic a, the first order theory of $D(\leq a)$, $Th(D(\leq a))$ has degree $\emptyset^{(\omega)}$.

The following corollary shows that if a is 1-generic, then not every degree $b < a$ is bounded by c of which a is a minimal cover.

Corollary 3.12 *Jockusch [11] If a is 1-generic, then there is a degree $b < a$ such that for every c with $b \leq c < a$, there exists d with $c < d < a$.*

Proof. Let a be 1-generic. Let $b < a$ be such that a is r.e. in b. Then for every c such that $b \leq c < a$, a is r.e. in c. The Corollary follows by relativizing the fact that any non-zero r.e. degree is not minimal.

Definition 3.3 *Let $\mathcal{A} = \langle U, \leq \rangle$ be a partially ordered set.*

   i. An automorphism of $\mathcal{A}$ is an isomorphism of $\mathcal{A}$ with $\mathcal{A}$.

   ii. A set $B \subseteq U$ is an automorphism base for $\mathcal{A}$ if every $f : B \longrightarrow B$ has at most one extension $\hat{f} : U \longrightarrow U$ such that $\hat{f}$ is an automorphism of $\mathcal{A}$.

So if $f : U \longrightarrow U$ is an automorphism of $\mathcal{A}$ such that $f(x) = x$ for all $x \in B$, then f is the identity.

Definition 3.4 *Let $\mathcal{A} = \langle U, \leq \rangle$ be a partially ordered set, and $B \subseteq U \subseteq D$ be given. We say B generates U if U is a subset of the smallest set $C \subseteq D$ which satisfies the following.*

   i. $B \subseteq C$,

   ii. $(\forall a, b \in C)[a \cup b \in C]$,

   iii. $(\forall a, b \in C)[\text{if } a \cap b \text{ exists}, a \cap b \in C]$.

Note if B generates U then B is an automorphism base for $\mathcal{A}$. Jockusch and Posner [13] showed that the set of 1-generic degrees $< 0'$ generates $D(\leq 0')$. The following is a key to prove it.

Lemma 3.2 *Jockusch and Posner [13]. Let $a \in D$, $b \geq a$, $b \notin GL_2(a)$ and $c \geq b \cup a'$ be given such that c is recursively enumerable in b. Then there are degrees $d, e \leq b$ such that $d' = e' = c$ and $d \cap e = a$.*

Lemma 3.3 *Jockusch and Posner [13]. $\bar{L}_2$ generates $D(\leq 0')$.*

Proof. It suffices to show that H_1 generates L_2. Fix $a \in L_2$. Applying the above lemma with $b = 0'$ and $c = 0^{(2)}$, we obtain $d, e \in H_1$ such that $d \cap e = a$.

Proposition 3.3 *Jockusch and Posner [13]. The set of 1-generic degrees $< 0'$ generates $D(\leq 0')$.*

Proof. By the above lemma, It suffices to show that the set of 1-generic degrees $< 0'$ generates $\bar{L}_2$. By modifying the proof of Theorem 2.2, it is easy to show that for each $a \in \bar{L}_2$, there are 1-generic degrees $b, c < 0'$ such that $b \cup c = a$.

Jockusch and Posner [13] showed that for any degree a, there are minimal degrees m_j, $0 \leq j \leq 3$, such that $a = (m_0 \cup m_1) \cap (m_2 \cup m_3)$. So the set of minimal degrees generates D. Also Jockusch and Posner [13] showed that any comeager set of degrees generates D, so the set of generic degrees generates D.

References

[1] Chong, C. T., and Downey, R. G. *On degrees bounding minimal degrees* Annals of Pure and Applied Logic 48, 1990, pp 215-225.

[2] Chong, C. T., and Jockusch, C. G. *Minimal degrees and 1-generic degrees below 0'*, Computation and Proof Theory, Lecture Notes in Mathematics 1104, Springer-Verlag, Berlin, Heidelberg, New York, Tokyo, 1983, pp 63-77.

[3] Cooper, S.B. *The strong anticupping property for recursively enumerable degrees*, Journal of Symbolic Logic 54, 1989, pp 527-539.

[4] Feferman, S. *Some application of the notion of forcing and generic sets*, Fundamenta Mathematicae 55, 1965, pp 325-345.

[5] Friedberg, R. M. *A criterion for completeness of degrees of unsolvability*, Journal of Symbolic Logic 22, 1957, pp 159-160.

[6] Harrington, L., and Kechris, A. *A basis result for Σ_3^0 sets of reals with an application to minimal covers*, Proc. Amer. Math. Soc. 53, 1975, pp 445-448.

[7] Haught, C. *The degrees below 1-generic degrees $< 0'$*, Journal of Symbolic Logic 51, 1986, pp 770-777.

[8] Hinman, P. G. *Some applications of forcing to hierarchy problems in arithmetic*, Z. Math. Logik Grundlagen Math 15, 1969, pp 341-352.

[9] Hinman, P. G. *Recursion-Theoretic Hierarchies*, Springer-Verlag, Berlin, Heiderberg, New York, 1977.

[10] Jockusch, C. G. *Simple proofs of some theorems on high degrees of unsolvability*, Canadian Journal of Math. 29, 1977, pp 1072-1080.

[11] Jockusch, C. G. *Degrees of generic sets*, Recursion Theory-Its General-izations and Applications-, London Mathematical Society Lecture Notes, Cambridge University Press, Cambridge, 1980, pp 110-139.

[12] Jockusch, C. G. and Posner,D. *Double jumps of minimal degrees*, Journal of Symbolic Logic 43, 1978, pp 715-724.

[13] Jockusch, C. G. and Posner,D. *Automorphism bases for degrees of un-solvability*, Journal of Symbolic Logic 40, 1981, pp 150-164.

[14] Jockusch, C. G., and Shore, R. A. *REA operators, R.E. degrees and min-imal covers*, Proceeding of Symposia in Pure Mathematics 42, American Mathematical Society Providence, Rhode Island, 1985, pp 3-11.

[15] Jockusch, C. G. and Soare, R. I. *Minimal covers and arithmetical sets*, Proceedings of the American Mathematical Society 25, 1970, pp 856-859.

[16] Kumabe, M. *A 1-generic degree which bounds a minimal degree*, Journal of Symbolic Logic 55, 1990, pp 733-743.

[17] Kumabe, M. *Relative recursive enumerability of generic degrees*, Journal of Symbolic Logic 56, 1991, pp 1075-1084.

[18] Kumabe, M. *Every n-generic degree is a minimal cover of an n-generic degree*, Journal of Symbolic Logic 58, 1993, pp 219-231.

[19] Kumabe, M. *Generic degrees are complemented*, Annals of Pure and Applied Logic 59, 1993, pp 257-272.

[20] Kumabe, M. *Minimal upper bounds for the arithmetical degrees*, Journal of Symbolic Logic 59, 1994, pp 516-528.

[21] Kumabe, M. *A 1-generic degree with a strong minimal cover*, in prepa-ration.

[22] Kumabe, M. *A homogeneity result on principal ideals of generic degrees*, in preparation.

[23] Kumabe, M. *On the structure of 1-generic generic degrees below 0'*, in preparation.

[24] Lerman, M. *Initial segments of degrees of unsolvability*, Annals of Math-ematics 93, 1971, pp 365-389.

[25] Lerman, M. *Degrees of Unsolvability*, Springer-Verlag, Berlin, Heidel-berg, New York, Tokyo, 1983.

[26] Lerman, M. *Degrees which do not bound minimal degrees*, Annals of Pure and Applied Logic 30, 1986, pp 249-276.

[27] Macintyre, J. M. *Transfinite extensions of Friedberg's completeness criterion*, Journal of Symbolic Logic 42, 1977, pp 1-10.

[28] Martin D. A. *Classes of recursively enumerable sets and degrees of unsolvability*, Z. Math. Logik Grundlagen Math. 12, 1966, pp 295-310.

[29] Odifreddi, P. *Forcing and reducibilities. I. Forcing in Arithmetic*, Journal of Symbolic Logic 48, 1983, pp 288-310.

[30] Posner, D. *The upper semilattice of degrees $\leq 0'$ is complemented*, Journal of Symbolic Logic 46, 1981, pp 705-713.

[31] Sacks, G. E. *The recursively enumerable degrees are dense*, Annals of Mathematics 80, 1964, pp 300-312.

[32] Sacks, G. E. *Recursive enumerability and the jump operator*, Trans. Amer. Math. Soc. 108, 1963, pp 223-239.

[33] Seetapun, D. and Slaman, T. *Minimal complements*, to appear.

[34] Selman, A. L. *Applications of forcing to the degree-theory of the arithmetical hierarchy*, Proc. London Math. Soc. 25, 1972, pp 586-602.

[35] Shoenfield, J. R. *A theorem on minimal degrees*, Journal of Symbolic Logic 31, 1966, pp 539-544.

[36] Shore, R. *On the $\forall\exists$-sentences of α-recursion theory*, Generalized recursion theory II, Studies in Logic and the foundation of mathematics 94, North-Holland, Amsterdam, 1978, pp 331-354.

[37] Slaman, T. A. and Steel, J. R. *Complementation in the Turing degrees*, Journal of Symbolic Logic 54, 1989, pp 160-176.

[38] Slaman, T. A. and Woodin, H. *Definability in the Turing degrees*, Illinois Journal of Mathematics 30, 1986, pp 320-334.

[39] Slaman, T. A. and Woodin, H. *Definability in degrees structures*, to appear.

[40] Soare, R. I. Recursively Enumerable Sets and Degrees, Springer-Verlag, Berlin, Heidelberg, New York, Tokyo, 1987.

[41] Spector, C. *On degrees of recursive unsolvability*, Annals of Mathematics 64, 1956, pp 581-592.

[42] Yates, C. E. *Initial segments of degrees of unsolvability, Part II, Minimal Degrees*, Journal of Symbolic Logic 35, 1970, pp 243-266.

[43] Yates, C. E. *Banach-Mazur games, comeager sets, and degrees of unsolvability*, Mathematical Proceeding of the Cambridge Philosophical Society 79, 1976, pp 195-220.

EMBEDDINGS INTO THE RECURSIVELY ENUMERABLE DEGREES[1]

MANUEL LERMAN

Department of Mathematics
University of Connecticut
Storrs, CT 06269-3009
e-mail: mlerman@math.uconn.edu

Section 1: Introduction

One of the most efficient methods for proving that a problem is undecidable is to code a second problem which is known to be undecidable into the given problem; a decision procedure for the original problem would then yield one for the second problem, so no such decision procedure can exist. Turing [1939] noticed that this method succeeds because of an inherent notion of *information content*, coded by a set of integers in the countable situation. This led him to introduce the relation of *relative computability* between sets as a way of expressing that the information content contained in one set was sufficient to identify the members of the second set.

Post [1944], and Kleene and Post [1954] tried to capture the notion of relative computability algebraically. They noticed that the pre-order relation induced on sets of integers by relative computability gave rise to an equivalence relation, and that the equivalence classes form a poset with least element. This structure, known as the *degrees of unsolvability* or just the *degrees* has since been intensively studied, and it is of interest whether the algebraic structure completely captures the notion of *information content*. This question reduces to the determination of whether the degrees are rigid, i.e., whether this algebraic structure has any non-trivial automorphisms, a question to which a positive result has recently been announced by Cooper.

One of the major problems one encounters in trying to produce, or rule out automorphisms of the degrees is that the structure is uncountable.

---

[1]We are grateful to the University of Leeds for their support and hospitality during the initiation of this project. Discussions with S.B. Cooper, S. Lempp, T.A. Slaman, and X. Yi have influenced the presentation of this paper. Research partially supported by NSF Grant DMS92-00539, and by E.P.S.R.C.(U.K.) Research Grant no. GR/J28018 ('Leeds Recursion Theory Year 1993/94').

Typeset by $\mathcal{AMS}$-TEX

However, Slaman and Woodin [ta] have shown that the rigidity of the (re-cursively) enumerable degrees (the degrees of sets which can be enumerated by computer) would imply the rigidity of the degrees. As the latter struc-ture is countable, this problem tends to be more accessible. Again, Cooper has recently announced the construction of non-trivial automorphisms of the enumerable degrees.

The enumerable degrees $\mathcal{R}$ were studied by Post [1944] even before the study of the structure of all degrees was begun. One way of trying to build automorphisms is through a back-and-forth argument. One starts with an isomorphism between (finite) substructures of the degrees, takes a finite extension of one of these structures, and tries to extend the other (within the degrees) in a way which also allows the extension of the isomorphism. Results which allow one to carry out such a procedure are called *extensions of embeddings* results. Sacks [1964] proved a result of this type for linear substructures by showing that the degrees are dense. Shoenfield [1965] then conjectured that such a result was also true in the non-linear case, with the only obstructions to extensions of embeddings given by the additional upper semi-lattice structure on the degrees (every pair of degrees have a least upper bound). The almost-immediate refutation of Shoenfield's Conjecture led to attempts, over the past thirty years, to try to understand the nature of the additional obstructions to extensions of embeddings. Recent results have brought us closer to such an understanding, and it is the goal of this paper to summarize both the status of this problem, and recent work which sheds new light on the problem.

Lachlan [1968] introduced a two-pronged approach for deciding $\forall\exists$-theories of certain structures; one must first characterize the obstructions to extensions of embeddings, and then prove an extensions of embeddings theorem covering all cases not ruled out by the obstructions. A typical $\forall\exists$-sentence in a relational language can usually be reduced to deciding a disjunctive extension of embeddings question: When is it true, given a fi-nite substructure P of the given structure, and finitely many extensions Q_i of P, that every realization (isomorphic copy) of P in the given structure can be extended to a realization of one of the Q_i? In trying to reduce this question to one with a single extension Q of P, one can view Lach-lan's method for characterizing obstructions to extensions of embeddings as one of adding additional predicates to the language to characterize the obstructions, and then deciding the existential theory of the structure in this expanded language. We will try to follow this approach by suggesting a language covering $\mathcal{R}$. We begin with a language containing a binary rela-tion symbol $\leq$ representing the ordering of $\mathcal{R}$, and constant symbols 0 and 1 representing, respectively, the least and greatest elements of $\mathcal{R}$. An early result of Sacks [1963] implies the decidability of the $\exists$-theory of $\mathcal{R}$ in this language, and Lempp, Nies, and Slaman [ta] have shown the $\exists\forall\exists$-theory of $\mathcal{R}$ to be undecidable. The decidability of the remaining level, the $\forall\exists$-theory of $\mathcal{R}$ requires the type of analysis we have already discussed. This leads us to a summary of the known obstructions to extensions of embeddings.

As $\mathcal{R}$ has a least and a greatest element, we can immediately preclude the extensions of embeddings statements for a single $Q_i = Q$ which specify the existence of an element p such that either $0 \not\leq p$ or $p \not\leq 1$. As $\mathcal{R}$ is an uppersemilattice, we can immediately preclude the extension of embeddings statement in which $P = \{p, q, r\}$ with $p < r$, $q < r$, and p and q are incomparable, and Q has four elements $\{p, q, r, s\}$ with $p < s$, $q < s$, and $s < r$. Thus we will want to add a 2-place function symbol $\vee$ to our language to represent the join, or least upper bound of a pair of degrees. We note that the closure of the join operator on a finite set is finite; so while the resulting expanded language (with still more predicates) will not be relational, it will still lend itself to a Lachlan-type analysis.

Other obstructions to extensions of embeddings were discovered through attempts to refute Shoenfield's conjecture. The first such result was the construction, by Lachlan [1966] and Yates [1966] of a minimal pair of degrees of $\mathcal{R}$, i.e., two incomparable degrees a and b with meet $\mathbf{0}$, the smallest element of $\mathcal{R}$. Such degrees a and b are called *cappable*. Alternatively, one can reformulate this theorem as affirming the existence of an embedding of the four-element boolean algebra into $\mathcal{R}$ preserving least element. The above result, known as the Minimal Pair Theorem, initiated a program whose goal was the characterization of the finite lattices which can be embedded into $\mathcal{R}$. We will summarize these results later in this paper, and propose an approach towards obtaining such a characterization; we merely note here that any such embedding imposes a restriction on extensions of embeddings, and additional restrictions are imposed if the embedding is to preserve least and/or greatest element. As $\mathcal{R}$ is not a lattice, we cannot express these obstructions by introducing a meet function; rather, we introduce infinitely many relations $M_n(a_0, , ..., a_{n-1}, b)$ defined by $\forall x (\wedge \{x \leq a_i : i \leq n-1\} \rightarrow x \leq b)$. We call a structure in the language containing $\leq$, the join function, and the meet relations M_n a *partial lattice*.

Additional obstructions to extensions of embeddings occur when we require the embedding to preserve both the least and the greatest element of $\mathcal{R}$. A manifestation of this type of obstruction was initially noticed by Lachlan [1966] (Non-Diamond Theorem) who proved that the four-element boolean algebra cannot be embedded into $\mathcal{R}$ preserving both least and greatest elements. This obstruction to extensions of embeddings was finally understood when Ambos-Spies, Jockusch, Shore, and Soare [1984] showed that the degrees of promptly simple sets comprise a strong filter of $\mathcal{R}$ (an upwards closed set with the property that any two elements of the filter have a common predecessor in the filter), whose complement is the ideal of cappable degrees. This obstruction can then be expressed by adding a one-place predicate C to our language representing this ideal. It was also shown by Ambos-Spies, Jockusch, Shore, and Soare [1984] that the elements of the filter are exactly those degrees which can be joined to the greatest element, $0'$, of $\mathcal{R}$, by some low degree (a degree $\mathbf{a}$ satisfying $\mathbf{a}' = 0'$), i.e., that the filter consists of the *low cuppable degrees*. This result will combine with the next type of obstruction to yield an additional obstruction to extensions of embeddings.

The final type of obstruction to extensions of embeddings was first discovered in the process of characterizing the degree of $\mathrm{Th}(\mathcal{R})$, the elementary theory of $\mathcal{R}$. Slaman has called these *saturation properties*, but the more common terminology has been to call them *strong embedding properties*. An example of such a result is the existence of a triple $\{\mathbf{a}, \mathbf{b}, \mathbf{c}\}$ of pairwise incomparable enumerable degrees such that for every enumerable degree $\mathbf{d} < \mathbf{a}$, either $\mathbf{d} < \mathbf{b}$ or $\mathbf{d} \cup \mathbf{b} \geq \mathbf{c}$. Another such result states that one can embed the diamond lattice (four-element boolean algebra) into $\mathcal{R}$ so that if $\mathbf{b}$ and $\mathbf{c}$ are the images of the two incomparable degrees, $\mathbf{a}$ is the image of the smallest degree, and $\mathbf{d}$ is the image of the largest degree, then for every degree $\mathbf{f}$ such that $\mathbf{a} < \mathbf{f} < \mathbf{b}$, $\mathbf{f} \cup \mathbf{c} = \mathbf{d}$. The obstructions introduced can be expressed by adding predicates $S_{m_0,\ldots,m_n}$ to our language, and defining

$$S_{m_0,\ldots,m_n}(a_{0,0}, \ldots, a_{0,m_0}, d_{0,0}, \ldots, d_{0,m_0}, \ldots, a_{n,0}, \ldots, a_{n,m_n}, d_{n,0}, \ldots, d_{n,m_n}, c, b)$$

to hold if for all $e_0, \ldots, e_n$ and $i \leq n$, if $e_i \leq a_{i,j}$ for all $j \leq m_i$ and $e_i \not\leq d_{i,j}$ for any $j \leq m_i$, then $e_0 \vee \ldots \vee e_n \vee c \geq b$.

Little progress was made in the extension of embeddings direction until Slaman and Soare [1995] decided the extension of embeddings problem in the language consisting just of $\leq$, and showed that all obstructions to extensions of embeddings in this language were either lattice embedding obstructions, or strong lattice embedding obstructions.

Motivated by the Slaman-Soare extension of embeddings theorem, Lerman conjectured, in the spirit of Shoenfield, that the lattice obstructions, the strong embedding obstructions, the obstructions imposed by the existence of least and greatest elements in $\mathcal{R}$, and the obstructions imposed by the ideal/filter decomposition of $\mathcal{R}$ arising from the cappable degrees were the only types of obstructions. The original conjecture was that there was a language which could express these obstructions such that the $\exists$-theory of $\mathcal{R}$ in this language is decidable, and that an extension of embeddings theorem could be proved to cover the cases not ruled out by the obstructions provided by the true $\exists$ sentences of this language. (It is no longer clear, however, that this language suffices.) Lerman also made specific conjectures about a decision procedure for the existential theory in the expanded language which have subsequently proved incorrect; however, additional information has led to a better understanding of what the existential theory can be. First, if one considers just the join function and meet relations, the negation of the condition NEC of Ambos-Spies and Lerman [1986], adjusted to this new language from the language of lattices, is conjectured to be necessary and sufficient for embeddability (an existential sentence in the expanded language). Next, if 0 and 1 are added to the language, the only additional obstruction is the one isolated by Ambos-Spies, Lempp, and Lerman [1994] dealing with interactions with the ideal C of cappable degrees. Work by Cooper, Slaman, and Yi has pointed out an additional obstruction to extensions of embeddings obtained when strong embeddings are required. This condition is obtained from the characterization of the complement of C as the filter of low-cuppable degrees. A typical manifestation of this latter obstruction occurs in the double-diamond lattice DD (see Figure 1.1), where we cannot require the strong embedding conditions simultaneously on both sides of the upper diamond. This observation follows easily from known results. We note that $\mathbf{a}$ and $\mathbf{b}$ are cappable, so lie in the ideal C of cappable degrees. As C is an ideal, $\mathbf{a} \cup \mathbf{b} = \mathbf{c} \cap \mathbf{d} \in C$. Now it cannot be the case that both $\mathbf{c}$ and $\mathbf{d}$ are in C, else we would have $\mathbf{c} \cup \mathbf{d} = \mathbf{0}' \in C$, which is not the case as $C \neq \mathcal{R}$. Also, it cannot be the case that both $\mathbf{c}$ and $\mathbf{d}$ are not in C, as the complement of C is a strong filter, so this would imply that $\mathbf{c} \cap \mathbf{d} \notin C$, contrary to what we have shown. Hence without loss of generality, we may assume that $\mathbf{c} \in C$ and $\mathbf{d} \notin C$. Now by the Sacks Splitting Theorem [1963], we can express $\mathbf{d}$ as the join of two incomparable low degrees $\mathbf{e}$ and $\mathbf{f}$; one of these, say $\mathbf{e}$, must satisfy $\mathbf{e} \not\leq \mathbf{c}$. Strong embeddability would now require that $\mathbf{e} \cup \mathbf{c} = \mathbf{0}'$; but as no degree in C is low-cuppable, this is impossible.

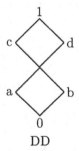

DD

Figure 1.1

Cooper, Slaman, and Yi [ta] have also recently obtained an extension of embeddings result which incorporates prompt simplicity in a way which explains the extent of the above example.

Yet more recently, Slaman has noted that NEC together with the Slaman-Soare extensions of embeddings theorem preclude the embeddability of DD (even with greatest element < 1) satisfying both $S_{0,0}(a_0, b_0, b_0, a_0)$ and $S_{0,0}(b_0, a_0, a_0, b_0)$. Even more strikingly, Slaman has proved that one cannot embed the partial ordering on $\{0, b, c, e, f\}$ where $0 < b, c < e$, $b < f$, b and c are incomparable, and c, and f are incomparable, if one also imposes the strong embedding conditions $S_{0,0}(c, b, b, c)$ and $S_{0,0}(e, f, f, e)$. The latter non-embedding result does not seem to follow from known results, and indicates that we still do not have enough of an understanding of the existential theory in the expanded language to conjecture a decision procedure for it.

Thus there are two areas where more information is needed to obtain a decision procedure. One is the understanding of the interactions of generalized strong embedding predicates with the other predicates. The other is the need to obtain a necessary and sufficient condition for partial lattice embeddings into $\mathcal{R}$. We turn to this latter question in the next section.

Section 2: Partial Lattice Embeddings

The techniques for embedding partial lattices into $\mathcal{R}$ are very similar to the embedding techniques for related structures. There are specific partial lattices which have been isolated as typical representatives of larger classes; when these specific representatives can be embedded, then the embedding techniques extend to the larger classes. The specific partial lattices which are considered are almost always lattices, and provide the embedding strategies; the passage to partial lattices is necessary later to counterbalance the obstructions to the embedding techniques.

First attempts to prove embedding results generally involve attempts at embedding all finite distributive lattices. The typical representative for this class is the diamond lattice M_2 (see Fig. 2.1) if there is no attempt to preserve both least and greatest element, and the double diamond lattice DD if there is such an attempt. One then passes to the non-distributive case. There are two smallest non-distributive lattices, N_5 and M_3 (see Fig. 2.1). Each has five elements, and any non-distributive lattice embeds one of these two lattices. The technique for embedding N_5 is simpler than that for M_3, so one normally passes to N_5 next. It is not easy to describe the class of finite (partial) lattices for which N_5 is a typical representative, as its description is not, on the surface, a description of a computable class. There is an intermediate class of lattices which is frequently studied, namely those with the *trace-probe property* (TPP); these lattices were introduced by Lerman, Shore, and Soare [1984] to obtain a categoricity result. Finally, one attempts to embed M_3; again the class of (partial) lattices represented by M_3 is not obviously computable.

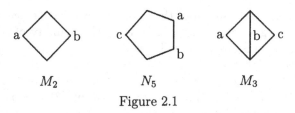

M_2 N_5 M_3

Figure 2.1

The first lattice embedding results dealt with embeddings into $\mathcal{R}$ preserving least element. (We will, henceforth, use the terminology *preserving* 0 in place of preserving least element, and *preserving* 1 in place of preserving greatest element.) As mentioned earlier, Lachlan [1966] and Yates [1966] succeeded in embedding the diamond lattice preserving 0, and the construction was generalized to embed all finite distributive lattices by Lachlan, Lerman, and Thomason [1971]. Lachlan [1972] then showed that N_5 and M_3 could also be embedded into $\mathcal{R}$ preserving 0. In studying the embedding techniques, Lerman noted that they were not sufficiently powerful to embed S_8 (Fig 2.2), and raised the possibility that S_8 might not be embeddable; the nonembeddability was confirmed by Lachlan and Soare [1980]. Ambos-Spies then generalized the non-embeddability techniques, and these efforts culminated with the formulation of the Non-Embedding Condition (NEC) of Ambos-Spies and Lerman [1986]. The condition will be presented in the next section, but for the purposes of this section, we would like to decompose it into two parts. The condition states that a lattice satisfying this condition can be decomposed into two parts, a bottom part which contains a *critical triple* (terminology later introduced by Downey [1990] which we will define in a later section), and an upper diamond lattice, with some amalgamation of the parts. Ambos-Spies and Lerman [1989] then tried to show that all finite lattices which fail to satisfy NEC can be embedded into $\mathcal{R}$ preserving 0. They introduced a complicated embedding condition, EC, which was an abstraction of properties which would ensure the success of a certain construction, and proceeded to show that every finite lattice with EC can be embedded into $\mathcal{R}$ (preserving 0). It is unknown whether EC and NEC are complementary, and we will propose a new approach for the investigation of this question.

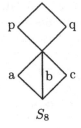

S_8

Figure 2.2

Ambos-Spies [1980] initiated the study of (partial) lattice embeddings into $\mathcal{R}$ preserving 1. He demonstrated the embeddability of M_2 and N_5, and then of the finite distributive lattices and the lattices with the trace-probe property. Subsequently, Ambos-Spies, Lempp, and Lerman [1994a] also embedded M_3 into $\mathcal{R}$ preserving 1. It is conjectured that the class of finite partial lattices which can be embedded into $\mathcal{R}$ preserving 1 is the same as the class of finite partial lattices which can be embedded into $\mathcal{R}$ preserving 0.

The first result about (partial) lattice embeddings into $\mathcal{R}$ preserving both 0 and 1 was the negative Non-Diamond Theorem of Lachlan [1966]. The study of promptly simple degrees by Ambos-Spies, Jockusch, Shore, and Soare [1984] which led to the ideal/filter decomposition of $\mathcal{R}$ induced by the cappable degrees was a natural generalization of the Non-Diamond Theorem. It was noted that the techniques for embedding lattices into $\mathcal{R}$ preserving 0 and those for preserving 1 could be straightforwardly combined if the lattice to be embedded could be separated into two disjoint lattices, so that all elements of the first lattice are less than all elements of the second lattice. New techniques were then introduced by Ambos-Spies, Lempp, and Lerman [1994] to obtain a characterization of the finite distributive lattices which can be embedded into $\mathcal{R}$ preserving both 0 and 1. These techniques seem to combine readily with those for embedding non-distributive partial lattices. No obstruction to such embeddings has been observed, other than those imposed by ordinary partial lattice embeddings and the ideal/filter decomposition induced by the cappable degrees.

The problem of characterizing the lattices which can be embedded into all intervals $[\mathbf{a}, \mathbf{b}]$ of degrees has also been studied, and no difference has been found if one imposes the additional restriction that $\mathbf{a} = \mathbf{0}$. Slaman [1991] observed that the diamond lattice can be embedded into all such intervals, and Ambos-Spies, Lempp, and Soare [ta] proved a similar result for TPP lattices. Downey [1990] studied the case in which $\mathbf{a} = \mathbf{0}$, and embedded all distributive lattices in this case. Weinstein [1988] and Downey [1990] also proved a non-embeddability result for lattices with critical triples. The permitting techniques needed in this situation irresolvably conflict with the techniques for embedding lattices with critical triples (M_3 is such a lattice). But for the lattices which can be embedded, the embedding techniques are a combination of the unrestricted embedding techniques with multiple permitting. A problem similar to that encountered in trying to show that EC and NEC are complementary seems to be the primary obstruction to obtaining a necessary and sufficient condition for embeddability in this case.

Another interesting structure for which lattice embeddings have been studied is the uppersemilattice of ideals of $\mathcal{R}$. Calhoun [1990] has shown that the availability of ideals allows one to use the techniques for embedding

N_5 to embed M_3. Calhoun has embedded a large class of lattices into this structure, and conjectures that all finite lattices are embeddable. The obstruction to proving such a result is similar to that encountered in proving that EC and NEC are complementary.

Some of the embedding results have been generalized to strong embeddings. In particular, Cooper, Sui, and Yi [ta] have strongly embedded the diamond lattice preserving 0, and Slaman and Soare [1995] have embedded the diamond partial lattice preserving 1, with one strong embedding condition.

The summary of results above lends credence to our belief that there are a handful of techniques which are relevant to constructions of lattice embeddings. It is our belief the failure of a partial lattice P to satisfy NEC implies that P can be embedded into $\mathcal{R}$, and that the techniques already discovered and utilized in the proofs of (strong) embedding and nonembedding results and their interaction with prompt simplicity are sufficient to decide the $\forall\exists$-theory of $\mathcal{R}$.

Section 3: Embeddings into $\mathcal{R}$

Fix a finite partial lattice $\mathcal{P}$ with universe P for the remainder of this paper. In order to embed $\mathcal{P}$ into $\mathcal{R}$ as a partial lattice, we must construct an enumerable set A_d of degree $\mathbf{a_d}$ for each $d \in P$ such that $\mathbf{a_0} = \mathbf{0}$, $\mathbf{a_1} = \mathbf{0'}$, and the following requirements are satisfied for all $a, b, c \in P$:

(3.1)   Comparability Requirements: $a \leq b \Rightarrow A_a \leq_T A_b$.

(3.2)   Incomparability Requirements: $a \nleq b \Rightarrow A_a \nleq A_b$.

(3.3)   Join Requirements: $a \vee b = c \Rightarrow A_c \leq_T A_a \oplus A_b$.

(3.4)   Meet Requirements: $a \wedge b = c \Rightarrow \forall W (W \leq_T A_a, A_b \rightarrow W \leq_T A_c)$

There seems to be no advantage to using nonstandard methods to satisfy comparability or incomparability requirements. Thus in isolation, we will try to employ the following strategies to satisfy requirements.

We try to satisfy (3.1) by constructing a global function $\Xi_{a,b}$ such that $\Xi_{a,b}(A_b) = A_a$. For the most part, we will force numbers x to enter A_b whenever they enter A_a. However, there may be times when we want to place x into A_b before deciding whether or not to place x into A_a; the use, $\xi_{a,b}(x)$ is then lifted, allowing us to place a larger y into A_b when we place x into A_a.

We employ the Friedberg-Mučnik diagonalization method to satisfy (3.2). Thus we will appoint a follower x for A_a, and prevent x from entering A_a until we find a computation $\Phi(A_b; x) \downarrow = 0$ (here, we have a separate requirement for each computable partial function Φ). We will then try to restrain A_b by preventing a number less than or equal to the use $\varphi(A_b; x)$ from entering A_b, and place x into A_a when it is safe to do so.

Join requirements in (3.3) will be satisfied through a method similar to that used for satisfying comparability requirements. Thus in order to place x into A_c in the face of a prior computation of $x \notin A_c$ via $\Delta_{a,b}(A_a \oplus A_b; x) = 0$, we must place some $y \leq$ the use $\delta_{a,b}(A_a \oplus A_b; x)$ of this computation into either A_a or A_b; the choice of *target set* will depend on the primary restraint in force at the time we want to place y into one of these sets. We may have to repeatedly change targets back and forth between A_a and A_b, as primary restraints change when x is subject to different meet requirements, in order to build a configuration which makes it safe to place x into A_c. If $\mathcal{P}$ has a critical triple, then the iterative trace-assignment process is potentially infinitary; otherwise, there are ways of organizing the hierarchy of traces so that the iterated trace-assignment process is finitary.

The meet requirements of (3.4) are handled by the standard branching degree methods. Given a pair of computable partial functionals Φ and Ψ, we define axioms for the functional $\Gamma(A_c)$ at expansionary stages for $\Phi(A_a) = \Psi(A_b)$. The axioms are defined for arguments which are less than or equal to the length of agreement we find for the equality between the functionals, and have large use so that if any trace in existence at the time the axiom is defined later enters A_c, we will be able to correct any axioms which were injured. Thus the use must be greater than or equal to any traces in existence at the time the axiom is defined.

We commented earlier on potential conflicts between requirements when join requirements generate an infinite string of traces. They may also arise when we change the primary meet requirement to which a given trace is subject. To see how the trace-generation process unfolds, let us study first the example of N_5 where the process is finitary, and then M_3 where the process is infinitary. In each case, we will be faced with a prohibition set, i.e., a set for which traces may not be targeted, as such targeting will cause injury to a higher priority diagonalization or meet requirement.

Example 3.1. Consider the lattice N_5 pictured in Figure 2.1. Suppose that we have targeted a number x for A_c, and the prohibition set A_b is imposed by a diagonalization requirement. As $b \vee a > c$, the entry of x into A_c must be affirmed by allowing a trace y of x to enter A_b or A_a; and as the prohibition set is A_b, the target set for y must be A_a. Note that y does not require a trace for any join computation. In the construction, once y enters A_a, a computation from oracle A_b has been found, and this computation will be protected from the appointment of any new large trace z targeted for A_b. z will not require traces for any join requirement, as we will arrange for x to enter A_c if and when z enters A_b.

Example 3.2. Consider the lattice M_3 pictured in Figure 2.1. Suppose that we have a number x targeted for A_c, and the prohibition set is A_b. As $b \vee a > c$, the entry of x into A_c must be affirmed by allowing a trace y of x to enter A_b or A_a; and as the prohibition set is A_b, the target set for y must be A_a. But now as $b \vee c > a$, the entry of y into A_a must be affirmed by allowing a trace z of y to enter A_b or A_c; and as the prohibition set is A_b, the target set for z must be A_c. z now behaves like x, and the process will continue. Note that the simultaneous entry of (different) numbers into A_a and A_c causes no injury to this meet requirement; and we can arrange the construction so that later appointment of traces to any number in this current set of traces does not affect the satisfaction of this meet requirement. Thus we will need to iteratively appoint traces hereditarily linked to x until (and if) x enters its target set. If the latter never occurs, we will have appointed an infinite set of traces during the course of the construction.

In order to see how the trace appointment procedure can conflict with the procedure to satisfy meet requirements, we discuss such potential conflicts in the context of a pinball machine argument. The use of pinball machines to model such constructions was introduced by Lerman [1971]; a good description of their use to embed M_5 appears in Downey and Shore [ta].

Example 3.3. Consider again the case of M_3 as described in Example 3.2. The pinball machine approach requires that once a number is assigned to a requirement and begins accumulating a string of traces, this string is subject to modification by higher priority meet requirements, one at a time. Thus suppose that we have a finite string of traces targeted, alternately, for A_a and A_c, which comes under the influence of a meet requirement generated by the lattice meet $a \wedge c = 0$. Such a minimal pair requirement cannot allow small numbers to enter A_a and A_c simultaneously, but can allow numbers to enter A_a and A_b simultaneously, and also to enter A_c and A_b simultaneously. Thus we take the largest trace of this string of traces, targeted, say, for A_a, and begins assigning traces (one at each stage) targeted, alternately, for A_b and A_a. Higher priority meet requirements may necessitate the appointment of traces targeted for A_c for some of the traces in this set, but we can arrange the construction so that this does not affect the original meet requirement. Once all of these traces enter their target sets, the next largest trace (targeted for A_c) becomes eligible, and we begin to assign traces (one at each stage) targeted, alternately, for A_b and A_c. The process continues until x is free to enter its target set.

In general, we will have trees of traces rather than sequences of traces. The traces will have to be linearly ordered each time a new meet requirement is encountered. A problem will occur if we need to declare an axiom from oracle A_c for a meet requirement corresponding to $a \wedge b = c$ with $c \neq 0$ at a time when no trace yet appointed and targeted for A_c will later enter its target set, yet traces x and y targeted separately for A_a and A_b will enter their respective target sets. Before either of the latter traces enters its target set, however, we may have to appoint a new trace z targeted for A_c for the sake of a join requirement, and this trace may be linked to a higher priority requirement than those to which x and y are linked. The entry of z into its target set after x and y enter their respective target sets may injure the axiom declared for meet requirement along the true path of the construction; yet the entry of z into its target set is required in order to satisfy the high priority requirement to which z is linked. Lerman observed that a problem of this sort for which a solution is not apparent occurs in the lattice S_8; subsequently, Lachlan and Soare [1980] showed that S_8 is not embeddable into $\mathcal{R}$ and so that the conflict between requirements cannot be resolved in this situation.

It is not known whether the late appointment of traces for high priority requirements which will cause injury to meet requirements is restricted to partial lattices for which join requirements must generate infinitely many traces (these will be seen to be the partial lattices which have weak critical triples). There is part of a typical construction where the late appointment of potentially destructive traces can occur. When traces become subject to a new minimal pair requirement, the decomposition process will lead to the appointment of new traces; however, the only cases known where this step creates a conflict are partial lattices with weak critical triples. Thus it is natural to turn to the simpler problem of determining whether all finite partial lattices with no weak critical triples are embeddable into $\mathcal{R}$. It is thus helpful to consider the remainder of the paper separately as it applies to weak critical triples.

Section 4: Principal Decomposability, Critical Triples, and NEC

Nonembeddability results for partial lattices will be of the form: If $\mathcal{L}$ is a finite partial lattice into which the finite partial lattice $\mathcal{P}$ can be embedded (as a partial lattice), then $\mathcal{L}$ cannot be embedded into $\mathcal{R}$.

Critical triples play a central role in nonembeddability results. They occur within every partial lattice satisfying NEC, and their existence precludes a partial lattice from being embedded into some intervals of $\mathcal{R}$ of the form $[0, d]$.

Definition 4.1. A *critical triple* in a partial lattice $\mathcal{P} = \langle P, \leq, \vee, \{M_n : n \in N\}\rangle$ consists of three distinct points $a, b, c \in P$ satisfying the following conditions:

(4.1) a, b, and c are pairwise incomparable.

(4.2) $a \vee b = c \vee b$.

(4.3) $M_2(a, c, b)$.

Weinstein [1988] noted that the Ambos-Spies and Lerman [1986] proof actually showed that a weaker condition precludes embeddability of a finite partial lattice P into $\mathcal{R}$; he defines *weak critical triples*, and shows that if P embeds a weak critical triple, then P cannot be embedded into some subinterval $[0, d]$ of $\mathcal{R}$. This condition is equivalent to the similar condition for critical triples when P is a lattice. Nonembeddability proofs using the weaker condition require only minor modifications. We say that a, b, c form a *weak critical triple* if (4.1) and (4.2) hold, and in addition, we have:

(4.4) For all $d \in P$, if $d < a$ and $d < c$, then $d \vee b < a \vee c$.

The difficulty in working with partial lattices which embed weak critical triples lies in the fact that there will be requirements of our construction whose aim is to preserve joins, and which seem to require potentially infinite action. Satisfaction of such requirements requires only finitary action when weak critical triples are not present.

The nonembeddability condition for $\mathcal{R}$ requires the embeddability of a partial lattice which contains a critical triple with additional properties.

Definition 4.2. A partial lattice $\mathcal{P} = \langle P, \leq, \vee, \{M_n : n \in N\} \rangle$ has NEC if there are five distinct points $a, b, c, p, q \in P$ satisfying the following conditions:

(4.5) $\langle a, b, c \rangle$ forms a critical triple.

(4.6) p and q are incomparable, and $q \not\leq a \vee b$.

(4.7) There is an $r \in P$ such that $a \leq r \leq a \vee b$ and $p \wedge q = r$.

Again this condition can be weakened while preserving nonembeddability, to a condition which is equivalent in the context of lattices. We say that $\mathcal{P}$ has WNEC if we have $a, b, c, p, q \in P$ which satisfy (4.6) and

(4.8) $\langle a, b, c \rangle$ forms a weak critical triple.

(4.9) $a \leq p$ and $a \leq q$ and $M_2(p, q, a \vee b)$.

It was shown by Ambos-Spies and Lerman [1986] that no partial lattice with NEC can be embedded into $\mathcal{R}$, and we note that a minor modification of that proof allows the replacement of NEC with WNEC.

Both embedding and extensions of embeddings proofs are best analyzed by looking at ideals and filters in partial lattices and their sub-partial lattices. Given a finite partial lattice $\mathcal{P}$ with universe P and $c < d \in P$, we let $\mathcal{P}[c, d]$ be the sub-partial lattice of $\mathcal{P}$ obtained by restricting P to $P[c, d] = \{b \in P : c \leq b \leq d\}$. We present some definitions, and then present conditions equivalent to the existence of weak critical triples, NEC, and WNEC, in terms of properties of filters of sub-partial lattices of $\mathcal{P}$.

Definition 4.3. An *ideal* I of the partial lattice $\mathcal{P}$ is a non-empty subset of P with the following properties:

(4.10) If $a \in I$, $b \in P$ and $b < a$, then $b \in I$.

(4.11) If $a, b \in I$, then $a \vee b \in I$.

A *filter* F of $\mathcal{P}$ is a non-empty subset of P with the following properties:

(4.12) If $a \in F$, $b \in P$ and $b > a$, then $b \in F$.

An ideal I (filter F, resp.) of $\mathcal{P}$ is *principal* if it has a greatest (least, resp.) element. The filter F is *prime* if for all $a, b \in P$, if $a \vee b \in F$ then either $a \in F$ or $b \in F$.

Ambos-Spies and Lerman [1986] showed that, from an algebraic point of view, the lattices which fail to have critical triples are exactly those lattices which are principally decomposable in the following sense.

Definition 4.4. A partial lattice $\mathcal{P}$ with universe P is *principally decomposable* if for all $d, b \in P$ such that $b < d$, there is a non-trivial principal prime filter F of $P[0, d]$ such that $b \notin F$.

It thus follows that $\mathcal{P}$ has a weak critical triple iff there are $b < d \in P$ such that no nontrivial prime filter F of $\mathcal{P}[0, d]$ for which $b \notin F$ is principal; and if $\mathcal{P}$ is a lattice, then $\mathcal{P}$ has NEC if there are $b < d \in P$ such that for every nontrivial prime filter F of $\mathcal{P}[0, d]$ for which $b \notin F$, F is non-principal and there are $p \neq q \in P$ such that $q \notin F$ and $p \wedge q \in F$. A simple modification of the proof of Ambos-Spies and Lerman [1986] will yield the equivalence: $\mathcal{P}$ has WNEC if there are $b < d \in P$ such that for every nontrivial prime filter F of $\mathcal{P}[0, d]$ for which $b \notin F$, F is non-principal and there are $p \neq q \in P$ such that $q \notin F$, $M_2(p, q, d)$, and there is a $r \in F$ such that $r < p$ and $r < q$.

Section 5: The Pinball Machine Approach

A typical construction utilizes positive action to satisfy comparability, incomparability, and join requirements, and this action consists of appointing followers and traces which must enter target sets of possibly incomparable degree in a prescribed order; but negative action to satisfy meet requirements will not allow traces to simultaneously enter sets of incomparable degree unless the requirement has prepared, in advance, for the specific instances where such action occurs. The pinball machine method for constructing embeddings of finite lattices into $\mathcal{R}$ presents an organized approach for resolving conflicts of this nature. The method orders the meet requirements of higher priority than the diagonalization requirement in reverse order of their priority, and demands an algorithm which requires followers and traces to obtain permission from the higher priority meet requirements to enter their target sets. Such an algorithm will usually separate the traces, allow some of these to enter their target sets, and then appoint new traces for the remaining followers and traces which, together with the numbers for which they are traces, are permitted by the next meet requirement to simultaneously enter their respective target sets. As mentioned in the previous section, a vague description of the problem to be avoided for a meet requirement $p \wedge q = r$ is to prevent the late appointment of a trace x targeted for A_r when traces targeted for A_p and A_q may precede x into their respective target sets, but no trace precedes x into A_r. We again refer the reader to Lerman [1971] or Downey and Shore [ta] for a precise description of the operation of pinball machine constructions.

Ambos-Spies and Lerman [1989] presented a dynamic sufficient condition, EC, for embeddability. In essence, that condition anticipated the obstacles which might arise in a pinball machine construction, and incorporated the possible ways they found for circumventing those obstacles. The resulting condition is an $\forall \exists \forall$-condition which states that the construction can successfully be executed. More specifically, the condition, in words, states: For all possible configurations of active meet requirements of higher priority than a given diagonalization requirement, there is a uniformly (in the configuration) computable method (satisfying specified conditions) for assigning traces and determining the order in which they will be released by the meet requirements, so that for all possible scenarios for the evolution of the trace assignment process and the release of traces by meet requirements, no injury will occur to any of these meet requirements, and the diagonalization requirement will eventually be satisfied. A rigorous statement of this condition can be found in Ambos-Spies and Lerman [1989], but we avoid stating the condition here because of its technical nature. We note, however, that this condition involves two alternations of quantifiers, so is not, on the surface, effective. It is not known whether this condition is complementary to NEC.

We now suggest a new approach, resulting from the analysis of examples of particular embeddable lattices (undertaken jointly with Lempp) rather than from the embedding construction. This approach has yielded an effective embeddability condition, EEC, which ensures embeddability into $\mathcal{R}$ for finite partial lattices which fail to embed critical triples. The condition has undergone many modifications based on investigations of its applicability to particular lattices, and we anticipate that it could undergo more modifications as new examples are investigated and attempts are made to prove such a condition. We are stating this condition here, not because we have confidence that it is complementary to NEC (as indicated above, we would not be surprised to see a need for further modifications), but because we feel that it provides a promising direction for finding an effective condition complementary to NEC. As a first attempt, we would suggest considering such a condition for finite partial lattices which fail to embed weak critical triples. If and when such a condition is obtained, we believe that it will give insight into how to find a condition (perhaps the same one) which also applies to the class of all finite partial lattices. The effectiveness of EEC will make it easier to work with than EC, should there, in fact, be a necessary and sufficient condition of this type. It also makes investigating complementarity of NEC with such a condition through the use of a computer program more tractable. EEC implies that the procedure for appointing traces will have the property that no traces appointed (hereditarily) for a follower at some late stage can cause injury to a meet requirement of high priority. Before formulating EEC, we present some definitions. Below, $\mathcal{P}$ will always represent a finite partial lattice with universe P.

Definition 5.1. Given $S \subseteq P$, then (S) denotes the upward closure of S, i.e., the filter generated by S. If $S = \{a\}$, then we write (a) in place of $(\{a\})$. Similarly, for a sequence $H = \langle h_0, \ldots, h_k \rangle$, we write (H) for $\cup \{(h_i) : i \leq k\}$.

The upward closure of a set S represents the elements of P which can record the entry of numbers into S. As indicated earlier, in order for a pinball construction to succeed, the upward closure of a full sequence of traces must be a prime filter of P. We keep track of the sequences of traces through the next definition.

Definition 5.2. Given $S, T \subseteq P$, we say that S *covers* T if both (S) and $(S \cup T)$ are prime filters of P.

When a sequence of traces reaches a gate, we must ensure that the target sets for this sequence receive numbers in an order which allows us to preserve one side of each high priority meet computation. This will be accomplished by allowing traces to enter only one of the target sets determining the meet at crucial portions of the construction. We will use the following terminology to keep track of the target sets avoided by filters of P.

Definition 5.3. Fix $b \in P$ and $S \subseteq P$. We say that S is *b-free* if $S \cap P[0, b] = \emptyset$.

The Effective Embedding Condition will try to replace properties of sequences of elements of P with properties of sets of elements of P to the extent possible. It will require that we can gradually modify a prime filter by breaking it up into pieces so that we preserve the covering condition required to satisfy join requirements without introducing traces targeted for elements of P which are not in the current prime filter, unless such traces can be safely introduced (without causing irreparable injury to any meet requirement); furthermore, the last such modification of the filter must be safe for a given meet requirement, i.e., it must be able to be released, intact, from the gate without causing injury to the corresponding meet requirement. We define the notion of *safety*, and then introduce the Effective Embedding Condition.

Definition 5.4. Let G and H be subsets of P. We say that H is *safe for G* if given any $a, b, c \in P$ such that $a \mid b$, $a \wedge b = c$, and $c \in H$, then either $a \notin G$ or $b \notin G$.

Definition 5.5. We say that P has the *Effective Embedding Condition (EEC)* if P fails to have NEC and given any $p, q, r \in P$ such that $p \mid q$ and $p \wedge q = r$, a prime filter F_0 of P, a sequence $F_1 = \langle f_0, ..., f_r \rangle$ of elements of P such that for all $i \leq r$, $F_0 \cup (f_0) \cup \cdots \cup (f_i)$ is a prime filter of P and (F_1) is q-free, and a subset $K \subseteq F_0$ such that $K \cup (F_1)$ is a prime filter of P, then there is a q-free $K_1 \subseteq K$, a finite sequence $G_0, G_1, ..., G_k$ of prime filters of P, a sequence $H_0, H_1, ..., H_k$ of sequences of elements of P (set $H_i = \langle h_{i,0}, ..., h_{i,r(i)} \rangle$, and for $j \leq r(i)$, set $H_{i,j} = \langle h_{i,0}, ..., h_{i,j} \rangle$), and embeddings $e_i : H_i \to H_{i+1}$ for all $i < k$ such that:

(5.1) $G_0 = F_0$, $(H_0) = (F_1) \cup K_1$, F_1 is a subsequence of H_0 and $G_k = \emptyset$.

(5.2) For all $i < k$ and $r < s \leq r(i)$,
    (i) $e_i(h_{i,r}) \leq h_{i,r}$; and
    (ii) if $e_i(h_{i,r}) = h_{i+1,u}$ and $e_i(h_{i,s}) = h_{i+1,v}$, then $u < v$; and
    (iii) if $H_{i+1} \neq H_i$ then $(H_{i+1}) \neq (H_i)$.

(5.3) For all $i \leq k$ and $r \leq r(i)$, $G_i \cup (H_{i,r})$ is a prime filter of P.

(5.4) For all $i < k$, $G_{i+1} - G_i$ is safe for G_i, and for all $a, b, c \in P$ such that $a|b$ and $a \wedge b = c$
    and all $i < k$ and $u \leq r(i)$, if $h_{i+1,u} \leq c$, then either:
    (i) $a \notin G_i \cup (H_i)$; or
    (ii) $b \notin G_i \cup (H_i)$; or
    (iii) there are r and $v \leq u$ such that $e_i(h_{i,r}) = h_{i+1,v}$ and $h_{i,r} \leq c$.

(5.5) For some $i \leq k$, $G_i \cup (H_i)$ is q-free.

Suppose that $\langle G_0, G_1, ..., G_k \rangle$ and $\langle H_0, H_1, ..., H_k \rangle$ satisfy (5.1)-(5.5). There are only $2^{|K|}$ many possible choices for K_1. Furthermore, by (5.2)(i),(ii), $\langle (H_0), ..., (H_k) \rangle$ is non-decreasing; thus by (5.2)(iii),

$$|\{i : H_{i+1} \neq H_i\}| \leq |P|.$$

Now if there are numbers i and j such that $i < j$, $G_i = G_j$, $G_{i+1} = G_{j+1}$, and for all $m \in [i, j+1]$, $H_m = H_i$, then the subsequences obtained by removing $G_i, ..., G_{j-1}$ and $H_i, ..., H_{j-1}$ also satisfy (5.1)-(5.5); hence if a pair of sequences satisfying (5.1)-(5.5) for K_1 exists, then there must be such a pair of sequences of length $\leq (2^{|P|})(|P|+1)$. It thus follows that we can effectively determine whether P has EEC.

We use pinball machine terminology to relate EEC to a pinball machine embedding construction. The aim of the condition is to localize each decision about the appointment of new traces to a single gate corresponding to a meet requirement for $p \wedge q = r$. Thus when a set of traces arrives at this gate, we want to be able to treat all traces in this set as if they were appointed at the same time (we call such a set of traces a *contemporaneous set*).

Suppose that a contemporaneous set of traces targeted for the elements in the prime filter $S \subset P$ arrives at a gate corresponding to a meet requirement for $p \wedge q = r$. If $r \in S$, then we can allow all these traces past the gate at the next expansionary stage, as we will be able to use an A_r oracle to identify any injury to computations from A_p and A_q on the same argument. Otherwise, we will organize the elements of P coding targets of traces into blocks (which will be the union of sequences of principal filters of P) such that no block contains both p and q, and every block is covered by the union of the preceding blocks. These blocks will alternately contain exactly one of p and q, and the contemporaneity will allow us to add elements of earlier blocks to later blocks withput destroying the p-freeness or q-freeness of a block, so that the set K as specified in Definition 5.5 exists. Thus we can reduce the problem of avoidance of injury to the consideration of two successive blocks F_0 and F_1, and a subset K of F_0. K will usually contain p, and F_1 will contain q. EEC now allows us to consecutively define subsets of P in such a way that we can safely allow the final subset, H_k, by the gate without injuring the corresponding meet requirement, and treat H_k as a contemporaneous set. Each G_m and H_m will be appointed at an expansionary stage preceding the stage at which the traces targeted for sets coded by G_{m-1} are allowed past the gate. Conditions (5.4) and (5.5) allow us to ensure that no injury will occur to meet requirements represented by a higher priority gate for $a \wedge b = c$, as any injury to computations from both A_a and A_b via traces y and z respectively on a fixed argument will be identified through the prior or simultaneous entry of some trace x into A_c, and x will trace its heredity for A_c back to a stage at least as early as the heredity of y for A_a or z for A_b.

There are other ways of arranging a pinball machine construction, some of which do not allow one to easily keep track of contemporaneous traces. We have proposed EEC as stated, as it explains the examples we have examined, and seems to be somewhat easier to work with than other conditions. We hope that this, or some closely related condition can be shown to be complementary to NEC.

REFERENCES

K. Ambos-Spies [1980], *On the structure of the recursively enumerable degrees*, Doctoral Dissertation, University of Munich.

K. Ambos-Spies, C.G. Jockusch, Jr., R.A. Shore, and R.I. Soare [1984], *An algebraic decomposition of the recursively enumerable degrees and the coincidence of several degree classes with the promptly simple degrees*, Trans. Amer. Math. Soc. **281**, 109-128.

K. Ambos-Spies, S. Lempp, and M. Lerman [1994], *Lattice embeddings into the r.e. degrees preserving 0 and 1*, J. London Math. Soc. (2)**49**, 1-15.

K. Ambos-Spies, S. Lempp, and M. Lerman [1994a], *Lattice embeddings into the r.e. degrees preserving 1*, Logic, Methodology and Philosophy of Science IX, D. Prawitz, B. Skyrms and D. Westerståhl, eds. Elsevier Science, Amsterdam, New York.

K. Ambos-Spies, S. Lempp, and R.I. Soare [ta], *Intervals of recursively enumerable degrees: Lattice embeddings and non-$\aleph_0$-categoricity of the partial order*, To appear.

K. Ambos-Spies and M. Lerman [1986], *Lattice embeddings into the recursively enumerable degrees*, J. Symbolic Logic **51**, 257-272.

K. Ambos-Spies and M. Lerman [1989], *Lattice embeddings into the recursively enumerable degrees, II*, J. Symbolic Logic **54**, 735-760.

W.C. Calhoun [1990], *The lattice of ideals of recursively enumerable degrees*, Doctoral Dissertation, University of California at Berkeley.

S.B. Cooper, T.A. Slaman, and X. Yi [ta], *An extension theorem for the recursively enumerable degrees*, To appear.

S.B. Cooper, R. Sui, and X. Yi [ta], *Minimal pairs and the Slaman-Soare phenomenon*, To appear.

R.G. Downey [1990], *Lattice nonembeddings and initial segments of the recursively enumerable degrees*, Ann. Pure Appl. Logic **49**, 97-119.

R.G. Downey and R.A. Shore [ta], *Lattice embeddings below a Nonlow$_2$ recursively enumerable degree*, To appear.

S.K. Kleene and E.L. Post [1954], *The upper semi-lattice of degrees of recursive unsolvability*, Ann. of Math. (2) **59**, 379-407.

A.H. Lachlan [1968], *On the lattice of recursively enumerable sets*, Trans. Amer. Math. Soc. **130**, 1-37.

A.H. Lachlan [1966], *Lower bounds for pairs of recursively enumerable degrees*, Proc. London Math. Soc. **16**, 537-569.

A.H. Lachlan [1972], *Embedding nondistributive lattices in the recursively enumerable degrees*, Conference in Mathematical Logic - London '70, W. Hodges ed., Lecture Notes in Mathematics 255, Springer-Verlag, Berlin, Heidelberg, New York, pp. 149-177.

A.H. Lachlan and R.I. Soare [1980], *Not every finite lattice is embeddable in the recursively enumerable degrees*, Advances in Math. **37**, 74-82.

S. Lempp, A. Nies, and T. Slaman [ta], *The undecidability of the Π_3-theory of the r.e. Turing degrees*, To appear.

M. Lerman [1973], *Admissible ordinals and priority arguments*, Cambridge Summer School in Mathematical Logic - Proceedings 1971, A.R.D. Mathias and H. Rogers, Jr. eds., Lecture Notes in Mathematics 337, Springer-Verlag, Berlin, Heidelberg, New York, pp. 311-344.

M. Lerman, R.A. Shore, and R.I. Soare [1984], *The elementary theory of the recursively enumerable degrees is not $\aleph_0$-categorical*, Advances in Math. **53**, 301-320.

E.L. Post [1944], *Recursively enumerable sets of positive integers and their decision problems*, Bull. Amer. Math. Soc. **50**, 284-316.

G.E. Sacks [1963], *Degrees of Unsolvability*, Annals of Mathematics Studies No. 55, Princeton University Press, Princeton, NJ.

G.E. Sacks [1964], *The recursively enumerable degrees are dense*, Ann. Math. **80**, 300-312.

J.R. Shoenfield [1964], *Applications of model theory to degrees of unsolvability*, Symposium on the Theory of Models, North-Holland Pub. Co., Amsterdam, pp. 359-363.

T.A. Slaman [1991], *The density of infima in the recursively enumerable degrees*, Ann. Pure Appl. Logic **52**, 1-25.

T.A. Slaman and R.I. Soare [1995], *Algebraic aspects of the computably enumerable degrees*, Proc. Nat. Acad. Sci., USA **92**, 617-21.

T.A. Slaman and W.H. Woodin [ta], *Definability in the Turing degrees*.

S.K. Thomason [1971], *Sublattices of the recursively enumerable degrees*, Z. Math. Logik Grundlag. Math. **17**, 273-280.

A.M. Turing [1939], *Systems of logic based on ordinals*, Proc. London Math. Soc. (2)**45**, 161-228.

B. Weinstein [1988], *On embedding of the 1-3-1 lattice into the recursively enumerable degrees*, Doctoral Dissertation, University of California at Berkeley.

C.E.M. Yates [1966], *A minimal pair of recursively enumerable degrees*, J. Symb. Logic **31**, 159-168.

On a Question of Brown and Simpson

Michael E. Mytilinaios
Theodore A. Slaman*

1 Introduction

Brown and Simpson (1993) considered two versions of the Baire Category Theorem, which they called BCT-I and BCT-II. They showed that BCT-I is provable in RCA_0, the subsystem of second order arithmetic specifying some basic algebraic properties of the numbers, induction for Σ_1^0 sets and the recursive comprehension axiom. Additionally, they showed that BCT-II is not provable in RCA_0. Brown and Simpson then introduced RCA_0^+, the extension of RCA_0 which for each n includes the statement that for each subset X of the natural numbers $\mathbb{N}$, there is a $G \subseteq \mathbb{N}$ such that G is Cohen generic for n-quantifier arithmetic relative to X. They showed that BCT-II is provable in RCA_0^+ and posed the question whether BCT-II and its consequence the Open Mapping Theorem are provable in a system weaker than RCA_0^+.

In this paper, we introduce an equivalent formulation of BCT-II, which we denote BCT-Π_1^0. It states that the intersection of every sequence of dense open subsets of Baire space which has a uniformly Π_1^0 representation is dense. In the language of forcing, the set of reals which are generic for the sequence $\langle O_n : n \in \mathbb{N} \rangle$ is dense. Using this recursion theoretic characterization, we show that there is an ω-model of BCT-Π_1^0 which is not a model of RCA_0^+ (Corollary 4.3, thereby answering the question raised by Brown and Simpson.)

Brown and Simpson compared BCT-II with WKL_0, which consists of RCA_0 and the assertion (of compactness) that every infinite binary tree has an infinite branch. They showed that neither of these two systems is a subtheory of the other, over the base system RCA_0. We continue in this line by comparing BCT-Π_1^0 with the assertions of the various cases of Ramsey's Theorem. See (Ramsey 1930).

*Slaman was partially supported by NSF Grant DMS 91-06714 and SERC Visiting Fellowship Research Grant ("Leeds Recursion Theory Year 1993/94") No. GR/H 91213. The authors would like to thank S. Simpson for bringing the Brown and Simpson question to their attention.

Definition 1.1 Let $[\mathbb{N}]^n$ denote the size n subsets of $\mathbb{N}$. Suppose that n and m are positive integers and F is a function from $[\mathbb{N}]^n$ to $\{1, \ldots, m\}$. We say that $H \subseteq \mathbb{N}$ is *homogeneous* for F if F is constant on H^n.

Theorem 1.2 (Ramsey's Theorem) *For all positive integers n and m, if $F : [\mathbb{N}]^n \to \{1, \ldots, m\}$ then there is an infinite set H such that H is homogeneous for F.*

If we fix n and m, we represent the above conclusion as $\mathbb{N} \to [\mathbb{N}]_m^n$. To state the principle for all n and m, we write $\mathbb{N} \to [\mathbb{N}]_{<\mathbb{N}}^{<\mathbb{N}}$.

Jockusch (1972) showed that there is a recursive partition of $[\mathbb{N}]^3$ into 2 pieces such that $0'$ is recursive in any infinite homogeneous set. The same proof shows that $RCA_0 + \mathbb{N} \to [\mathbb{N}]_2^3 \vdash ACA_0$, where ACA_0 is RCA_0 with a scheme for arithmetic comprehension. Seetapun showed that if F is a partition of $[\mathbb{N}]^2$ into finitely many pieces and X is not recursive in F then there is an infinite set H such that H is homogeneous for F and X is not recursive in H. Consequently, $RCA_0 + \mathbb{N} \to [\mathbb{N}]_2^2 \nvdash ACA_0$. Seetapun's theorem suggests that $RCA_0 + \mathbb{N} \to [\mathbb{N}]_2^2$ is a relatively weak subtheory of Z_2. On the other hand, Slaman showed that there is an arithmetic statement φ such that $RCA_0 + \mathbb{N} \to [\mathbb{N}]_2^2 \vdash \varphi$ and $RCA_0 \nvdash \varphi$. Slaman's theorem suggests that the first order consequences of $RCA_0 + \mathbb{N} \to [\mathbb{N}]_2^2$ is a relatively strong subtheory of PA. Seetapun and Slaman asked whether $RCA_0 + \mathbb{N} \to [\mathbb{N}]_2^2$ proves PA; as of the writing of this paper, the Seetapun and Slaman question is open. The results of Seetapun and Slaman may be found in (Seetapun and Slaman 1994).

We will show that $RCA_0 + \mathbb{N} \to [\mathbb{N}]_{<\mathbb{N}}^2 \vdash I\Sigma_2^0$ and that $RCA_0 + \mathbb{N} \to [\mathbb{N}]_2^2 + I\Sigma_2^0 \vdash BCT\text{-}\Pi_1^0$. We will also observe that $RCA_0 + BCT\text{-}\Pi_1^0 \nvdash \mathbb{N} \to [\mathbb{N}]_2^2$.

2   Separable metric spaces within second order arithmetic

We reproduce the Brown and Simpson definitions.

The language of second order arithmetic Z_2 consists of two sorted variables, the *number variables* $i, j, k, l, m, n, \ldots$ and the *set variables* $X, Y, Z, \ldots$. The *numerical terms* are built up from number variables, the constant symbols 0 and 1, and the binary operations of addition and multiplication. *Atomic formulas* are $t_1 = t_2$, $t_1 < t_2$ and $t_1 \in X$ where t_1 and t_2 are numerical terms. *Formulas* are built up from atomic formulas by means of the propositional connectives, number quantifiers and set quantifiers.

A formula is Σ_0^0 or Π_0^0 if it has only number quantifiers, all of which are bounded. Φ is Σ_{n+1}^0 if Φ is of the form $(\exists n_1) \ldots (\exists n_k)\varphi$ and φ is Π_n^0. Similarly, Φ is Π_{n+1}^0 if Φ is of the form $(\forall n_1) \ldots (\forall n_k)\varphi$ and φ is Σ_n^0. A formula is Σ_0^1 or Π_0^1 if it only has number quantifiers. Φ is Σ_{n+1}^1 if Φ is of the form $(\exists X_1)\ldots(\exists X_k)\varphi$ and φ is Π_n^1. Finally, Φ is Π_{n+1}^1 if Φ is of the form $(\forall X_1)\ldots(\forall X_k)\varphi$ and φ is Σ_n^1. Note that in each case above, φ may have free real variables.

RCA_0 is the subsystem of Z_2 which includes the *ordered semiring axioms* for $\langle \mathbb{N}, +, \cdot, 0, 1, < \rangle$ together with schemes for Σ_1^0 induction $(I\Sigma_1^0)$ and recursive comprehension $(\Delta_1^0 CA)$.

WKL_0 is the subsystem of Z_2 which consists of RCA_0 plus an additional axiom which states that every infinite binary tree has an infinite path and ACA_0 is the subsystem of Z_2 which consists of RCA_0 plus the scheme for arithmetic comprehension.

Within RCA_0, Brown and Simpson represent metric spaces as follows.

A (code for a) *complete separable metric space* consists of a set $A \subseteq \mathbb{N}$ together with a distance function d. A point in the *completion* $\widehat{A}$ (of A) is a function f from $\mathbb{N}$ to A such that $(\forall n)(\forall i)[d(f(n), f(n+i)) < 2^{-n}]$. That is, f is a Cauchy sequence which converges geometrically. The pseudometric d on A extends to a pseudometric $\widehat{d}$ on $\widehat{A}$. Within RCA_0, we can form the infinite product space $\widehat{A} = \prod_{i=0}^{\infty} \widehat{A}_i$ from the sequence of (codes for) complete separable metric spaces $\widehat{A}_i$, $i \in \mathbb{N}$. Thus, we can represent Cantor space $2^{\mathbb{N}}$ and Baire space $\mathbb{N}^{\mathbb{N}}$ inside RCA_0 and we can identify their points with functions f from $\mathbb{N}$ to $\{0,1\}$ and g from $\mathbb{N}$ to $\mathbb{N}$, respectively.

Let $\widehat{A}$ be a complete separable metric space. An *open ball* $B(x, \epsilon)$ in $\widehat{A}$ is a set associated with the ordered pair (x, ϵ) (the center and the radius of the ball, respectively) with $x \in \widehat{A}$ and $\epsilon \in \mathbb{R}^+$ such that $B(x, \epsilon)$ is equal to $\{y : \widehat{d}(x,y) < \epsilon\}$. An open ball with center $a \in A$ and radius $r \in \mathbb{Q}^+$ is called a *basic open* set. A code for an *open* set U is a sequence of basic open sets $\langle B(a_n, r_n) : n \in \mathbb{N} \rangle$ and we say that $x \in \widehat{A}$ is in U if there is a basic open set $B(a, r)$ in the sequence such that $x \in B(a, r)$. A *closed* set is one which is the complement of an open set.

A (code for a) *separably closed* set C is a sequence $S = \langle x_n : n \in \mathbb{N} \rangle$ of points in A whose closure is C, namely $C = \overline{S}$. In other words, a separably closed set is one which is represented by some countable set which is dense within it. A set is *separably open* if it is the complement of a separably closed set.

Suppose that $\widehat{A}$ and $\widehat{B}$ are complete separable metric spaces with codes A and B. Working within RCA_0, we define a (code for a) *continuous partial function from $\widehat{A}$ to $\widehat{B}$* to be a function $F : \mathbb{N} \to A \times \mathbb{Q}^+ \times B \times \mathbb{Q}^+$ such that for all $m, n \in \mathbb{N}$, $a, a' \in A$, $b, b' \in B$ and $r, r', s, s' \in \mathbb{Q}^+$:

$$F(m) = (a, r, b, s) \ \& \ F(n) = (a, r, b', s') \implies d(b, b') < s + s';$$
$$F(m) = (a, r, b, s) \ \& \ B(b, s) \subseteq B(b', s') \implies (\exists k)[F(k) = (a, r, b', s')];$$
$$F(m) = (a, r, b, s) \ \& \ B(a', r') \subseteq B(a, r) \implies (\exists k)[F(k) = (a', r', b, s)].$$

We will use some basic consequences of RCA_0 and some standard results about complete separable metric spaces provable within RCA_0. The facts that we will use without proof include the following.

Every Σ_1^0 set has a least element, every bounded Σ_1^0 set has a bounded enumeration, for every natural number m there is no Σ_1^0 bijection between $\{n \in \mathbb{N} : n < m\}$ and $\{n \in \mathbb{N} : n < m+1\}$, if Φ is $\boldsymbol{\Pi}_1^0$ then $(\exists i \le j)\Phi$ is equivalent to a $\boldsymbol{\Pi}_1^0$ formula. (See (Kirby and Paris 1977).)

Working within a complete separable metric space, the countable union of open sets is open, the finite intersection of open sets is open, if U is open and $\{a_1, a_2, a_3, \ldots, a_n\}$ a finite set of points which are in U then the set $U - \{a_1, a_2, a_3, \ldots, a_n\}$ is open. (See (Brown 1987).)

Let $\widehat{A}$ be a complete separable metric space. We say that an open set U is *dense* if for every basic open set $B(a, r)$ there exists an element $x \in \widehat{A}$, such that $x \in U \cap B(a, r)$. A closed set is said to be *nowhere dense* if its complement is a dense set. Equivalently over RCA_0 if C contains no open ball.

Brown and Simpson defined RCA_0^+ to be the subsystem of second order arithmetic $\boldsymbol{Z}_2$ whose axioms are those of RCA_0 plus a scheme which states that given a sequence of arithmetically defined dense subsets of $2^{<\mathbb{N}}$ there exists an element of $2^{\mathbb{N}}$ which meets them all.

3 $\boldsymbol{\Pi}_1^0$-Open Sets

We recall the Brown and Simpson statement of *BCT-II*.

Definition 3.1 *BCT-II* is the following statement. Let $\widehat{A}$ be a complete separable metric space, and let $\langle\langle x_{n,k} : k \in \mathbb{N}\rangle : n \in \mathbb{N}\rangle$ be a sequence of (codes for the complements of) dense separably open subsets O_n of $\widehat{A}$. If U is a (code for a) nonempty open set in $\widehat{A}$, then there is a point $x \in \widehat{A}$ such that $x \in U$ and $x \in O_n$ for all $n \in \mathbb{N}$ (*i.e.*, $x \notin \overline{\{x_{n,k} : k \in \mathbb{N}\}}$ for any $n \in \mathbb{N}$).

We give a recursion theoretic formulation of *BCT-II*.

Definition 3.2   • A set O is a $\boldsymbol{\Pi}_1^0$-*open set* if there is a (code for a) sequence of basic open sets $\langle B(a_i, r_i) : i \in \mathbb{N}\rangle$ and a $\boldsymbol{\Pi}_1^0$ formula Φ such that O is equal to $\bigcup\{B(a_i, r_i) : \Phi(i)\}$.

   • A sequence of open sets $\langle O_n : n \in \mathbb{N}\rangle$ is *uniformly* $\boldsymbol{\Pi}_1^0$ if there are (codes for) a sequence $\langle B(a_{n,i}, r_{n,i}) : n \in \mathbb{N} \,\&\, i \in \mathbb{N}\rangle$ and a sequence of $\boldsymbol{\Pi}_1^0$ formulas $\langle \Phi_n : n \in \mathbb{N}\rangle$, such that for each n, O_n is equal to $\bigcup\{B(a_{n,i}, r_{n,i}) : \Phi_n(i)\}$.

Definition 3.3 $BCT\text{-}\mathit{\Pi}_1^0$ is the following statement. Let $\langle O_n : n \in \mathbb{N} \rangle$ be a uniformly $\mathit{\Pi}_1^0$ sequence of dense open subsets of $\mathbb{N}^{\mathbb{N}}$. If U is a (code for a) nonempty open subset of $\mathbb{N}^{\mathbb{N}}$, then there is a point $x \in \mathbb{N}^{\mathbb{N}}$ such that $x \in U$ and, for each n in $\mathbb{N}$, $x \in O_n$.

Note that $BCT\text{-}\mathit{\Pi}_1^0$ is written as a hypothesis on Baire space rather than on all complete separable metric spaces. In order to prove the equivalence of $BCT\text{-}II$ and $BCT\text{-}\mathit{\Pi}_1^0$, we check that the usual propositions establishing the universal role of $\mathbb{N}^{\mathbb{N}}$ among all complete separable metric spaces are true at the effective level.

To fix some notation, for $\sigma \in \mathbb{N}^{<\mathbb{N}}$ and $k \in \mathbb{N}$, let $\sigma * k$ be the sequence obtained by appending k to σ. Let $lh(\sigma)$ denote the length of σ. For n in $\mathbb{N}$ and x in $\mathbb{N}^{\mathbb{N}}$, we let $x \lceil n$ be the sequence obtained by restricting x to its first n values.

Fix a representation of a complete separable metric space $\widehat{A}$. We construct a continuous map $\pi : \mathbb{N}^{\mathbb{N}} \to \widehat{A}$ which we will use to reduce instances of the Baire Category Theorem for $\widehat{A}$ to ones involving $\mathbb{N}^{\mathbb{N}}$.

Let $\mathcal{B}_A$ be the collection of basic open subsets of $\widehat{A}$ and fix an enumeration $\langle B(a_i, r_i) : i \in \mathbb{N} \rangle$ of $\mathcal{B}_A$ such that every element of $\mathcal{B}_A$ appears infinitely often. We define π_0 mapping the nontrivial elements σ in $\mathbb{N}^{<\mathbb{N}}$ to finite sequences from $\mathcal{B}_A$ by recursion. If σ is a sequence $\langle k \rangle$ of length 1 then $\pi_0(\sigma)$ is the kth element $B(a, r)$ of $\mathcal{B}_A$ such that r is less than or equal to 1. When we speak of the kth element, we are referring to the kth element in the fixed enumeration of $\mathcal{B}_A$. Given that $\pi_0(\sigma)$ is defined and equal to $B(a(\sigma), r(\sigma))$ let $\pi_0(\sigma * k)$ be the kth element $B(a, r)$ of $\mathcal{B}_A$ such that r is less than or equal to the minimum of $\{1/2^{lh(\sigma)}, r(\sigma)\}$ and such that $d(a(\sigma), a) \leq r(\sigma) - r$.

We have defined π_0 so that for each σ, $\pi_0(\sigma)$ is an element of $\mathcal{B}_A$ of radius less than or equal to $1/2^{lh(\sigma)-1}$ and so that for each k, $\pi_0(\sigma * k) \subseteq \pi_0(\sigma)$. Note that π_0 is defined by a Σ_1^0 recursion relative to the presentation of A and, as such, is well defined within RCA_0.

We define $\pi : \mathbb{N}^{\mathbb{N}} \to \widehat{A}$ by setting $\pi(x)$ equal to $\langle a(x\lceil n) : n \in \mathbb{N} \rangle$, where $B(a(x\lceil n), r(x\lceil n))$ is $\pi_0(x\lceil n)$. We defined π_0 so that $\langle a(x\lceil n) : n \in \mathbb{N} \rangle$ would be a suitable Cauchy sequence. The surjectivity of π is built into the representation of elements of $\widehat{A}$ as Cauchy sequences with a geometric convergence rate. The continuity of π is ensured in the specification of π_0.

Lemma 3.4 (RCA_0) *For every complete separable metric space $\widehat{A}$ the following conditions hold.*

- *If O is an open dense subset of $\widehat{A}$ then $\pi^{-1}(O)$ is dense in $\mathbb{N}^{\mathbb{N}}$. Note, this claim does not depend on the presentation of O and holds if O is an arbitrary union of basic open subsets of $\widehat{A}$.*

- *If O is a $\boldsymbol{\Pi}_1^0$ open dense subset of $\widehat{A}$ then there is an O^* such that O^* is a $\boldsymbol{\Pi}_1^0$ open dense subset of $\mathbb{N}^\mathbb{N}$ and $O^* \subseteq \pi^{-1}(O)$. Further, the $\boldsymbol{\Pi}_1^0$ index for O^* is obtained uniformly recursively from the $\boldsymbol{\Pi}_1^0$ index for O.*

Proof: Suppose $\widehat{A}$ is a complete separable metric space. Recall our notation from above: $\langle \sigma_i : i \in \mathbb{N} \rangle$ is an enumeration of $\mathbb{N}^{<\mathbb{N}}$, $\pi_0 : \sigma_i \mapsto B(a(\sigma_i), r(\sigma_i))$, and π is the continuous function derived from π_0.

Suppose that O is an arbitrary union of basic open subsets of $\widehat{A}$ and is dense in $\widehat{A}$. To check that $\pi^{-1}(O)$ is dense, let U be an open subset of $\mathbb{N}^\mathbb{N}$ and let i be given so that $B(\sigma_i, 1/2^{lh(\sigma_i)})$ is contained in U. Since O is dense, there is a basic open subset of $\widehat{A}$, $B(a, r)$, such that $B(a, r) \subseteq B(a(\sigma_i), r(\sigma_i)) \cap O$ and r is less than $1/2^{lh(\sigma_i)}$. In particular, a in an element of O. Now, consider a sequence x such that for each n, if n is less than or equal to $lh(\sigma_i)$ then $x(n) = \sigma_i(n)$ and if n is greater than $lh(\sigma_i)$ then $B(a(x \lceil n), r(x \lceil n))$ is of the form $B(a, r(x \lceil n))$. (For example, the leftmost such sequence x can be constructed recursively in the presentation of A.) This x is an element of $B(\sigma_i, 1/2^{lh(\sigma_i)})$, is hence an element of U and satisfies $\pi(x) = a$. In particular, $x \in \pi^{-1}(O) \cap U$, as required.

Now, suppose that O is a $\boldsymbol{\Pi}_1^0$ open dense subset of $\widehat{A}$. Say that O is $\bigcup \{ B(a_i, r_i) : \Phi(i) \}$ where $\langle B(a_i, r_i) : i \in \mathbb{N} \rangle$ is an enumeration of elements from $\mathcal{B}_A$ and $\Phi(i)$ is $\boldsymbol{\Pi}_1^0$. We must find a $\boldsymbol{\Pi}_1^0$ open dense subset O^* of $\mathbb{N}^\mathbb{N}$ such that $O^* \subseteq \pi^{-1}(O)$.

For each σ in $\mathbb{N}^{<\mathbb{N}}$, $B(\sigma, 1/2^{lh(\sigma)})$ is the basic open set, consisting of all x in $\mathbb{N}^\mathbb{N}$ such that $x \lceil lh(\sigma) = \sigma$. Fix an enumeration $\langle B(\sigma_i, 1/2^{lh(\sigma_i)}) : i \in \mathbb{N} \rangle$ of all such sets. Suppose that $\Phi(i)$ is $(\forall t)\varphi(t, i)$, where φ is $\boldsymbol{\Pi}_0^0$. We let $\Psi(k)$ be

$$(\exists i \leq k) \left[\pi_0(\sigma_k) \subseteq B(a_i, r_i) \ \& \ (\forall t)\varphi(t, i) \right].$$

Note that $\pi_0(\sigma_k) \subseteq B(a_i, r_i)$ can be rewritten recursively as $r_i \geq r(\sigma_k)$ and $d(a_i, a(\sigma_k)) \leq r_i - r(\sigma_k)$. It follows from $I\boldsymbol{\Sigma}_1^0$ that Ψ is equivalent to a $\boldsymbol{\Pi}_1^0$ formula. Of course, Ψ holds of k if and only if there is at least one i less than or equal to k such that $\pi_0(\sigma_k)$ is contained in $B(a_i, r_i)$ and Φ holds of i. We define

$$O^* = \bigcup \{ B(\sigma_k, 1/2^{lh(\sigma_k)}) : \Psi(k) \}.$$

O^* is explicitly presented as a $\boldsymbol{\Pi}_1^0$ open subset of $\mathbb{N}^\mathbb{N}$.

To see that $\pi(O^*) \subseteq O$, let x be an element of O^*. There is a k such that $x \in B(\sigma_k, 1/2^{lh(\sigma_k)}) \ \& \ \Psi(k)$. Fixing such a k, we have $x \lceil lh(\sigma_k) = \sigma_k$. Then, $\pi_0(\sigma_k)$ is contained in an open ball, $B(a_i, r_i)$, contained in O and π maps every element of $B(\sigma_k, 1/2^{lh(\sigma_k)})$ to an element of $\pi_0(\sigma_k)$ and hence to an element of O. In particular, $\pi(x) \in O$.

To see that O^* is dense, let σ be given; we will show that there is a k such that $B(\sigma * k, 1/2^{lh(\sigma*k)})$ is contained in O^*. Now, $\pi_0(\sigma)$ is a basic open subset $B(a(\sigma), r(\sigma))$ of $\widehat{A}$. Since O is dense there is an i such that $\Phi(i)$ (and so $B(a_i, r_i)$ is contained in O) and $B(a_i, r_i) \cap B(a(\sigma), r(\sigma))$ is not empty. But then there are a^* in A and r^* in $\mathbb{Q}$ such that $B(a^*, r^*)$ is contained in $B(a_i, r_i) \cap B(a(\sigma), r(\sigma))$. Let k be an index greater than i such that $\pi_0(\sigma * k)$ is equal to $B(a^*, r^*)$. There is such an index because $B(a^*, r^*)$ is an acceptable value for π_0 on an extension of σ and every basic open subset of $\widehat{A}$ has infinitely many indices. ∎

Proposition 3.5 (RCA_0) *BCT-II is equivalent to BCT-$\mathit{\Pi}_1^0$.*

Proof: Let $\widehat{A}$ be a complete separable metric space. Let π be the continuous map from $\mathbb{N}^{\mathbb{N}}$ to $\widehat{A}$ defined above; let π_0 be its corresponding map from $\mathbb{N}^{<\mathbb{N}}$ to basic open subsets of $\widehat{A}$; let $\pi_0(\sigma)$ be denoted by $B(a(\sigma), r(\sigma))$.

$(BCT\text{-}\mathit{\Pi}_1^0 \implies BCT\text{-}II)$ Assume that $\mathbb{N}^{\mathbb{N}}$ satisfies the Baire Category Theorem for uniformly $\mathit{\Pi}_1^0$ sequences of dense open sets $(BCT\text{-}\mathit{\Pi}_1^0)$.

Let $\langle \langle x_{n,k} : k \in \mathbb{N} \rangle : n \in \mathbb{N} \rangle$ be a sequence of (codes for the complements of) dense separably open subsets O_n of $\widehat{A}$ and let U be a (code for a) nonempty open set in $\widehat{A}$. We must show that there is a point $x \in \widehat{A}$ such that $x \in U$ and $x \in O_n$ for all $n \in \mathbb{N}$. As indicated above, this is equivalent to our showing that for each n, $x \notin \overline{\{x_{n,k} : k \in \mathbb{N}\}}$.

Consider O_n; we claim that $\pi^{-1}(O_n)$ is a $\mathit{\Pi}_1^0$ open dense subset of $\mathbb{N}^{\mathbb{N}}$. By Lemma 3.4, $\pi^{-1}(O_n)$ is dense. It has a $\mathit{\Pi}_1^0$ presentation as follows. Let $\langle \sigma_i : i \in \mathbb{N} \rangle$ be a recursive enumeration of $\mathbb{N}^{<\mathbb{N}}$ (*i.e.*, of the basic open subsets of $\mathbb{N}^{\mathbb{N}}$). Then, $\pi_0(\sigma_i)$ is contained in the complement of $\overline{\{x_{n,k} : k \in \mathbb{N}\}}$ if and only if for every k, $x_{n,k}$ is not an element of $B(a(\sigma_i), r(\sigma_i))$. This property of σ_i is $\mathit{\Pi}_1^0$ relative to the sequence $\langle x_{n,k} : k \in \mathbb{N} \rangle$. Thus, the sequence $\langle B(\sigma_i, 1/2^{lh(\sigma_i)}) : i \in \mathbb{N} \rangle$ and the formula stating that $B(a(\sigma_i), r(\sigma_i)) \cap \overline{\{x_{n,k} : k \in \mathbb{N}\}}$ is empty is a $\mathit{\Pi}_1^0$ representation of $\pi^{-1}(O_n)$.

This is a uniform representation of $\pi^{-1}(O_n)$ and so $\langle \pi^{-1}(O_n) : n \in \mathbb{N} \rangle$ is uniformly $\mathit{\Pi}_1^0$. We may apply $BCT\text{-}\mathit{\Pi}_1^0$ to obtain an x such that $x \in \pi^{-1}(U)$ and $x \in \pi^{-1}(O_n)$ for all $n \in \mathbb{N}$. Then $\pi(x)$ is in U and in the intersection of the O_n, as required.

$(BCT\text{-}II \implies BCT\text{-}\mathit{\Pi}_1^0)$ Assume that $\mathbb{N}^{\mathbb{N}}$ satisfies $BCT\text{-}II$. Fix a uniformly $\mathit{\Pi}_1^0$ sequence, $\langle O_n : n \in \mathbb{N} \rangle$, of dense open subsets of $\mathbb{N}^{\mathbb{N}}$. Let U be a (code for a) nonempty open subset of $\mathbb{N}^{\mathbb{N}}$. We must show that there is a point $x \in \mathbb{N}^{\mathbb{N}}$ such that $x \in U$ and $x \in O_n$ for all $n \in \mathbb{N}$.

Consider O_n; we claim that O_n has a dense separably open subset O_n^*, whose code is obtained uniformly recursively from the $\mathit{\Pi}_1^0$ presentation of O_n. Note that we will not show that O_n is itself separably open.

Let $\langle B(\sigma_{n,j}, 1/2^{lh(\sigma_{n,j})}) : j \in \mathbb{N}\rangle$ and $(\forall t)\varphi_n$ be the components of a $\mathbf{\Pi}_1^0$ presentation of O_n. Let $\langle B(\sigma_i^*, 1/2^{lh(\sigma_i^*)}) : i \in \mathbb{N}\rangle$ be a recursive enumeration of all the basic open sets in $\mathbb{N}^{\mathbb{N}}$. We define a sequence $\langle x_k : k \in \mathbb{N}\rangle$ by recursion as follows. At stage s, we say that u *requires attention* if u is less than or equal to s, for all $t < s$, $x_t \notin B(\sigma_u^*, 1/2^{lh(\sigma_u^*)})$ and $B(\sigma_u^*, 1/2^{lh(\sigma_u^*)})$ is not a subset of $\bigcup\{B(\sigma_{n,v}, 1/2^{lh(\sigma_{n,v})}) : v \le u \,\&\, (\forall t \le s)\varphi_n(t, v)\}$.

Note that in $\mathbb{N}^{\mathbb{N}}$ one open ball is contained in the union of finitely many others is a recursive condition. Thus, whether u requires attention during stage s is uniformly recursive in the parameters of the $\mathbf{\Pi}_1^0$ presentation of O_n. Further, if u does require attention during stage s then σ_u^* is not in $\bigcup\{B(\sigma_{n,v}, 1/2^{lh(\sigma_{n,v})}) : v \le u \,\&\, (\forall t \le s)\varphi_n(t, v)\}$.

Our action during stage s, is as follows. If there is a number which requires attention then let u be the least such (u is well defined by $I\Sigma_1^0$) and set x_s equal to σ_u^*. Otherwise, let x_s be σ_0^*. (Here we view $\mathbb{N}^{<\mathbb{N}}$ as a subspace of $\mathbb{N}^{\mathbb{N}}$.) The sequence $\langle x_s : s \in \mathbb{N}\rangle$ is uniformly recursive in the parameters of the $\mathbf{\Pi}_1^0$ representation of O_n and hence well defined within RCA_0.

If u requires attention during stage s then either some number smaller than u requires attention or x_s is an element of $B(a_u^*, r_u^*)$ and u never requires attention during any later stage. Consequently, if u requires attention during stage s then the set of stages within $[s, s+u]$ during which a number smaller than u requires attention is a Δ_1^0 subset of $[s, s+u]$ with at most u many elements. By $I\Sigma_1^0$ (in particular, the pigeon hole principle for Δ_1^0 functions), there is a stage s^* in $[s, s+u]$ such that no number less than u requires attention. By construction, either u does not require attention during stage s^* because there is a t less than s^* such that x_t is in $B(\sigma_u^*, 1/2^{lh(\sigma_u^*)})$ or x_{s^*} is chosen to be an element of $B(\sigma_u^*, 1/2^{lh(\sigma_u^*)})$. In either case, if u ever requires attention then there is an x_{s^*} such that $x_{s^*} \in B(a_u^*, r_u^*)$.

Let O_n^* be complement of $\overline{\{x_s : s \in \mathbb{N}\}}$.

First, we show that $O_n^* \subseteq O_n$. Suppose that x is an element of O_n^*. Then there is a basic open set, say $B(\sigma_j^*, 1/2^{lh(\sigma_j^*)})$, such that $B(\sigma_j^*, 1/2^{lh(\sigma_j^*)})$ is a subset of O_n^* and $x \in B(\sigma_j^*, 1/2^{lh(\sigma_j^*)})$. Since u's ever requiring attention is a Σ_1^0 property, we may apply $I\Sigma_1^0$ to conclude that there is a stage, call it s_1, such that, for every subsequent stage t, if u requires attention during stage t then u is greater than or equal to j. Similarly, there is a stage, call it s_2, such that for every larger stage s,

$$\{B(\sigma_v^*, 1/2^{lh(\sigma_v^*)}) : v \le u \,\&\, (\forall t \le s)\varphi_n(t, v)\} =$$
$$\{B(\sigma_v^*, 1/2^{lh(\sigma_v^*)}) : v \le u \,\&\, (\forall t)\varphi_n(t, v)\}.$$

If $B(\sigma_j^*, 1/2^{lh(\sigma_j^*)}) \not\subseteq \bigcup\{B(\sigma_v^*, 1/2^{lh(\sigma_v^*)}) : v \le u \,\&\, (\forall t)\varphi_n(t, v)\}$, s is a stage greater than the maximum of s_1 and s_2 and there is no t less than s such that x_t is an element of $B(\sigma_j^*, 1/2^{lh(\sigma_j^*)})$ then j will require attention during stage s. But then there will be a t such that $x_t \in B(\sigma_j^*, 1/2^{lh(\sigma_j^*)})$, a contradiction to our choice of j. Consequently, $B(a_j^*, r_j^*)$ must be a subset of $\bigcup\{B(a_v, r_v) : v \leq u \ \& \ (\forall t)\varphi_n(t, v)\}$ and so x is an element of this union as well.

Second, we show that O_n^* is dense. Let U^* be an open set. Since O_n is dense, there is an x in $O_n \cap U^*$. But then, there must be a $B(\sigma_j, 1/2^{lh(\sigma_j)})$ such that $x \in B(\sigma_j, 1/2^{lh(\sigma_j)}) \cap U^*$. By construction, for all s greater than j, $x_s \notin B(\sigma_j, 1/2^{lh(\sigma_j)})$. Thus, the set $\{x_s : s \in \mathbb{N} \ \& \ x_s \in B(\sigma_j, 1/2^{lh(\sigma_j)})\}$ is finite. Thus, the intersection of $\overline{\{x_s : s \in \mathbb{N}\}}$ with $B(\sigma_j, 1/2^{lh(\sigma_j)})$ is also finite. Since $B(\sigma_j, 1/2^{lh(\sigma_j)}) \cap U^*$ is nonempty and open, it is not finite. Thus, U^* has nonempty intersection with O_n^*, which is the complement of $\overline{\{x_s : s \in \mathbb{N}\}}$, as required.

We may now apply $BCT\text{-}II$ to U and the sequence $\langle O_n^* : n \in \mathbb{N}\rangle$ to obtain an element x of $U \cap \bigcap_{n \in \mathbb{N}} O_n^*$. Then, x is an element of $U \cap \bigcap_{n \in \mathbb{N}} O_n$, as required by $BCT\text{-}\boldsymbol{\Pi}_1^0$. ∎

Henceforth, we will speak of $BCT\text{-}\boldsymbol{\Pi}_1^0$ rather than $BCT\text{-}II$.

4   A model of $RCA_0 + BCT\text{-}\boldsymbol{\Pi}_1^0$

Definition 4.1 A subset G of $\mathbb{N}$ is *low* if the complete Σ_1^0 predicate relative to G is Δ_2^0 (equivalently, G' has the same Turing degree as $0'$).

Proposition 4.2 *There is an ω-model $\mathfrak{M}$ of $RCA_0 + BCT\text{-}\boldsymbol{\Pi}_1^0$ in which every real is low.*

Proof: We begin with the standard model of arithmetic and construct $\mathfrak{M}$ by iteratively adding reals to meet uniformly $\boldsymbol{\Pi}_1^0$ sequences of dense open sets and closing under relative computability.

Our induction step goes as follows. At step s, we will have added all of the sets recursive in G_s. We view $\mathbb{N}^{<\mathbb{N}}$ as the (Cohen) partial order for adding an element H_{s+1} to $\mathbb{N}^{\mathbb{N}}$ using finite conditions, ordered by extension.

Let B be a basic open subset of $\mathbb{N}^{\mathbb{N}}$ and let $\langle O_{s,n} : n \in \mathbb{N}\rangle$ be a uniformly $\Pi_1^0(G_s)$ sequence of dense open subsets of $\mathbb{N}^{\mathbb{N}}$.

We also let $\langle D_n : n \in \mathbb{N}\rangle$ be the sequence of dense open subsets of $\mathbb{N}^{\mathbb{N}}$ deciding the $\Sigma_1^0(G_s)$ theory of H_{s+1}. Explicitly, if $(\exists t)\theta_n(t, G_s \restriction t, H_{s+1} \restriction t)$ is the nth Σ_1^0 formula relative to G_s and H_{s+1} then D_n is equal to the union of the collection of $B(\sigma, 1/2^{lh(\sigma)})$ such that either $\theta(lh(\sigma), G_s \restriction lh(\sigma), \sigma)$ or for all τ, if τ is compatible with σ then $\neg\theta(lh(\tau), G_s \restriction lh(\tau), \tau)$. By meeting the set D_n, we can ensure the $\Pi_1^0(G_s, H_{s+1})$ condition $(\forall t)\neg\theta_n(t, G_s \restriction t, H_{s+1} \restriction t)$ is $\Sigma_1^0(G_s' \oplus H_{s+1})$. Thus, by meeting all the sets D_n, we can ensure that $(G_s \oplus H_{s+1})'$ is recursive in $G_s' \oplus H_{s+1}$. A set in the intersection of $\{D_n : n \in \mathbb{N}\}$ is said to be *1-generic relative to G_s*.

We build H_{s+1} to be in B, generic with respect to meeting each $O_{s,n}$ and also 1-generic relative to G_s. We let G_{s+1} be the recursive join of G_s with

H_{s+1}. The steps of meeting B, of meeting a basic open subset of $O_{s,n}$ or of deciding the next Σ_1^0 statement about $G_s \oplus H_{s+1}$ are uniformly recursive in G_s' and hence, by induction, uniformly recursive in $0'$. Thus, G_{s+1}' is recursive in $0'$.

By fixing a reasonable pairing function and considering the eth index for uniformly $\Pi_1^0(G_s)$ sequence of open sets during stage $\langle e, s \rangle$, we may ensure that every $\langle O_n : n \in \mathbb{N} \rangle$ which is uniformly Π_1^0 relative to some parameter in $\{G : (\exists s)(G \leq_T G_s)\}$ appears as $\langle O_{s,n} : n \in \mathbb{N} \rangle$ during some stage s.

Let $\mathfrak{M}$ be the ω-model whose reals are the sets which are recursive in some G_s. Clearly, $\mathfrak{M}$ is a model of RCA_0; we claim that it is also a model of BCT-Π_1^0. Let $\langle O_n : n \in \mathbb{N} \rangle$ be a uniformly Π_1^0 sequence of dense open subsets of $\mathbb{N}^{\mathbb{N}}$, let U be an open subset of $\mathbb{N}^{\mathbb{N}}$ and let B be a basic open subset of U. By construction, there is an s such that H_{s+1} is an element of the intersection of B with all of the O_n. But, then H_{s+1} is an element of U and for each $n \in \mathbb{N}$, of O_n. Thus, the conclusion of BCT-Π_1^0 is satisfied. ∎

Corollary 4.3 *There is a model of $RCA_0 + BCT$-Π_1^0 which is not a model of RCA_0^+.*

Proof: Let $\mathfrak{M}$ be the model from Proposition 4.2.

Fixing an arithmetic enumeration, let X_n be the nth Δ_2^0 element of $2^{\mathbb{N}}$, let O_n be the dense open subset of $2^{\mathbb{N}}$ given by $\{\sigma : (\exists m \leq lh(\sigma))[\sigma(m) \neq X_n(m)]\}$. Then $\langle O_n : n \in \mathbb{N} \rangle$ is an arithmetic sequence of dense open subsets of $2^{\mathbb{N}}$ which has no Δ_2^0 element in its intersection.

Every element of $\mathfrak{M}$ is low and hence Δ_2^0 so there is no element in the intersection of $\{D_n : n \in \mathbb{N}\}$ in $\mathfrak{M}$. Consequently, $\mathfrak{M}$ is not a model of RCA_0^+. ∎

5   Ramsey's Theorem

We begin by observing a second corollary to Proposition 4.2.

Corollary 5.1 $RCA_0 + BCT$-$\Pi_1^0 \nvdash \mathbb{N} \to [\mathbb{N}]_2^2$.

Proof: Corollary 5.1 follows from Proposition 4.2 and a theorem of (Jockusch 1972). Jockusch showed that there is a recursive partition of pairs for which there is no infinite homogeneous set which is recursive in $0'$. Let $\mathfrak{M}$ be the ω-model found in Proposition 4.2. Then, $\mathfrak{M}$ is a model of $RCA_0 + BCT$-Π_1^0 and every real in $\mathfrak{M}$ is low, hence recursive in $0'$. But, then there is a partition of pairs in $\mathfrak{M}$ (the recursive one provided by Jockusch) which has no infinite homogeneous set in $\mathfrak{M}$. ∎

Proposition 5.2 $RCA_0 + \mathbb{N} \to [\mathbb{N}]_{<\mathbb{N}}^2 \vdash I\Sigma_2^0$.

Proof: Let $\mathfrak{M}$ be a model of RCA_0 and suppose that $I\Sigma_2^0$ fails in $\mathfrak{M}$.

By Kirby and Paris (1977), we may fix a and a Π_1^0 formula Φ, with two free variables, such that there is a proper cut I contained in the numbers less than a for which Φ defines a strictly increasing function f by $f(b) = y$ if and only if $\Phi(b,y)$. Suppose that Φ has the form $(\forall t)\varphi$, where φ has no unbounded quantifiers.

We define a function F from M^2 to $\{0, \ldots, a\}$, where M denotes the numbers in $\mathfrak{M}$. We use the function f to determine the partition; we ensure that for each x, for all but boundedly many s, $F(x,s)$ is equal to the least b such that $f(b) > x$. For each s and x less than s, let $F(x,s)$ be the least number b less than a such that $(\forall y < x)(\exists t < s)\neg\varphi(b,y)$, if there is such a b; let $F(x,s)$ be equal to a, otherwise.

Fixing x, consider the eventual behavior of $F(x,s)$, as s increases. By $B\Sigma_1^0$ in $\mathfrak{M}$, if b is such that $(\forall y < x)(\exists t)\neg\varphi(b,y)$ then there is an s such that $(\forall y < x)(\exists t < s)\neg\varphi(b,y)$. By $I\Sigma_1^0$, if there is any b with this property then there is a least such. Of course, there is such a b by the choice of f. Thus, F has the desired property: $F(x,s)$ is eventually equal to the least b such that $f(b)$ is greater than x.

But now, for each x there are only boundedly many y's such that the set $\{s : F(x,s) = F(y,s)\}$ is unbounded. Thus, there is no unbounded homogeneous set for F. Consequently, $\mathfrak{M}$ is not a model of $\mathbb{N} \to [\mathbb{N}]^2_{<\mathbb{N}}$. ∎

Proposition 5.3 $RCA_0 + \mathbb{N} \to [\mathbb{N}]^2_2 + I\Sigma_2^0 \vdash BCT\text{-}\Pi_1^0$.

Proof: We will use $\mathbb{N}$ to denote the numbers and $\mathbb{N}^\mathbb{N}$ to denote Baire space. However, our proof will be completely within the system $RCA_0 + \mathbb{N} \to [\mathbb{N}]^2_2 + I\Sigma_2^0$.

Let $\langle O_n : n \in \mathbb{N} \rangle$ be a uniformly Π_1^0 sequence of dense open subsets of $\mathbb{N}^\mathbb{N}$. To fix some notation, let O_n be the union of the set $\{B(a_{n,i}, r_{n,i}) : \Phi(n,i)\}$, where Φ is the Π_1^0 formula $(\forall x)\varphi(n,i,x)$ and φ is Π_0^0.

Recursively in the parameters defining $\langle O_n : n \in \mathbb{N} \rangle$, we construct the partition $F : [\mathbb{N}]^2 \to \{0,1\}$ and the functional Γ so that for any infinite set H, if H is F-homogeneous then $\Gamma(H)$ is an element of $\bigcap\{O_n : n \in \mathbb{N}\}$.

We build F and enumerate Γ in the context of a Σ_2 priority construction. Specifically, each of our strategies will act only finitely often but the number of actions it may take will depend on the stage at which all of the strategies of higher priority reach their limit state. Without having a recursive bound on the number of actions of each strategy, we have not been able to prove that all of the requirements are satisfied without invoking $I\Sigma_2^0$.

We will proceed by recursion on stages s. During stage s, we will enumerate computations $\langle x, y, P, N \rangle$ into Γ; here, $\langle x, y, P, N \rangle \in \Gamma$ indicates that if $X \subseteq \mathbb{N}$, $P \subseteq X$ and $N \cap X = \emptyset$ then $\Gamma(x, X) = y$. We will label each number less than s with either 0 or 1. We will define $F(x, s)$ for each x less than s so that $F(x, s)$ is equal to i if and only if x is labeled by i at the end of stage s. We shall satisfy the following requirements.

- For each x, there is a stage after which the label of x does not change. Let $\mathcal{L}(x)$ denote the particular requirement on the label of x.

- For each O_n, there is a t such that for each σ, if σ is F-homogeneous with value i and each element of σ is eventually labeled i and σ has an element greater than t then $B(\Gamma(\sigma), 1/2^{lh(\Gamma(\sigma))})$ is contained in O_n. In short, the amount of Γ which can be computed relative to the finite amount of information in σ is enough to ensure being an element of O_n. Let $\mathcal{D}(n, i)$ denote the particular requirement on O_n and i.

Our strategies will take two types of action: enumerating computations into Γ and setting the labels of various numbers. When a strategy sets the label of x then that label can only be set to a new value by a strategy of greater than or equal priority.

During stage s, we will approximate whether a finite set H_0 has an infinite homogeneous extension by the examining whether H_0 is homogeneous with value i and each element of H_0 is labeled i. The first requirement is that this approximation is eventually correct. By satisfying this requirement, if H_0 is F homogeneous with value i then we can also ensure that H_0 has no infinite F homogeneous extension by labeling at least one element of H_0 with $1 - i$.

Ensuring that labels reach a limit. The strategy $L(x)$ to ensure the satisfaction of $\mathcal{L}(x)$ operates as follows. If no strategy of higher priority than $L(x)$ has set the label of x then $L(x)$ defines the label of x to be 0. Thus, $L(x)$ fixes the label of x except for the action of strategies of higher priority than $L(x)$. Since the strategies of higher priority will only change the label of x finitely often, this simple form of $L(x)$ is sufficient to ensure that $\mathcal{L}(x)$ is satisfied.

Ensuring that $\Gamma(H) \in O_n$. The strategy $D(n, i)$ to ensure the satisfaction of $\mathcal{D}(n, i)$ operates as follows. Suppose that s_0 is the last stage during which $D(n, i)$ is injured (by the action of some $D(m, j)$ of higher priority during stage s_0).

$D(n, i)$ must respect the strategies of higher priority: to respect $L(x)$ for $x < n$, $D(n, i)$ will not change the label of any number less than n; to respect $D(m, j)$ for $m < n$ or $m = n$ and $j < i$, $D(n, i)$ will not change the label of any number less than s_0 and will only enumerate computations into Γ which are applied to sets which are F homogeneous with value i and are compatible with those enumerated by $D(m, i)$.

Let $\mathcal{H}$ be the collection of finite subsets H_0 of $[0, s_0)$ such that H_0 is F homogeneous with value i. For each H_0 in $\mathcal{H}$, let $\Gamma^*(H_0)[s_0]$ be the finite binary string determined by the computations in Γ during stage s_0 relative to H_0 and those computations that are being enumerated on the one point extensions of H_0 by the strategies, $D(m, i)$, for $m < n$. During stage s greater than or equal to s_0, $D(n, i)$ acts as follows.

> For each H_0 in $\mathcal{H}$, let $i(H_0)[s]$ be the least i such that for all t less than or equal to s, $\varphi(n, i, t)$ and $\Gamma^*(H_0)[s_0]$ can be extended to an element σ of $B(a_{n,i}, r_{n,i})$. That is, $i(H_0)[s]$ is the least i which during stage s appears to be the index for a basic open subset of O_n containing an extension of $\Gamma^*(H_0)[s_0]$.

> 1. If there is an H_0 in $\mathcal{H}$ such that $i(H_0)[s]$ is not equal to $i(H_0)[s-1]$ then label every number in $[s_0, s)$ with $1 - i$. This action by $D(n, i)$ injures every strategy of lower priority. *By changing the labels of these numbers, $D(n, i)$ ensures that none of the computations enumerated during stages in the interval $[s_0, s)$ are relevant to an infinite extension of H_0 which is F homogeneous with value i.*

> 2. For each m less than or equal to the length of σ, enumerate $\langle m, \sigma(m), H_0 \cup \{s-1\}, [0, s-1) - H_0 \rangle$ into Γ. *By enumerating these computations into Γ, we ensure that if H is an infinite extension of H_0 then $\Gamma(H)$ is an element of $B(a_{n,i}, r_{n,i})$.*

Since O_n is dense, for each H_0 in $\mathcal{H}$ there is an i such that some extension of $\Gamma^*(H_0)[s_0]$ is in $B(a_{n,i}, r_{n,i})$. By $I\Pi_1$ (an equivalent form of $I\Sigma_1$), there is a least such i. By $B\Sigma_1$, for each H_0 in $\mathcal{H}$, $D(n, i)$ can only execute Step 1 for the sake of H_0 finitely often. By $B\Sigma_2$, there is a stage s_1 such that $D(n, i)$ does not execute Step 1 after stage s_1. In other words, $D(n, i)$ only injures the strategies of lower priority finitely often.

Suppose that H is an infinite set which is F-homogeneous with value i. Let H_0 be the restriction of H to $[0, s_0)$. By $I\Sigma_1$, H_0 is an element of $\mathcal{H}$. Let s be the least element of H which is greater than or equal to s_0. During stage s, $D(n, i)$ enumerates a computation into Γ so that $\Gamma(H)$ is an element of $B(a_{n,i(H_0)[s]}, r_{n,i(H_0)[s]})$. Since H is F-homogeneous with value i, $D(n, i)$ never set the label of s equal to $1 - i$. Thus, $B(a_{n,i(H_0)[s]}, r_{n,i(H_0)[s]})$ is contained in O_n and so $\Gamma(H) \in O_n$.

We omit the description of the construction of F, except to say that it is an application of the Σ_2-priority method similar to the proof of the Sacks splitting theorem. We note that combinatorial arguments behind constructions of this sort can be implemented within the theory $P^- + I\Sigma_2$.

brief218 *Michael E. Mytilinaios and Theodore A. Slaman*

Now apply $\mathbb{N} \to [\mathbb{N}]_2^2$ to find an H such that H is F-homogeneous. By the above remarks, $\Gamma(H)$ is an element of $\bigcap_{n \in \mathbb{N}} O_n$, as required to verify $BCT\text{-}\Pi_1^0$. ∎

References

Brown, D. K. (1987). *Functional analysis in weak subsystems of second order arithmetic.* Ph. D. thesis, The Pennsylvania State University, University Park, PA, USA.

Brown, D. K. and S. G. Simpson (1993). The Baire category theorem in weak subsystems of second-order arithmetic. *Journal of Symbolic Logic 58*(2), 557–578.

Jockusch, Jr., C. G. (1972). Ramsey's theorem and recursion theory. *Journal of Symbolic Logic 37*, 268–280.

Kirby, L. A. and J. B. Paris (1977). Initial segments of models of Peano's axioms. In *Set theory and hierarchy theory V (Bierutowice, Poland, 1976)*, Berlin, pp. 211–226. Lecture Notes in Mathematics, vol. 619, Springer-Verlag.

Ramsey, F. P. (1930). On a problem in formal logic. *Proceedings of the London Mathematical Society 30*, 264–286.

Seetapun, D. and T. A. Slaman (1994). On Ramsey's theorem for pairs. unpublished.

RELATIVIZATION OF STRUCTURES
ARISING FROM COMPUTABILITY THEORY

ANDRÉ NIES[1]

Cornell University

ABSTRACT: We describe a general method to separate relativizations of structures arising from computability theory. The method is applied to the lattice of r.e. sets, and the partial orders of r.e. m-degrees and T-degrees. We also consider classes of oracles where all relativizations are elementarily equivalent. We hope that the paper can serve as well as an introduction to coding in these structures.

1. Introduction. The relativization of a concept from computability theory to an oracle set Z is obtained by expanding the underlying concept of computation in a way such that, at any step of the computation procedure, tests of the form "$n \in Z$", where n is some number obtained previously in the computation, are allowed. For instance, the relativization of the concept of r.e. sets to Z is "set r.e. in Z". In this paper, we study to what extent the isomorphism type and the theory of the relativization A^Z of a structure A from computability theory depend on the oracle set Z. We consider mainly the case that A is the structure $\mathcal{E}$ of r.e. sets under inclusion or a degree structure on r.e. sets, but first discuss the case that A is the structure of $\mathcal{D}_T$ all T-degrees or $\mathcal{D}_m$ of all m-degrees. In this case, $\mathcal{D}_m^Z$ is the structure of degrees of subsets of ω under many-one reductions via (total) functions recursive in Z, while $\mathcal{D}_T^Z$ is simply the upper cone of $\mathcal{D}_T$ above the T-degree of Z.

It is a common phenomenon in computability theory that the proof of a result is actually a proof of all relativized forms of the result. Thus, the proof that there is a minimal T-degree below $0''$ actually shows that each degree $\mathbf{z}$ has a minimal cover below $\mathbf{z}''$, and the construction of a maximal r.e. set actually gives an index i such that (W_i^Z) is a coatom in $(\mathcal{E}^Z)^*$.

This observation led to the "strong homogeneity conjecture" [Rogers 67] that, for each Z, $\mathcal{D}_T^Z \cong \mathcal{D}_T$. Yates [Ya 70] speculated, based on results of

---

[1]Research partially funded by Mathematical Sciences Institute.

Martin, that the conjecture and also its weaker form asserting that $\mathcal{D}_T^Z$ is elementarily equivalent to $\mathcal{D}_T$ for each Z is independent of ZFC. Even the weaker form of the conjecture was refuted by Shore [Sh 82]: if $\mathcal{D}_T \equiv \mathcal{D}_T^Z$, then Z must be of arithmetical degree. Here already some of the ideas occur which will be exploited in the present paper.

Surprisingly, the analog of the homogeneity conjecture holds for $\mathcal{D}_m$. Ershov [Er 75], with an addendum by Paliutin gave a characterization of $\mathcal{D}_m$ which is purely algebraic: $\mathcal{D}_m$ is the only distributive upper semilattice with 0 that has cardinality 2^ω, the countable predecessor property and a certain extension property for ideals of cardinality $< 2^\omega$. Relativizations of the proofs that these properties hold give exactly the same properties for $\mathcal{D}_m^Z$, so $\mathcal{D}_m^Z \cong \mathcal{D}_m$.

There are several reasons to study relativizations of structures. One is that, as mentioned above, relativized versions of results are often already implicitly obtained. Moreover, in some cases the relativized structures arise naturally in some other way. For instance, if $Z = \emptyset^{(n-1)}$, then $\mathcal{E}^Z$ is the lattice of Σ_n^0–sets, and for any Z, if $\mathbf{z} = \deg_T(Z)$, the relativization of the Δ_2^0–Turing degrees to Z is the interval $[\mathbf{z}, \mathbf{z}']$.

The way to prove $A^Z \not\cong A^W$ if Z, W are sufficiently different oracle sets is to show that, to some accuracy, the complexity of the oracle set X can be recovered from the isomorphism type of A^X. To make this precise, we need the notion of (uniform) coding of *extended standard models of arithmetic* (extended SMA). An extended SMA is a structure (M, U), where $M \cong \mathbb{N}$ and $U \subseteq M$. In general, a coding with parameters of a relational structure C of finite signature in a structure $\mathcal{D}$ is given by a scheme S of formulas $\varphi_S(x, \bar{p})$ and $\varphi_R(x_1, \dots, x_n; \bar{p})$ for each n–ary relation symbol R in the language of C (including equality) such that, for an appropriate list $\bar{d}$ of parameters in D, $\varphi_=$ defines an equivalence relation on $\{x : D \models \varphi_S(x, \bar{d})\}$ and the structure defined on equivalence classes by the remaining formulas φ_R is isomorphic to C.

¿From now on, we focus on arithmetical structures A of finite signature. Such a structure is determined by a scheme of arithmetical formulas without parameters, which gives a representation of A in terms of natural numbers ("indices"). For instance, the scheme for $\mathcal{E}$ contains a Π_2^0 formula defining $\{\langle i, j \rangle : W_i \subseteq W_j\}$. Suppose the ground level Δ_1^0 of the arithmetical hierarchy is defined in terms of the Kleene T–predicate. Then we obtain relativizations of each arithmetical formula to an "oracle predicate" Z by replacing the computations the definition of T is based on by oracle computations.

In the terminology of Hodges [Ho 94], there is an interpretation Γ of structures in the language of A in the extended SMA $(\mathbb{N}, Z)$ and A^Z can be defined as $\Gamma(\mathbb{N}, Z)$. We call the least number r such that each arithmetical formula needed in defining A is a boolean combination of $\Sigma_r -$ formulas the *arithmetical complexity* of A. This complexity is 2 for $\mathcal{E}$, 3 for $\mathcal{E}^*$ and $\mathcal{R}_m$, and 4 for $\mathcal{R}_T$.

Note that, for each Z, there is a representation of the diagram of A^Z which is recursive in $Z^{(r)}$.

Now suppose that in the converse direction there is a coding scheme S for coding the extended SMA (M, X) in A^X with parameters. This coding condition, which is satisfied e.g. by $\mathcal{R}_m$, $\mathcal{R}_T$ and $\mathcal{E}$ (see below), is a crude form of expressing that the complexity of the oracle X is reflected in A^X, the isomorphism type of A^X. We abbreviate the coding condition by CO ("coded oracle").

We will always assume that, if (M, X) is coded by a certain list of parameters $\bar{p}$, M is a model of a finitely axiomatized fragment PA^- of Peano arithmetic (say Robinson arithmetic R) which implies M is an end extension of $\mathbb{N}$. This can be expressed by a first order condition on $\bar{p}$.

Whenever an extended SMA (M, V) is coded in A^X, then by combining this coding with the coding of A^X in $(\mathbb{N}, X)$ we obtain that V is $\Sigma_d^0(X)$ for some natural number d. Thus, if A satisfies CO, then $A^Z \cong A^W$ implies that Z, W have the same arithmetical degree. If we can in addition recognize standardness of coded models M by a first–order condition on parameters (call this coding condition $\mathrm{CO}_{\mathrm{st}}$), then we obtain an elementary difference between A and A^Z, for $Z \notin \Sigma_d^0$: the first–order sentence expressing

$(*)$ "Whenever (M, V) is a coded extended SMA, then V

(as a subset of M) is Σ_d^0 "

holds in A, but not in A^Z.

In Section 2, we use a still stronger coding condition $\mathrm{CO}_{\mathrm{st}}(k)$, which depends on $k \geq 1$, to refine these separations of isomorphism types and of theories. (A somewhat similar idea was used first in [Sh 81] for the special case $A = \mathcal{D}_T(\leq 0')$.) In the central Section 3, we explain why such a coding condition is satisfied for $\mathcal{R}_m$, $\mathcal{R}_T$ and $\mathcal{E}$. For $\mathcal{R}_T$ and $\mathcal{E}$, the full proofs are in [Ha, N ta] and [N, Sh, Sl ta], respectively, and we review them here in survey style.

In Section 4 we discuss "large" classes of oracles where relativizations of A are all elementarily equivalent. Finally, in Section 5, we show that the subset of $\mathrm{Th}\,(A)$ of relativizing sentences is much more complex than $\mathrm{Th}\,(A)$, assuming $\mathrm{CO}_{\mathrm{st}}$. This fact was obtained in collaboration with T. Slaman. It implies that there is no way to give an effective relativizability criterion C such that $\mathrm{Th}\,(A) \cap C$ is the set of relativizable sentences, i.e. the sentences which hold in every relativization of A. In other words, it is not possible to distinguish, say, in $\mathcal{R}_m$ a relativizing sentence like "each incomplete degree has a minimal cover" from a sentence like $(*)$ above (assuming the sentences are true). (For the *particular* sentence $(*)$ it is easy to grasp why it does not relativize: to say that V as a subset of M is Σ^0_d keeps the same meaning in all relativizations.)

2. Separating relativizations. We first list the hierarchy of coding conditions used. In saying A satisfies a certain coding condition, we view A as an interpretation in extended SMA's.

CO   In a uniform way, it is possible to code $(\mathbb{N}, X)$ in A^X.

CO_{st} In the underlying scheme s to code structures M, $M \models PA^-$, in a relativization A^X, one can recognize standardness of M by a fixed first order condition on parameters.

CO(k) $(k \geq 1)$ Suppose the arithmetical complexity of A is r, and let $c = r + k - 1$. The extended SMA $(\mathbb{N}, \overline{X^{(c)}})$ can be uniformly coded in A^X using a scheme of Σ_k-formulas with parameters.

$CO_{st}(k)$ CO(k), and (as in the conditions CO_{st}) standardness can be recognized.

2.1 Separation Theorem. *Suppose A satisfies CO(k). Then, if $Z^{(c)} \not\equiv_T W^{(c)}$, $A^Z \not\cong A^W$, where $c = r + k - 1$ and r is the arithmetical complexity of A.*

Proof: If M is a model of PA^- coded in A^X via the scheme s, then there is an $f \leq_T X^{(c)}$ such that $f(n)$ is an index of n^M in the canonical representation of A^X. For, the successor relation S of M, viewed as a relation on indices, is r.e. in the $(k-1)$-th jump of the atomic diagram of A^X (since the scheme is Σ_k), so it is r.e. in $X^{(c)}$. To compute f inductively, let $f(0)$ be an index of 0^M, and let $f(n+1)$ be an index j such that $Sf(n)j$ holds. Then, since M is an end extension of $\mathbb{N}$, f is total, and f is recursive in $X^{(c)}$.

Now we can obtain an upper bound on the complexity of U, for an extended SMA (M, U) coded in A^X: U is r.e. in $X^{(c)}$ via the enumeration procedure which enumerates n into U iff the Σ_k-formula defining U (with a fixed list of parameters in A^X) holds for $f(n)$.

Suppose $Z^{(c)} \not\leq_T W^{(c)}$. Then $Z^{(r+k)}$ is not r.e. in $W^{(r+k-1)}$. By hypothesis, the extended SMA $(M, Z^{(c)})$ can be coded in A^Z. But if $(M, \overline{Z^{(c)}})$ can be coded in A^W, then $Z^{(r+k)}$ is r.e. in $W^{(c)}$, contradiction. So $A^Z \not\cong A^W$. $\square$

Recall that a set $U \subseteq \mathbb{N}$ is *implicitly definable in arithmetic* (i.d.) if there is a first-order description ψ in the language of extended SMA's such that $(\mathbb{N}, X) \models \psi \Leftrightarrow X = U$. For instance, all recursive jumps $\emptyset^{(\alpha)}$, $\alpha <$

ω_1^{CK}, are i.d. Implicit definability of U only depends on the arithmetical degree of U, and can only hold if U is hyperarithmetical.

2.2 Theorem (Separation Theorem for Elementary Equivalence).

Suppose $\mathrm{CO}_{st}(k)$ *is satisfied for* A. *Let* $c = r + k - 1$. *If* $Z^{(c)} \not\leq_T W^{(c)}$ *and* Z *or* W *is implicitly definable in arithmetic, then* $A^Z \not\equiv A^W$.

Note that this includes the case that $W = \emptyset$ and $Z \notin \mathrm{Low}_c$. Thus for sufficiently complex Z, the theory of the relativization to Z differs from the theory of the unrelativized structure.

Proof: We attempt to express the fact which led to $A^Z \not\equiv A^W$ in the first–order language of A. First suppose that Z is implicitly definable. Then the statement

"Some (M, U) can be coded such that M is standard and there is $e \in M$, where $\{e\}^U$ satisfies the description of Z, such that $\overline{U} = (\{e\}^U)^{(c)}$"

is expressible in that language, holds in A^Z but fails in A^W.

If W is implicitly definable, we distinguish two cases. If $Z^{(c)} \not\geq_T W^{(c)}$, then $A^Z \not\equiv A^W$ by the argument above. Else $Z^{(c)} >_T W^{(c)}$, and there is an index e such that $\{e\}(\overline{Z^{(c)}}) = W$. So the first–order sentence expressing

"there is a coded extended SMA (M, U) and an $e \in M$ such that $\{e\}^U$ satisfies the description of W and $U \notin \Sigma_{c+1}^0(\{e\}^U)$"

is true in A^Z via a coding of $(M, \overline{Z^{(c)}})$, but not in A^W. $\qquad\square$

3. The structures $\mathcal{R}_m$, $\mathcal{E}$ and $\mathcal{R}_T$.

We sketch proofs that $\mathcal{R}_m$ and $\mathcal{E}$ satisfy the condition $\mathrm{CO}_{st}(k)$ used in the Separation Theorem for elementary equivalence. In the cases $\mathcal{R}_m$ and $\mathcal{R}_T$ the coding condition holds with $k = 1$. The full proofs for $\mathcal{R}_T$ and $\mathcal{E}$ are implicit in the results in [N, Sh, Sl ta] and [Ha, N ta], respectively. In all proofs, it is sufficient to consider the unrelativized structure and note the relativizability of the proof techniques used.

3.1 R.e. many–one degrees.

In five steps, we build up a coding scheme of Σ_1-formulas for coding an extended SMA $(M, \overline{X^{(3)}})$ in $\mathcal{R}_m^X$ with parameters. This proves the condition $\mathrm{CO}(1)$, since $r = 3$ for $\mathcal{R}_m$.

We use two auxiliary structures: first a bipartite graph and then a distributive lattice. This makes it necessary to apply a transitive version of coding with Σ_1-formula: as in [N ta1], a relational structure A is Σ_1-e.d.(p) in a structure B if there is a coding scheme of Σ_1 formulas (with parameters) for defining the universe of A, the relations of A *and their complements*.

Step 1. $\mathbb{N}$ is Σ_1–e.d. in a recursive bi–partite graph $G = (Le, Ri, E)$, using the coding given in the proof of Theorem 4.2 in [N ta1]. The class of vertices representing numbers is a recursive Σ_1–definable subset of the left domain Le of G.

Step 2. G is Σ_1–e.d.p. in a recursive distributive lattice L_G, viewed as a p.o. This step is carried out in [N ta1] for finite bipartive graphs, in order to show that the Π_3–theory of the class of finite distributive lattices (as p.o.) is hereditarily undecidable. An obvious modification of the proof yields L_G. For instance, to define a sequence of infinitely many independent elements A_i (representing the left domain of G) in an appropriate recursive distributive lattice L by a quantifier free formulas with one parameter, consider copies $\mathbf{B}_1$, $\mathbf{B}_2$ of the boolean algebra of finite or cofinite subsets of CO, put $\mathbf{B}_2$ on top of $\mathbf{B}_1$ where $P :=$ greatest element of $\mathbf{B}_1 =$ least element of $\mathbf{B}_2$. For each i, insert the new element A_i between the i–th coatom of $\mathbf{B}_1$ and the i–th atom of $\mathbf{B}_2$. In this way, obtain L. Now

$$\{A_i : i \in \omega\} = \{X \in L : X, P \text{ incomparable}\},$$

and no A_i is below a finite supremum $\bigvee_{j \in F} A_i$ for $i \notin F$ (i.e., the elements A_i are independent).

The coding of G in a lattice L_G is obtained by an extension of this: take another copy of L, such that elements $\{\widetilde{A}_j : j \in \omega\}$ represent the right domain, and add further parameters C_E, $C_{\overline{E}}$ such that $Eij \Leftrightarrow C_E \not\leq A_i \vee \widetilde{A}_j$, and similarly for $\overline{E}$ and $C_{\overline{E}}$. In [N ta1] it is described how the further parameters can be introduced without interfering with the Σ_1–definition of the set of elements representing the left and right domain.

Step 3. By a theorem of Lachlan [La 72], $L_G \cong [\mathbf{0}, \mathbf{a}]$ for some $\mathbf{a} \in \mathcal{R}_m$, by an effective map on indices. This gives a scheme S_M with parameters $\overline{p}$ (including the upper bound $\mathbf{a}$) to code SMA's M. Note that, by effectivity, for the particular $\mathbf{a}$ above, there is a uniformly r.e. sequence $(\mathbf{c}_i)$ of m–degrees such that $\mathbf{c}_i$ represents i^M: the sequence $(\mathbf{c}_i)$ is a subsequence of the degrees representing the elements A_i of L_G. Thus, also $(\mathbf{c}_i)$ is an independent sequence.

Step 4. Given a Π_3^0–complete (or in fact, any Π_3^0–) set S, by the Exact Degree Theorem for structures of arithmetical complexity 3 in [N ta2], there is a $\mathbf{b} \in \mathcal{R}_m$ such that $i \in S \Leftrightarrow \mathbf{c}_i \not\leq \mathbf{b}$. Including $\mathbf{b}$ as a parameter, we obtain the desired scheme S in the unrelativized case.

Step 5. Since all the proof techniques used are relativizable, via the same scheme, CO(1) is satisfied: for each X, there is a list of parameters in $\mathcal{R}_m^X$ coding $(\mathbb{N}, \overline{X^{(3)}})$ via S.

To recognize standardness, we argue as in [N 94], where an interpretation of true arithmetic in Th $(\mathcal{R}_m)$ is given. For any model M coded in $\mathcal{R}_m^X$ by the scheme s_M, if M satisfies PA^-, then $\{i : \deg(W_i^X)$ *is a standard number of* $M\}$ is $\Sigma_k^0(X)$ for some fixed k and bounded from above by **a**. By the relativized form of the Definability Lemma in [N 94] we can quantify over such sets in the first order language of $\mathcal{R}_m$ and therefore, we can express that M is standard.

Applying 2.2, we now obtain the following result:

3.1 Theorem. *If* $Z^{(3)} \not\leq_T W^{(3)}$ *and* Z *or* W *is implicitly definable in arithmetic, then* $\mathcal{R}_m^Z \not\equiv \mathcal{R}_m^W$. $\qquad\qquad\square$

3.2 The lattice of r.e. sets.

We review the necessary facts about $\mathcal{E}$ to prove that $\mathcal{E}$ satisfies $CO_{st}(k)$ for some k. As in [Ha ta] and [Ha, N ta], for any r.e. set E, $\mathcal{B}(E)$ is the boolean algebra of r.e. subsets X of E such that $E - X$ is r.e. and $\mathcal{R}(E)$ is the ideal of recursive subsets of E. *The variables* R, S *range over recursive sets.* If $X \in \mathcal{B}(E)$, we write $X \sqsubset E$. An ideal I of $\mathcal{B}(E)$ is k–acceptable if $\mathcal{R}(E) \subseteq I$ and I has a Σ_k^0 index set. I is *acceptable* if it is k–acceptable for some k.

A class $\mathcal{C}$ of subsets of a structure S is *uniformly definable* if, for some formula $\varphi(x; \bar{p})$, $\mathcal{C}$ is the class of sets defined by this formula as $\bar{p}$ varies over tuples of parameters in S. (Sometimes in the literature it is only required that $\mathcal{C}$ be included in such a class, e.g., in [N 94].)

Ideal Definability Lemma [Ha ta]. *For each nonrecursive r.e. set* E *and each* $n \geq 1$, *the class of* $2n + 1$–*acceptable ideals of* $\mathcal{B}(E)$ *is uniformly definable by a formula* φ_{2n+1}.

The formula used for the 3–acceptable ideals is

$$\varphi_3(X; E, C) \equiv X \sqsubset E \wedge (\exists R \subseteq E)[X \subseteq C \cup R]$$

which clearly can only define 3–acceptable ideals. The formula φ_{2n+3} for $2n + 3$–acceptable ideals has an $\exists\forall$ quantifier prefix in front of an instance of φ_{2n+1} with different parameters and therefore only defines Σ_{2n+3}–ideals. More precisely,

$$\varphi_{2n+3}(X; E, \overline{C}, C_n)$$
$$\equiv X \sqsubset E \wedge (\exists R \subseteq E)(\forall S \subseteq E - R)[\varphi_{2n+1}(X \cap S \cap C_n; \overline{C})].$$

The general framework to use induction on k in this way for obtaining uniform definability of objects with Σ_k–index set is adapted from [N 94]. Note that φ_{2n+1} is a Σ_{2n-1}–formula in the language of $\mathcal{E}$, as a lattice with least and greatest element.

In [Ha, N ta], we use the Ideal Definability Lemma to establish the hypothesis $\mathrm{CO}_{st}(k)$ (some k) of the Separation Theorem for elementary equivalence. As for $\mathcal{R}_m$, here we describe the coding process in several steps.

Step 1. If E is an r.e. nonrecursive set, let $(P_k)_{k\in\omega}$ be any u.r.e. partition of E into nonrecursive sets. Such a partition can be obtained by the method of the Friedberg Splitting Theorem (see [So 87]). Modulo some ideal I, the sets P_k will be the elements of the SMA to be coded.

Step 2. For each r.e. nonrecursive set D, one can obtain uniformly in an index of D a maximal ideal $I(D)$ of $\mathcal{B}(D)$ which contains $\mathcal{R}(D)$ and has Δ_4^0 index set. Apply this process to each P_k, and let $I = \{X \sqsubset A : (\forall k)[X \cap P_k \in I(P_k)]\}$. Then $(P_k/I)_{k\in\omega}$ is a uniformly r.e. listing of the atoms in $\mathcal{B}(E)/I$ without repetitions.

Step 3. To be able to code ternary relations corresponding to the arithmetical operations $+$, $\times$ on the atoms of $\mathcal{B}(E)/I$ for an appropriate I, we require that E is hh–simple, where the lattice $\mathcal{L}^*(E)$ is isomorphic to the boolean algebra of finite and cofinite subsets of ω. Then, with the right choice of the ideals $I(P_k)$, the atoms of $\mathcal{L}^*(E)$ can be used to represent 3–tuples of atoms. (This, however, increases the arithmetical complexity of the index set of the ideals $I(P_k)$ and hence of I.) Then any recursive ternary relation on the atoms P_k/I can be defined in terms of three further acceptable ideals.

Thus we obtain a scheme with parameters to code a SMA M in $\mathcal{E}$.

Step 4. To be able to uniformly define subsets S of the standard part of a model of $PA^- \ M$ coded by this scheme which have an arithmetical index set, we first proceed as in the proof of the Separation Theorem: for some fixed c depending only on the coding formulas, there is an $f \leq \emptyset^{(c)}$, such that

$$i^M = W_{f(i)}/I$$

for each $i \in \omega$.

Moreover, since atoms of a boolean algebra are independent, S can be recovered from the ideal of $\mathcal{B}(E)$ it generates: let P, Q range over $\{X \sqsubset E : X/I \text{ atom in } B(E)/I\}$. If P/I is a standard number of M, then

$$P/I \in S \Leftrightarrow P \in I_S,$$

where I_S is the ideal generated by I and those Q such that $Q/I \in S$.

Clearly I_S is acceptable if S has an arithmetical index set (in the sense that $\{Q : Q/I \in S\}$ has one). Then, since the standard part of M is such a set, we can quantify over the possible subsets of M which can be the standard part and thus express that M is standard.

The same first–order condition for recognizing standardness works in every relativization $\mathcal{E}^Z$, since the proof of the Ideal Definability Lemma relativizes.

Step 5. To define extended SMA's of the type required to satisfy $CO_{st}(k)$, note that, using the function $f \leq \emptyset^{(c)}$ obtained in Step 4, if M is standard then, for some sufficiently large odd $d > c$,

$S \subseteq M$ is Σ_d^0 as a subset of $M \Leftrightarrow S$ has Σ_d^0 index set $\Leftrightarrow I_S$ has Σ_d^0 index set.

Then, for d large enough, because of the remarks following the Ideal Definability Lemma, an extended SMA (M, S) can be coded in $\mathcal{E}$ using a scheme of Σ_{d-2}–formulas (in the language of lattices with 0, 1), for any Σ_d^0–set S.

Moreover, by relativizability of the proof techniques, the same scheme can be used to code (M, S) in $\mathcal{E}^X$, if S is $\Sigma_d^0(X)$. We can conclude that $CO_{st}(d-2)$ is satisfied: recall that $r = 2$ for $\mathcal{E}$, so with the value $k = d-2$ there is a scheme of Σ_k–formulas such that each $\Sigma_{r+k}^0(X)$ set and hence $X^{\overline{(c)}}$ is coded, where $c = d - 1$. We have obtained the following.

3.2 Theorem. *[Ha, N ta] For some c, if $Z^{(c)} \not\leq_T W^{(c)}$ and Z or W is implicitly definable, then $\mathcal{E}^Z \not\equiv \mathcal{E}^W$. In particular, if $Z \notin Low_c$ then $\mathcal{E}^Z \not\equiv \mathcal{E}$.*

3.3 R.e. T–degrees, and r.e. m–degrees revisited.

The coding methods developed in [N, Sh, Sl ta] suffice to satisfy $CO_{st}(1)$ for $\mathcal{R}_T$, viewed as an u.s.l. Since $r = 4$ for $(\mathcal{R}_T, \vee)$, this gives the separation of $\text{Th}(\mathcal{R}_T^Z)$ and $\text{Th}(\mathcal{R}_T^W)$ for $Z^{(4)} \not\leq_T W^{(4)}$ if Z or W is implicitly definable.

We now describe a way to give, for both $\mathcal{R}_m$ and $\mathcal{R}_T$, a first–order condition $R(\overline{p})$ on parameters $\overline{p}$ coding an extended SMA (M, U) which, in each relativization $\mathcal{R}_T^X$, holds only if $U^{(3)} \equiv_T X^{(3)}$. Since some parameters will satisfy the condition, this can be interpreted by saying that we can, in a uniform first order way, recover the T–degree of $X^{(3)}$ from $\mathcal{R}_T^X$ and $\mathcal{R}_m^X$. Then, if Z is implicitly definable, there is a formula φ which holds in $\mathcal{R}_m^X$ $(\mathcal{R}_T^X)$ iff $Z^{(3)} \equiv_T X^{(3)}$.

We use that, with a suitable scheme s_M, $\mathcal{R}_m$ and $\mathcal{R}_T$ satisfy the coding condition

"for each M_1, M_2, the isomorphism between the standard parts of M_1, M_2 is uniformly definable",

i.e., there is a formula $\varphi(x, y, \overline{q})$ which, uniformly with parameters defines all these isomorphisms. This coding condition makes it possible to recognize standardness, and to code a SMA in the degree structure without parameters. For $\mathcal{R}_m$, we can use the scheme s_M introduced in 3.1. So it is part of a scheme for defining extended SMA's such that in $\mathcal{R}_m^X$, an extended SMA $(M, \overline{X^{(3)}})$ can be coded with appropriate parameters. Note that, if an extended SMA (M, Z) is coded, we can express that $Z = \overline{U^{(3)}}$ for some U, since any such U must satisfy $U = \{e\}^Z$ for some e, so U is represented within (M, Z). Now consider the property of a parameter list

"$\overline{p}$ codes an extended SMA $(M, \overline{U^{(3)}})$ such that, for each coded extended SMA $(N, \overline{V^3})$, $V^{(3)} \leq_T U^{(3)}$.

By the uniform definability of the isomorphism $h : M \leftrightarrow N$ and the remark above, this property is equivalent to a first–order property $R(\overline{p})$, since we can compare the T–degrees of $V^{(3)}$ and $h(U^{(3)})$ inside N.

It was proved above that an extended SMA $(M, \overline{X^{(3)}})$ can be coded in $\mathcal{R}_m^X$. Now, whenever $(N, \overline{V^3})$ is coded, then $V^3 \leq_T X^3$, because $\overline{V^{(3)}}$ is r.e. in $X^{(3)}$ by the argument used in the proof of the Separation Theorem. So the property $R(\overline{p})$ holds in $\mathcal{R}_m^X$ for any list of parameters coding an extended SMA $(M, \overline{X^{(3)}})$.

In $\mathcal{R}_T$ one can argue similarly to decode the degree of $X^{(4)}$ from $\mathcal{R}_T^X$. For decoding the degree of $X^{(3)}$, we use the fact, proved in [N, Sh, Sl ta] that

for each r.e. nonrecursive A, if $\mathbf{a} = \deg_T(A)$, the extended SMA $(M, \overline{A^{(3)}})$ is coded in the u.s.l. $[0, \mathbf{a}]$ using a fixed scheme of Σ_1–formulas.

Now consider the first–order property $R(\overline{p})$ expressing

"$\overline{p}$ codes a model $(M, \overline{U^{(3)}})$, M standard, such that $U^{(3)}$ is $\leq_T$–maximal with respect to the property that in each u.s.l. $[0, \mathbf{a}]$, $\mathbf{a} \neq \mathbf{0}$, a structure $(M, \overline{V^{(3)}})$ is coded such that $V^{(3)} \equiv_T U^{(3)}$"

If this property holds in $\mathcal{R}_T^X$ for $\overline{p}$, then $\overline{p}$ codes $(M, \overline{U^{(3)}})$, $U^{(3)} \equiv_T X^{(3)}$, since there is a nonzero $\mathbf{a} \in \mathcal{R}_T^X$ such that $\mathbf{a}' = X'$. The following theorem is now almost immediate.

Theorem. *Suppose $A = \mathcal{R}_m$ or $A = \mathcal{R}_T$.*

(i) If $Z^{(3)} \not\equiv_T W^{(3)}$, then $A^Z \not\cong A^W$.

(ii) If Z is implicitly definable in arithmetic, then there is a sentence φ such that, for each W,

$$A^W \models \varphi \Leftrightarrow W^{(3)} \equiv_T Z^{(3)}.$$

In particular, there is φ which holds precisely in the relativizations to Low_3 oracles.

4. Elementarily equivalent relativizations. We consider several results of the form

$(*)$ $\qquad\qquad\qquad Z, W \in \mathcal{C} \Rightarrow A^Z \equiv A^W,$

where C is an in some sense a large class of subsets of ω (reals). It is reasonable to assume that the isomorphism type of A^Z depends only on the T–degree of Z. Then, for any φ in the language of A

$$\{Z : A^Z \models \varphi\}$$

is an arithmetical class of reals closed under $\equiv_T$.

By arithmetic determinacy, $(*)$ holds for a class $C = \{Z : Z \geq_T F\}$, for some real F which is an upper bound for sets encoding winning strategies. This was observed in [Sh 81] for $A = \boldsymbol{D}_T(\leq 0')$.

In the following, we derive $(*)$ for the classes of ω–generic and ω–random sets Z. Recall that Z is ω–generic iff Z is in every comeager arithmetical class of reals, and Z is ω–random iff Z is in every arithmetical class of measure 1. Both classes can be defined in terms of forcing notions — see [Od ta] for the first and [Kau 91] for the second.

4.1 Proposition. *If Z, W are ω–generic, then $A^Z \equiv A^W$. Thus $(*)$ holds for a comeager class.*

Proof: Since $A^Z \models \varphi$ does not depend on finite variations of Z, the following equivalences hold:

$A^Z \models \varphi$ for some ω–generic Z

$\Leftrightarrow (\exists \theta) \subseteq Z \, [\theta \Vdash \text{``} A^X \models \varphi\text{''}]$, i.e., for all ω–generic $X \supset \theta, A^X \models \varphi$

$\Leftrightarrow A^X \models \varphi$ for all ω–generic X.

Note that the class of ω–generic reals G is radically different from the class of implicitly definable reals Z considered in Sections 2 and 3: If Z is arithmetical in G, then Z is an arithmetical set.

4.2 Proposition. *If Z, W are ω–random, then $A^Z \equiv A^W$. Thus $(*)$ holds for a class of measure 1.*

Proof: It follows from Kolmogorov's 0–1 law that each measurable degree–invariant (or even $=^*$–invariant) class of reals has measure 0 or 1. Thus, for Z, W ω–random

$$A^Z \models \varphi \Leftrightarrow \{X : A^X \models \varphi\} \text{ has measure 1}$$
$$\Leftrightarrow A^W \models \varphi.$$

For the rest of this section we assume that A satisfies the coding conditions CO_{st}.

Let G be some ω–generic and R some ω–random set.

4.3 Proposition. $Th(A^G) \equiv_T Th(A^R) \equiv_T \emptyset^{(\omega)}$.

Proof: We can assume $R, G \leq_T \emptyset^{(\omega)}$. By the hypothesis CO_{st}, true arithmetic can be interpreted in both theories. Conversely, to obtain $\emptyset^{(\omega)}$

as an upper bound, first note that, for any X, $\mathrm{Th}(A^X) \leq_T X^{(\omega)}$. But for $X = G$ and $X = R$, $X^{(n)} = X \oplus \emptyset^{(n)}$ where the T-reductions are obtained uniformly in n, by results in [Kur 81] and [Kau 91], respectively, so $X^{(\omega)} \equiv_T X \oplus \emptyset^{(\omega)} \equiv_T \emptyset^{(\omega)}$.

We now show that, assuming CO_{st} the three theories $\mathrm{Th}(A)$, $\mathrm{Th}(A^G)$ and $\mathrm{Th}(A^R)$ are all different. Thus the theory of the unrelativized structure behaves typically neither in the sense of category nor in the sense of measure.

4.4 Theorem. *The theories $Th(A)$, $Th(A^G)$ and $Th(A^R)$ are pairwise distinct.*

Proof: We first prove that the structures are nonisomorphic. If $A \cong A^X$ for $X = G$ or $X = R$, then an extended SMA (M,X) is coded in A. Hence X is Σ_c^0 for some c, which is impossible.

If $A^R \cong A^G$, then an extended SMA (M,R) can be coded in A^G, so R is in $\Sigma_c^0(G)$ for some sufficiently large c, and $R \leq_T \emptyset^{(c+1)} \oplus G$. This is impossible by the following

Fact: If R is $n+1$-random and G is $n+1$-generic, then $R \not\leq_T \emptyset^{(n)} + G$. (See [Kau 91] for definitions of k-random and k-generic. Here it is enough to know that these are arithmetical classes of reals whose intersection is the class of ω-random respectively ω-generic reals.) The proof of this fact is obtained in a straightforward way by adapting Kurtz's proof (Theorem 4.2 in [Kur 81]) that the downward closure of the class of 1-generic degrees has measure 0.

We now obtain the stronger facts that the structures are not elementarily equivalent: for $X = R$, G, A^X, but not A, satisfies

"an extended SMA (M,U) can be coded such that U is not Σ_c^0".

Moreover, A^R, but not A^G, satisfies

"an extended SMA (M,U) can be coded such that U is $c+2$-random."

For the second, use the above fact for $n = c+1$, together with Kautz's result [Kau 91] that $R^{(k)} \equiv_T R + \emptyset^{(k)}$ for $k+1$-random R. $\square$

5. The set of relativizable sentences. Assuming the coding conditions CO_{st} on A as in the preceding section, we investigate the theory

$$T = \bigcap_{X \subseteq \omega} \mathrm{Th}(A^X),$$

i.e., the class of sentences which hold in all relativizations of A. Note that it suffices to take the intersection over all hyperarithmetical X. Both of the facts we prove show that T is complicated in some sense.

5.1 Proposition. *If S is a consistent theory containing T, then $\emptyset^{(\omega)} \leq_m S$.*

Proof: Given $\varphi \in L(+, \times)$, let $F(\varphi)$ be the sentence expressing "φ holds in some coded SMA". Then $\varphi \in \mathrm{Th}(\mathbb{N})$ implies $\widetilde{\varphi} \in T$, so $F(\varphi) \in S$, and $\neg\varphi \in \mathrm{Th}(\mathbb{N})$ implies $F(\neg\varphi) \in S$, so $F(\varphi) \notin S$ since S is consistent. So $\emptyset^{(\omega)} \leq_m S$. $\qquad\square$

5.2 Proposition. T is Π_1^1-complete.

Proof: Since A^X is given as $\Gamma(\mathbb{N}, X)$ for an interpretation Γ, there is a fixed recursive function f such that $\mathrm{Th}(A^X) \leq_m X^{(\omega)}$ via f. Then

$$T = \{\varphi : (\forall X)(\forall Y)[Y = X^{(\omega)} \Rightarrow f(\varphi) \in Y]\}.$$

The matrix of this expression is arithmetical so T is Π_1^1.

To show completeness, we give a reduction of the Π_1^1-complete set

$$\{\psi \in L(+, \times, U) : \text{no extended SMA satisfies } \psi\}.$$

Let $g(\psi)$ be the negation nof the sentence in the language of A expressing "for some extended SMA (M, U), $M \models \psi(U)$".

Then, if no extended SMA satisfies ψ, $g(\psi) \in T$, and if some extended SMA does, then $g(\psi)$ fails in any A^U such that $(\mathbb{N}, U) \models \psi$ holds, so $g(\psi) \notin T$.

References

[Er 75]   Y. Ershov, *The u.s.l. of enumerations of a finite set* (transl), Alg. Log. **14**, 159–175.

[Ha, N ta]   L. Harrington, A. Nies, *Coding in the lattice of r.e. sets* (to appear).

[Ho 94]   W. Hodges, *Model Theory, Encyclopedia of Mathematics and its Applications*, vol. 42, Cambridge University Press, 1994.

[Kau 91]   S. Kautz, *Degrees of random sets*, Ph.D. Thesis, Cornell Unviersity (1991).

[Kur 81]   S. Kurtz, *Randomness and genericity in the degrees of unsolvability*, Ph.D. thesis, Univ. of Illinois at Urbana-Champaign (1981).

[La 72]   A. Lachlan, *R.e. m-degrees*, Alg. Log. (translation) **11** (1972), 186–202.

[N ta1]   A. Nies, *Undecidable fragments of elementary theories*, to appear in Algera Universalis.

[N ta2]   A. Nies, *On a uniformity in degree structure*, to appear in JSL.

[N,Sh,Sl ta]   A. Nies, R. Shore, T. Slaman, *Definability in the r.e. degrees*, to appear.

232 *André Nies*

[Od ta]   P. Odifreddi, *Classical Recursion Theory II*, to appear.
[Ro 67]   H. Rogers, Jr., *Theory of Recursive Functions and Effective Computability*, McGraw–Hill, 1967.
[Sh 81]   R. Shore, *The theory of degrees below* **0′**, J. London Math. Soc. (2) **24** (1981), 1–14.
[Sh 82]   R. Shore, *On homogeneity and definability in the first–order theory of T–degrees*, JSL **47**, 8–16.
[Ya 70]   C.E.M. Yates, *Initial segments of the degrees of unsolvability, part I*, North Holland 1970 (Bar-Hillel, editor), 63-83.

A Hierarchy of Domains With Totality, but Without Density*

Dag Normann

University of Oslo

Abstract

We construct a hierarchy of domains with totality, based on the canonical domains for the empty set, the set of boolean values and the set of natural numbers, and closed under dependent sums and products of continuously parameterised families of domains with totality. We show that all hereditarily total objects will be extensional, and that extensional equality between total objects is an equivalence relation.

1 Introduction

This paper is a continuation of the papers [4, 5] where we give a pure semantical analysis of some of the constructors of pure type theory. The constructors we have focused on so far, and will focus on in this paper, are the dependent sums $\sum(x \in A)B_x$ and dependent products $\prod(x \in A)B_x$.

It is important that we discuss possible interpretations of the constructors without having particular type theories in mind, because this is the only way to construct semantics that are robust with respect to extensions or alterations of theories.

The naive interpretation will be to let

$$\sum(x \in A)B_x = \{(x,y) \mid x \in A \land y \in B_x\}$$

and

$$\prod(x \in A)B_x = \{f \mid dom(f) = A \land \forall x \in A(f(x) \in B_x)\}$$

*The research for this paper has partly been supported by EU science plan, contract no. SCI*CT91-724

233

and then to construct a full hierarchy over some base types where we include sums and products of all possible parameterisations of types.

There are two problems with this naive approach.

One is that we cannot construct this hierarchy using standard set theory ZFC without a minimum of large cardinal axioms.

The other, and more serious problem, is that the types in a hierarchy like this will have little internal structure beyond what can be obtained from general set theory, and that there is thus no real information to be obtained from the naive approach.

We have to find a ballance between the need for generality, i.e. the construction lends itself to the implementation of the relevant fragment of any type theory, and the need for restrictions in order to get structured interpretations.

To us it is also important that the interpretation is as close to the set-theoretical intuition as possible. This means that in the end a sum-type should be interpreted as a set of ordered pairs, while a product-type should be interpreted as a set of choice functions.

There are two principles for restriction that suggest themselves, we require that all functions or parameterisations are continuous (locally finitely based), or we may even require that they are effective. We focus on continuity.

In the previous papers we have constructed hierarchies based on interpretations for the set of natural numbers and the set of boolean values. The elements of the hierarchy will in each case be a two-level structure; we have the underlying domain together with a canonical set of total objects. Consistency will be an equivalence relation on the set of total elements, and the equivalence classes will form structures that share many properties with the Kleene-Kreisel continuous functionals. The hierarchy of equivalence classes will be isomorphic to a hierarchy of sets of ordered pairs and choice functions.

In this paper we will construct a transfinite hierarchy of domains where we will use the canonical domains for the empty set, the set of boolean values and the set of natural numbers as the base of the hierarchy.

The construction of the hierarchy will be in two steps. First we construct one domain S of *syntactic forms* together with an interpretation $I(s)$ as a domain for each $s \in S$. Similar constructions can be found in Palmgren [12] and Palmgren and Stoltenberg-Hansen [13, 14], and this part of our construction is mainly covered by the methods of those papers.

The second step will be to define the well founded syntactic forms S_{wf} and the total elements $I(s)_{\mathrm{TOT}}$ for $s \in S_{\mathrm{wf}}$. It is important that this definition is accepted as canonical, since we do want to investigate the total elements of natural interpretations of typed expressions.

We will work in the context of domains. We will assume knowledge to the basic theory of domains, see e.g. Stoltenberg-Hansen, Lindström and

Griffor [15]. Following [12, 14] we will represent a domain as the set of ideals over a preordering with an explicit least upper bound construction. We will strengthen the concept from [14] slightly. This will be discussed in section 2.

Though we will work with total elements in a domain, our work will not fit into the axiomatisation of totality suggested by Berger [1]. In section 2 we will discuss this.

Since the total elements are in general not dense in the underlying domains, it will not always be the case that a continuous function on the total objects can be extended to a continuous function on the underlying domain. Since we want to be able to represent all continuous functions defined on the total objects, we introduce continuous multivalued functions. The total functions will be singlevalued when restricted to total input. In order to give a decent treatment of multivalued functions, we introduce a domain-constructor ^mD, the domain of multivalued elements of D, and we characterize the compact elements of ^mD. This will be done in section 3.

In section 4, we give an explicit construction of the domains D and S together with the interpretation of $s \in S$ as a subdomain of D.

In section 5, we define the well founded hierarchy of domains with totality, and we prove that all total elements will respect extensional equality. The fact that this hierarchy can be constructed, is the main result of the paper.

In section 6 we first show that a hierarchy of domains of singlevalued functions, where we systematically avoid the empty type, can be seen as a subhierarchy of the one constructed in this paper. A consequence will be that the hight of the hierarchy will be the first ordinal not recursive in ^3E and any real. We then give a simpler proof of this fact, using the presence of the empty type.

2   Domains and totality

It is well known that any domain can be represented as the set of ideals over some preordering where all finite bounded sets have least upper bounds. It appears, however, that it is useful to represent domains as sets of ideals over partial preorderings with an explicit least upper bound operator sup on bounded pairs. This was observed by Palmgren and Stoltenberg-Hansen [14], where they called these structures *conditional upper semilattices with least element, cusl*. In [4, 10] we also used such preorderings, but with the extra property that for each p there are only finitely many pairs p_1, p_2 with $p = sup\{p_1, p_2\}$ (in the explicit sense).

We call a cusl with this extra property an *iei-structure*, iei is short for *implicit-explicit-information*. In a preprint version of this paper we analysed this concept further. All constructions of domains in this paper will be explicit constructions of iei-structures.

The concept of totality is important in this paper. This concept was first discussed in abstract terms in Normann [8], and Berger [1] defined a general concept of sets of total objects for domains. In both cases an object will be total when it gives a well defined total output to some given set of input material. The total objects in this paper will not satisfy the axioms from [8] or [1]. One simple reason is that both assume that the procedures modelled in the domain will be singlevalued, while we will use multivalued objects, see section 3. In addition, Berger's axioms are implicitly based on the assumption that the input material relevant for totality is dense in the set of all input material.

In this paper we will not need an abstract notion of totality. All our constructions are explicit, and the total elements of the domains constructed are given canonically. Still it is important to understand the concept of totality at a more abstract level, and it would be an improvement if the constructions of this paper could be seen to be less ad hoc than they appear. As a part of the renewed analysis of totality that ought to come, let us discuss the conceptual analysis of the central properties of the total objects appearing in this paper. We will be very informal.

Let us assume that the total elements of a domain X are total because they answer a set Q of questions q that are not neccessarily dense in the underlying set of possible input material. Assume further that for an object to be total, we require that the answer to a question $q \in Q$ is single valued. Define the relation E on the set of total objects by

$$x E y \text{ if } x \sqcap y \text{ is total.}$$

We claim that it is reasonable to assume that E is an equivalence relation by the following argument:

E is obviously reflexive and symmetric.

Let xEy and yEz. We explain why $x \sqcap z$ should be total.

Let q be a question. Then $x \sqcap y$ will use some finite information q_0 from q in order to produce the answer a, and $y \sqcap z$ will use some finite information q_1 from q to produce the answer b.

Because y then gives both answers a and b to q, we must have that $a = b$.

Let q_2 be the union of q_0 and q_1. Then, assuming that x and z are ideals of pieces of information coded in the standard way, the information

$$\text{'we obtain the answer } a \text{ from } q_2\text{'}$$

will be in both x and z, i.e. in $x \sqcap z$. Thus $x \sqcap z$ will answer q.

We will prove that our total objects will satisfy that the relation E is an equivalence relation, and that it in a natural sense corresponds to hereditarily extensional equality between total elements. This property of the hierarchy will be the main technical result of the paper. We will further prove that each equivalence-class will have a unique maximal element.

3   Multiobjects

As mentioned in the introduction, we want to consider multivalued functions, and in order to give a clean account for this, we introduce the domains of multiobjects.

Definition 1 Let D be a domain. A *multiobject* in D is a nonempty set $X \subseteq D$ such that

i) If $\alpha \in X$ and $\beta \sqsubseteq \alpha$, then $\beta \in X$.

ii) If $Y \subseteq X$ is a bounded subset of D, then $\sqcup Y \in X$.

We let $^m D$ be the set of multiobjects in D ordered by inclusion.

Lemma 1 $^m D$ *is a domain.*

Proof:
The compact elements of $^m D$ will be represented by finite sets $\{C_1, \ldots, C_n\}$ of compact elements from D such that if C_i and C_j are consistent, then for some k, $C_i \sqcup C_j = C_k$. (This is one possible representation of the compact sets, which makes the preordering simple.) The corresponding compact set will be

$$X_{C_1, \ldots, C_n} = \{\alpha \mid \alpha \sqsubseteq C_i \text{ for some } i \leq k\}$$

The preordering on the representation of compact sets is determined from the following fact:
$X_{C_1, \ldots, C_n} \sqsubseteq X_{D_1, \ldots, D_m}$ if and only if for each $i \leq n$ there is a $j \leq m$ with $C_i \sqsubseteq D_j$.
We leave the formal definition of the correspondence between the ideals of compacts and $^m D$ to the reader.

There will be multiobjects that are pairwise bounded. These will be exactly the multiobjects of the form

$$\{\beta \mid \beta \sqsubseteq \alpha\}$$

for some $\alpha \in D$.
We will call an object of this form *single*, and we will identify it with α.
When X is a multiobject in D, and we write $X \in D$, or $X \in D'$ for some

subset D' of D, we mean that X is single, and without further mentioning, we will identify X with its corresponding element in D.

We will of course represent the domain ${}^m D$ of multiobjects using ideals of compacts. Thus, if D is the set of ideals in the iei-structure $(|D|, \leq, sup_D)$, the elements of ${}^m D$ will be sets of sets $\{C_1, \ldots, C_n\}$ where $C_1, \ldots, C_n$ are elements of $|D|$ closed under sup_D, and $sup_{{}^m D}\{X, Y\}$ is the closure of $X \cup Y$ under sup_D.

If $\alpha \in {}^m D$ we may identify α with

$$\{C \mid \exists X \in \alpha (C \in X)\}.$$

With this identification, we see D as a subset of ${}^m D$. Since we want ${}^m D$ to be based on an iei-structure, we will not make this identification explicitly.

4   The domains S and D

We will now define two domains D and S.

D will be a domain of all possible objects that we might be interested in, and we might have used a universal domain instead, see [15] for a construction.

The domain S will represent possible typeterms, we call the elements "syntactic forms" for types. As an integrated part of the definition of S, we have an interpretation $I(s)$ of $s \in S$ as a subdomain of D.

We find an analogue construction in the domain of types in [14], and we have used the same main idea in [4, 5, 10].

For technical reasons we will use another representation of the compact elements of a cartesian product than what is standard. We will explain later exactly where we need this seemingly unneccessarily complex representation. In order to give some intuition we can say that since our aim is to use pairs as elements in Σ-types, it will be an advantage to have a representation of the objects where the dependence of the second component from the first component is taken into account.

Definition 2 Let X and Y be two domains generated from iei-structures $|X|$ and $|Y|$ resp.

We represent the compact elements in $X \times Y$ by sets $\{(p_1, q_1), \ldots, (p_n, q_n)\}$ where $\{p_1, \ldots, p_n\}$ is a bounded set of compacts in $|X|$ and $\{q_1, \ldots, q_n\}$ is a bounded set of compacts in $|Y|$.

We let $\{(p_1, q_1), \ldots, (p_n, q_n)\} \leq \{(u_1, v_1), \ldots, (u_m, v_m)\}$ if $sup\{p_1, \ldots, p_n\} \leq sup\{u_1, \ldots, u_m\}$ and $sup\{q_1, \ldots, q_n\} \leq sup\{v_1, \ldots, v_m\}$

We may then use union as the sup-operator on the representations for compact sets in a cartesian product.

If $\gamma \in |X \times Y|$ we let

$$\pi_0(\gamma) = \{p \mid (\exists q \in |D|)(\exists C \in \gamma)(p,q) \in C\}$$
$$\pi_1(\gamma) = \{q \mid (\exists p \in |D|)(\exists C \in \gamma)(p,q) \in C\}$$

Definition 3 By recursion we define the set $|D|$ with the preordering $\leq$, and then the induced domain D, as follows:

$\perp \in |D|$ and $\perp$ will be the minimal object in $|D|$.

The Boolean values tt and ff will be atomic elements of $|D|$.

Each natural number n will be an atomic element of $|D|$.

If $r = \{(p_1, q_1), \ldots, (p_n, q_n)\}$ represents a compact set in $D \times D$, then $(P, r) \in |D|$. P is just a syntactic entity, meaning "pair".

We let $(P, r_1) \leq (P, r_2)$ if $r_1 \leq r_2$ in the preordering on the representations of compacts in cartesian products.

If C represents a compact in $D \to {}^m D$, we let $(\lambda, C) \in |D|$. λ is just a syntactic entity.

We define the preordering between the (λ, C)'s in the obvious way.

$(|D|, \leq)$ is extended to an iei-structure, using the obvious interpretation of *sup*.

Essentially, D will be a solution to the domain equation

$$D = Boolean \oplus Nat \oplus (D \times D) \oplus (D \to {}^m D)$$

The domain S will contain atomic objects B and N which will be interpreted as the subdomains of D representing Boolean and Nat. It will also contain the atomic element O representing the empty set.

If $s \in S$ with interpretation $I(s)$ and $F : I(s) \to {}^m S$ is continuous, then the pair s, F represents the multivalued parameterization $\{I(F(x))\}_{x \in I(s)}$, and S will essentially contain the elements $\sum(s, F)$ and $\prod(s, F)$.

We define the set of compacts $|S|$ with a preorder, and the interpretation $|I(\sigma)|$ as a subset of $|D|$ for each $\sigma \in |S|$ in a simultanous definition:

Definition 4 Let $\perp_S \in |S|$ with $|I(\perp_S)| = \{\perp\}$.

Let O be an atomic element of $|S|$ with $|I(O)| = \{\perp\}$.

Let B be an atomic element of $|S|$ with $|I(B)| = \{\perp, tt, ff\}$.

Let N be an atomic element of $|S|$ with $|I(N)| = \{\perp, 0, 1, \cdots\}$.

If $\sigma \in |S|$, $p_1, \ldots, p_n \in |I(\sigma)|$ and $C_1, \ldots, C_n \in |{}^m S|$, then

$$(\prod, \sigma, \{(p_1, C_1), \ldots, (p_n, C_n)\})$$

and

$$(\sum, \sigma, \{(p_1, C_1), \ldots, (p_n, C_n)\})$$

are in $|S|$.

We let $(\prod, \sigma, X) \leq (\prod, \tau, Y)$ and $(\sum, \sigma, X) \leq (\sum, \tau, Y)$ if $\sigma \leq \tau$ as elements of $|S|$, and $X \leq Y$ as elements of $|D \to {}^m S|$.

Now let $C_i = \{\tau_{i,1}, \ldots, \tau_{i,n_i}\}$ for $i = 1, \ldots, n$.

$|I((\Pi, \sigma, \{(p_1, C_1), \ldots, (p_n, C_n)\}))|$ contains $\perp$ together with all objects of the form

$$(\lambda, \{(q_1, Y_1), \ldots, (q_m, Y_m)\})$$

such that for each j we have

i) $q_j \in |I(\sigma)|$.

ii) Let Z_j be the subdomain of D with compacts $|Z_j|$ where $|Z_j|$ is the closure of

$$\cup\{|I(\tau_{i,k})| \mid p_i \leq q_j \wedge 1 \leq k \leq n_i\}$$

under sup.

We then in addition demand that $Y_j \in |{}^d(Z_j)|$.

Let $|I((\Sigma, \sigma, \{(p_1, C_1) \ldots (p_n, C_n)\}))| =$
$\{(P, \{(q_1, v_1), \ldots, (q_m, v_m)\}) \in |D| \mid q_j \in |I(\sigma)| \wedge v_j \in |I(Z_j)|$
for $j = 1, \ldots, m\}$
where Z_j is as constructed above.

If $s \in S$, we let $|I(s)| = \cup\{|I(\sigma)| \mid \sigma \in s\}$.

$(|S|, \leq)$ is extended to an iei-structure in the obvious way.

Lemma 2 a) *The function* $\sigma \to |I(\sigma)|$ *is monotone*

b) *If* $p \in |I(s)|$ *for some* $s \in S$, *there is a unique minimal* $\sigma(p) \in |S|$ *with* $p \in |I(\sigma(p))|$ *such that* $sup(\sigma(p), \sigma(q)) = \sigma(sup(p, q))$ *whenever the latter is defined.*

Proof
a) is trivial.
b) is proved by induction on the rank of p.
If $p = \perp$, then $\sigma(p) = \perp_S$.
If $p = tt$ or ff, then $\sigma(p) = B$.
If $p = n$ for some natural number n, then $\sigma(p) = N$.
Let $p = (\lambda, \{(q_1, Y_1), \ldots, (q_m, Y_m)\})$. Then
$\sigma(p) = (\Pi, sup\{\sigma(q_1), \ldots, \sigma(q_m)\}, \{(q_1, E_1), \ldots, (q_m, E_m)\})$ whenever this is defined, where $E_j = \{\sigma(q) \mid q \in Y_j\}$.
It is easy to see that $p \in |I(\sigma(p))|$ whenever $\sigma(p)$ is defined, and that the construction of $\sigma(p)$ in this case commutes with least upper bounds.
It remains to show that if

$$p \in |I(\tau)| \text{ with } \tau = (\textstyle\prod, \sigma, \{(p_1, C_1), \ldots, (p_n, C_n)\})$$

then $\sigma(p)$ is defined and $\sigma(p) \leq \tau$.

First we have that $q_j \in |I(\sigma)|$ for $1 \leq j \leq m$ so $\sigma(q_j) \leq \sigma$.
Thus $sup\{\sigma(q_1), \ldots, \sigma(q_m)\}$ exists by the induction hypothesis, and is bounded by σ.
Then we show that E_j is defined and that each element of E_j is bounded by a consistent subset of $\cup\{C_i \mid p_i \leq q_j\}$.
Let $q \in Y_j$. Then $\sigma(q)$ is a typical element of E_j.
Using the notation from the definition, we have $Y_j \in |^d(Z_j)|$ so $q \in |Z_j|$.
By the induction hypothesis, $\sigma(q)$ is defined, and by the definition of Z_j we have that $\sigma(q)$ is bounded by the sup of a consistent set of $\tau_{i,k}$ where $p_i \leq q_j$ and $1 \leq k \leq n_i$.
But this is exactly what we aimed to prove.

If $p = (P, \{(q_1, v_1), \ldots, (q_m, v_m)\})$ we let
$\sigma(p) = (\textstyle\sum, sup\{\sigma(q_1), \ldots, \sigma(q_m)\}, \{(q_1, E_1), \ldots, (q_m, E_m)\})$
where E_j is as above.
Here we use our special representation of compacts for pairs to see that the definition of $\sigma(p)$ commutes with sup.
The rest of the argument is like the $\prod$-case.

As a consequence we get

Corollary 1 *a) If $s, t \in S$ then $|I(s \cap t)| = |I(s)| \cap |I(t)|$*
b) If $s, t \in S$, $\alpha \in I(s)$ and $\beta \in I(t)$, then $\alpha \cap \beta \in I(s \cap t)$.

The proof is trivial.

If $s \in S$, either $s = \{\bot_S\}$ or s will be one of five sorts; s represents the empty type, the type of boolean values, the type of natural numbers, or it represents a $\sum$- or $\prod$-type.
If s contains an element of the form $(\prod, \sigma, X)$, we let

$$s_1 = \{\tau \mid (\textstyle\prod, \tau, Y) \in s \text{ for some } Y\},$$

and we let

$$F_s = \{Y \mid (\textstyle\prod, \tau, Y) \in s \text{ for some } \tau\}.$$

s_1 will be an element of S, and F_s will be the domain interpretation of a multivalued function from $I(s_1)$ to S. s_1 and $_sF$ will determine s completely. In this case we write $s = \prod(s_1, F_s)$

Likewise, if s contains an element of the form $(\sum, \sigma, X)$, we write
$s = \sum(s_1, F_s)$.

If $s, t \in S$ with $s \subseteq t$, we will use the following notation:
If $\alpha \in I(s)$, we let $[\alpha]_t$ be the closure of α to an ideal in $I(t)$.

Lemma 3 **a)** *If $s = \prod(s_1, F_s)$ and $t = \prod(t_1, F_t)$ with $r = s \cap t$, then*
$$r \in S , \ r_1 = s_1 \cap t_1 \text{ and } F_r(\alpha) = F_s([\alpha]_{s_1}) \cap F_t([\alpha]_{t_1})$$
for any $\alpha \in I(s_1 \cap t_1)$.

b) *If s , t and r are as in a),*
$$\alpha_1 \in I(s) , \ \alpha_2 \in I(t) \text{ and } \beta \in I(r_1),$$
then $(\alpha_1 \cap \alpha_2)(\beta) = \alpha_1([\beta]_{s_1}) \cap \alpha_2([\beta]_{t_1})$.

c) *If $s = \sum(s_1, F_s)$ and $t = \sum(t_1, F_t)$ with $r = s \cap t$, then*
$$r \in S , \ r_1 = s_1 \cap t_1 \text{ and } F_r(\alpha) = F_s([\alpha]_{s_1}) \cap F_t([\alpha]_{t_1})$$
for any $\alpha \in I(s_1 \cap t_1)$.

d) *If s , t and r are as in c), $\alpha \in I(s)$, $\beta \in I(t)$ and $\gamma = \alpha \cap \beta$ then*
$$\pi_0(\gamma) = \pi_0(\alpha) \cap \pi_0(\beta)$$
$$\pi_1(\gamma) = \pi_1(\alpha) \cap \pi_1(\beta) \cap |I(F_s(\pi_0(\alpha)) \cap F_t(\pi_1(\beta)))|.$$

We prove a) and leave the rest for the reader. The proof of c) is almost identical, the proof of b) follows the same pattern while the proof of d) follows the same idea, but is slightly more complicated.

Proof of a):
Clearly $r_1 \subseteq t_1$.
If $\sigma \in s_1 \cap t_1$, then $(\prod, \sigma, \emptyset) \in s \cap t = r$, so $\sigma \in r_1$.
We also clearly have

$$F_r(\alpha) \subseteq F_s([\alpha]_{s_1}) \cap F_t([\alpha]_{t_1}).$$

Let $X \in |^m S|$ be in $F_s([\alpha]_{s_1}) \cap F_t([\alpha]_{t_1})$.
Then for some $p_0 \in [\alpha]_{s_1}$ and $p_1 \in [\alpha]_{t_1}$ we have that

$$(\lambda, \{(p_0, X)\}) \in F_s \text{ and } (\lambda, \{(p_1, X)\}) \in F_t.$$

Then there is a $p_2 \in \alpha$ with $p_0 \leq p_2$ and $p_1 \leq p_2$.
It follows that $(\lambda, \{(p_2, X)\}) \in F_s \cap F_t$, so $X \in (F_s \cap F_t)(\alpha) = F_r(\alpha)$.
This complete the proof in case a).

5   Wellfounded types and total objects

We have introduced a system of multivalued functions and multivalued parameterisations. We are however mainly interested in a hierarchy of hereditarily singlevalued total functions and singlevalued, wellfounded parameterisations.

Definition 5 By simultanous recursion we define the set S_{wf} and $I(s)_{\text{TOT}}$ for $s \in S_{\text{wf}}$ as follows:

$s = \{\perp_S, O\} \in S_{\text{wf}}$ with $I(s)_{\text{TOT}} = \emptyset$.

$s = \{\perp_S, B\} \in S_{\text{wf}}$ with $I(s)_{\text{TOT}} = \{\{\perp, tt\}, \{\perp, ff\}\}$

$s = \{\perp_S, N\} \in S_{\text{wf}}$ with $I(s)_{\text{TOT}} = \{\{\perp, n\} \mid n \text{ is a natural number }\}$.

If $s = \prod(s_1, F_s)$ we let $s \in S_{\text{wf}}$ if $s_1 \in S_{\text{wf}}$ and if $F_s(\alpha) \in S_{\text{wf}}$ for all $\alpha \in I(s_1)_{\text{TOT}}$.

See the conventions on single objects in section 3 for the precise interpretation of this definition.

If $\gamma \in I(s)$ we let $\gamma \in I(s)_{\text{TOT}}$ if $\gamma(\alpha) \in I(F_s(\alpha))_{\text{TOT}}$ for all $\alpha \in I(s_1)_{\text{TOT}}$.

If $s = \sum(s_1, F_s)$ we let $s \in S_{\text{wf}}$ if $s_1 \in S_{\text{wf}}$ and if $F_s(\alpha) \in S_{\text{wf}}$ for all $\alpha \in I(s_1)_{\text{TOT}}$.

Let $\gamma \in I(s)$. Then $\gamma \in I(s)_{\text{TOT}}$ if $\pi_0(\gamma) \in I(s_1)_{\text{TOT}}$ and $\pi_1(\gamma) \in I(F_s(\pi_0(\gamma)))_{\text{TOT}}$.

Theorem 1 *Let* $s \subseteq t$ *,* $s, t \in S_{\text{wf}}$

a) *Let* $\alpha \in I(s)_{\text{TOT}}$. *Then* $[\alpha]_t \in I(t)_{\text{TOT}}$.

b) *Let* $\beta \in I(t)_{\text{TOT}}$. *Then* $\beta \cap |I(s)| \in I(s)_{\text{TOT}}$.

Proof:

We use induction on s in S_{wf}.

The cases $s = [O]_S$, $s = [B]_S$ and $s = [N]_S$ are trivial.

Case $s = \prod(s_1, F_s)$, $t = \prod(t_1, F_t)$.

a): Let $\alpha \in I(s)_{\text{TOT}}$. Let $\gamma \in I(t_1)_{\text{TOT}}$.

We will show that $[\alpha]_t(\gamma) \in I(F_t(\gamma))_{\text{TOT}}$.

Using the induction hypothesis we have

$$[\alpha(\gamma \cap |I(s_1)|)]_{F_t(\gamma)} \in I(F_t(\gamma))_{\text{TOT}}$$

so it is sufficient to prove

$$[\alpha]_t(\gamma) = [\alpha(\gamma \cap |I(s_1)|)]_{F_t(\gamma)}.$$

$\supseteq$ is trivial, so let $q \in [\alpha]_t$.

Then there is a $p \in \gamma$ with

$$(\lambda, \{(p, \{q\})\}) \in [\alpha]_t.$$

Further there is a $(\lambda, \{(p_1, X_1), \ldots, (p_m, X_m)\}) \in \alpha$ with

$(\lambda, \{(p, \{q\})\}) \leq (\lambda, \{(p_1, X_1), \ldots, (p_m, X_m)\})$.

Let $K = \{i \mid p_i \leq p\}$. Then there is a bounded set $Y \subseteq \cup\{X_i \mid i \in K\}$ with

$q \leq supY$.

Let $p' = sup\{p_i \mid i \in K\}$. Then $p' \in I(s_1)$ so $p' \in \gamma \cup |I(s_1)|$.

Moreover $q' = supY \in \alpha(\gamma \cap |I(s_1)|)$ and $q \leq q'$ so $q \in [\alpha(\gamma \cap |I(s_1)|)]_{F_t(\gamma)}$.

b): let $\gamma \in I(s_1)_{\text{TOT}}$. Then

$$[\gamma]_{t_1} \in I(t_1)_{\text{TOT}}$$

and

$$\beta([\gamma]_{t_1}) \in I(F_t([\gamma]_{t_1}))_{\text{TOT}}$$

so

$$\beta([\gamma])_{t_1} \cap |I(F_s(\gamma))| \in I(F_s(\gamma))_{\text{TOT}}$$

We will show that

$$\beta([\gamma]_{t_1}) \cap |I(F_s(\gamma))| = (\beta \cap |I(s)|)(\gamma).$$

$\supseteq$ is simple, and is left for the reader.

Let $q \in \beta([\gamma]_{t_1}) \cap |I(F_s(\gamma))|$.

Since $q \in \beta([\gamma]_{t_1})$ there is a $p \in [\gamma]_{t_1}$ with $(\Pi, \{(p, \{q\})\}) \in \beta$, and since $p \in [\gamma]_{t_1}$ there is a $p_1 \in \gamma$ with $p \leq p_1$.

Then $(\Pi, \{(p_1, \{q\})\}) \in \beta$.

Since $q \in |I(F_s(\gamma))|$ there is a $p_2 \in \gamma$ and a finite set X such that $(p_2, X) \in F_s$ and $q \in |I(X)|$

(where $|I(X)|$ is the closure of $\cup\{|I(r)| \mid r \in X\}$ under sup.)

X has to be singlevalued since $p_2 \in \gamma$, γ is total and $F_s(\gamma) \in S$.

Then $(\lambda, \{(p_2, \{q\})\}) \in |I(s)|$.

Let $p_3 = sup\{p_1, p_2\}$.

Then $(\lambda, \{(p_3, \{q\})\}) \in |I(s)| \cap \beta$ and $p_3 \in \gamma$ so $q \in (\beta \cap |I(s)|)(\gamma)$.

Case $s = \sum(s_1, F_s)$, $t = \sum(t_1, F_t)$.

a): Let $\alpha \in I(s)_{\text{TOT}}$

Then $[\pi_0(\alpha)]_{t_1} = \pi_0([\alpha]_t)$ so $\pi_0([\alpha])_t \in I(t_1)_{\text{TOT}}$.

It is sufficient to show

$$\pi_1([\alpha]_t) = [\pi_1(\alpha)]_{F_t([\pi_0(\alpha)]_{t_1})}.$$

$\supseteq$ is easy because $\pi_1([\alpha]_t) \in I(F_t(\pi_0(\alpha)))]_{t_1}$.

Let $q \in \pi_1([\alpha]_t)$. Then for some p we have that $(P, \{(p, q)\}) \in [\alpha]_t$ and for some p_1 and q_1 we have that $p \leq p_1$, $q \leq q_1$ and $(P, \{(p_1, q_1)\}) \in \alpha$.

Then $q_1 \in \pi_1(\alpha)$.

It remains to show that $q \in I([F_t(\pi_0(\alpha))]_{t_1})$.

But this is trivial, since $p \in \pi_0([\alpha]_t)$ and $(P, \{(p,q)\}) \in [\alpha]_t$ and a requirement for this last fact is that $q \in |I(F_t(p))|$.

Let $\beta \in I(t)_{\text{TOT}}$.
Then

$$\pi_0(\beta \cap |I(s)|) = \pi_0(\beta) \cap |I(s_1)|$$

and

$$\pi_1(\beta \cap |I(s)|) = \pi_1(\beta) \cap |I(F_s(\pi_0(\beta) \cap |I(s_1)|))|.$$

The second fact requires a small proof following the same line of argument as the proofs above. We leave this proof for the reader.

Corollary 2 a) *If* $s \subseteq t \subseteq r$, s *and* r *are in* S_{wf} *and* $t \in S$, *then* $t \in S_{\text{wf}}$.

b) *If* $s \subseteq t \subseteq r$, $\alpha \subseteq \beta \subseteq \gamma$, $s, r \in S_{\text{wf}}$, $t \in S$, $\alpha \in I(s)_{\text{TOT}}$, $\gamma \in I(r)_{\text{TOT}}$
and $\beta \in I(t)$, *then* $\beta \in I(t)_{\text{TOT}}$.

Proof:
This is proved by induction on $s \in S_{\text{wf}}$.
Using the above theorem, the steps are trivial.

Definition 6 By a simultaneous recursion we define the binary relation E on S_{wf} and the relation R on

$$\{(s, \alpha) \mid s \in S_{\text{wf}} \wedge \alpha \in I(s)_{\text{TOT}}\})$$

The relations are meant to represent extentional equality between types and extentional equality between objects in extentionally equal types.
We write $[O]$ instead of $\{\perp_s, O\}$ or $[O]_s$ etc. in this definition.
1. $[O] \; E \; [O]$

2. $[B] \; E \; [B]$, $([B], [tt]) \; R \; ([B], [tt])$ and $([B], [ff]) \; R \; ([B], [ff])$

3. $[N] \; E \; [N]$ and $([N], [n]) \; R \; ([N], [n])$ for all numbers n.

4. If $s = \prod(s_1, F_s)$ and $t = \prod(t_1, F_t)$, let
$s \; E \; t$ if $s_1 \; E \; t_1$ and for all $\alpha \in I(s_1)_{\text{TOT}}$ and $\beta \in I(t_1)_{\text{TOT}}$ we have that if $(s_1, \alpha) \; R \; (t_1, \beta)$ then $F_s(\alpha) \; E \; F_t(\beta)$.
If $s \; E \; t$ in this case, $\gamma_1 \in I(s)_{\text{TOT}}$ and $\gamma_2 \in I(t)_{\text{TOT}}$ then
$(s, \gamma_1) \; R \; (t, \gamma_2)$ if for all $\alpha \in I(s_1)_{\text{TOT}}$ and $\beta \in I(t_1)_{\text{TOT}}$ we have that if $(s_1, \alpha) \; R \; (t_1, \beta)$ then $(F_s(\alpha), \gamma_1(\alpha)) \; R \; (F_t(\beta), \gamma_2(\beta))$.

5. If $s = \sum(s_1, F_s)$ and $t = \sum(t_1, F_t)$, let

$s \ E \ t$ if $s_1 \ E \ t_1$ and for all $\alpha \in I(s_1)_{\mathrm{TOT}}$ and $\beta \in I(t_1)_{\mathrm{TOT}}$ we have that if $(s_1, \alpha) \ R \ (t_1, \beta)$ then $F_s(\alpha) \ E \ F_t(\beta)$.
If $s \ E \ t$ in this case, $\gamma_1 \in I(s)_{\mathrm{TOT}}$ and $\gamma_2 \in I(t)_{\mathrm{TOT}}$ then
$(s, \gamma_1) \ R \ (t, \gamma_2)$ if $(s_1, \pi_0(\gamma_1)) \ R \ (t_1, \pi_0(\gamma_2))$ and
$(F_s(\pi_0(\gamma_1)), \pi_1(\gamma_1)) \ R \ (F_t(\pi_0(\gamma_2)), \pi_1(\gamma_2))$.

We will characterise the relations and show that they are equivalence relations. The only obvious property the relations E and R have is symmetry.

Theorem 2 *Let E and R be the relations defined above.*

a) *Let s and t be elements of S_{wf}.*
      Then $s \ E \ t$ if and only if $s \cap t \in S_{\mathrm{wf}}$

b) *Let s and t be in S_{wf} with $s \ E \ t$.*
      Let $\alpha \in I(s)_{\mathrm{TOT}}$ and let $\beta \in I(t)_{\mathrm{TOT}}$.
      Then $(s, \alpha) \ R \ (t, \beta)$ if and only if $\alpha \cap \beta \in I(s \cap t)_{\mathrm{TOT}}$.

Proof:
We will prove the theorem by simultanous induction on the rank of s in S_{wf}. The induction start, when s represents the empty type, the type of boolean values or the type of natural numbers, is trivial, since E and R are the identity-relations in these cases.

Case $\prod$:
a): Let $s = \prod(s_1, F_s)$ and $t = \prod(t_1, F_t)$.
By Lemma 3

$$s \cap t = \prod(s_1 \cap t_1, F_{s \cap t})$$

where

$$F_{s \cap t}(\alpha) = F_s([\alpha]_{s_1}) \cap F_t([\alpha]_{t_1})$$

for $\alpha \in I(s_1 \cap t_1)$.
First assume that $s \ E \ t$.
By the induction hypothesis, $s_1 \cap t_1 \in S_{\mathrm{wf}}$ and by Theorem 1

$$[\alpha]_{s_1} \in I(s_1)_{\mathrm{TOT}} \text{ and } [\alpha]_{t_1} \in I(t_1)_{\mathrm{TOT}}$$

We further have that $\alpha \subseteq [\alpha]_{s_1} \cap [\alpha]_{t_1}$ so by Corollary 2

$$[\alpha]_{s_1} \cap [\alpha]_{t_1} \in I(s_1 \cap t_1)_{\mathrm{TOT}}.$$

By the induction hypothesis

$$(s_1, [\alpha]_{s_1}) \ R \ (t_1, [\alpha]_{t_1})$$

so by assumption

$$F_s([\alpha]_{s_1}) \; E \; F_t([\alpha]_{t_1}).$$

Then by the induction hypothesis

$$F_s([\alpha]_{s_1}) \cap F_t([\alpha]_{t_1}) \in S_{\text{wf}}.$$

This shows that $s \cap t \in S_{\text{wf}}$.

Now assume that $s \cap t \in S_{\text{wf}}$.
Then $s_1 \cap t_1 \in S_{\text{wf}}$ so $s_1 \; E \; t_1$ by the induction hypothesis.
Let $\alpha \in I(s_1)_{\text{TOT}}$ and $\beta \in I(t_1)_{\text{TOT}}$ such that $(s_1, \alpha) \; R \; (t_1, \beta)$.
Then $\alpha \cap \beta \in I(s_1 \cap t_1)_{\text{TOT}}$.
Clearly

$$F_{s \cap t}(\alpha \cap \beta) \subseteq F_s(\alpha) \cap F_t(\beta)$$

Since $s \cap t \in S_{\text{wf}}$ we have that $F_{s \cap t}(\alpha \cap \beta) \in S_{\text{wf}}$.
It follows from Corollary 2 that

$$F_s(\alpha) \cap F_t(\beta) \in S_{\text{wf}}$$

and then $F_s(\alpha) \; E \; F_t(\beta)$ by the induction hypothesis. This shows that $s \; E \; t$.

The proof of b) follows the same pattern, and is left for the reader.

Case $\sum$:
The proof of a) is exactly like the proof of a) in case $\prod$.

b): Let $s = \sum(s_1, F_s)$ and $t = \sum(t_1, F_t)$
Let $\alpha \in I(s)_{\text{TOT}}$ and let $\beta \in I(t)_{\text{TOT}}$.

Assume first that $(s, \alpha) \; R \; (t, \beta)$.
By Lemma 3 we have

$$\pi_0(\alpha \cap \beta) = \pi_0(\alpha) \cap \pi_0(\beta)$$
$$\pi_1(\alpha \cap \beta) = \pi_1(\alpha) \cap \pi_1(\beta) \cap |I(F_{s \cap t}(\pi_0(\alpha \cap \beta)))|$$

By the induction hypothesis

$$\pi_0(\alpha \cap \beta) \in I(s_1 \cap t_1)_{\text{TOT}}$$

and

$$\pi_1(\alpha) \cap \pi_1(\beta) \in I(F_s(\alpha) \cap F_t(\beta))_{\text{TOT}}.$$

Then by Theorem 1

$$\pi_1(\alpha) \cap \pi_1(\beta) \cap |I(F_{s \cap t}(\pi_0(\alpha \cap \beta)))| \in I(F_{s \cap t}(\pi_0(\alpha \cap \beta)))_{\text{TOT}}.$$

This shows that $\alpha \cap \beta \in I(s \cap t)_{\text{TOT}}$.

Conversly, assume that $\alpha \cap \beta \in I(s \cap t)_{\text{TOT}}$.
Then

$$\pi_0(\alpha) \cap \pi_0(\beta) \in I(s_1 \cap t_1)_{\text{TOT}}$$

so by the induction hypothesis

$$(s_1, \pi_0(\alpha)) \; R \; (t_1, \pi_0(\beta)).$$

Moreover

$$\pi_1(\alpha \cap \beta) \in I(F_{s \cap t}(\alpha \cap \beta))_{\text{TOT}}$$

and

$$\pi_1(\alpha \cap \beta) \subseteq \pi_1(\alpha) \cap \pi_1(\beta) \subseteq \pi_1(\alpha) \in I(F_s(\pi_0(\alpha)))_{\text{TOT}}$$

so

$$\pi_1(\alpha) \cap \pi_1(\beta) \in I(F_s(\pi_0(\alpha)) \cap F_t(\pi_0(\beta)))_{\text{TOT}}.$$

This shows

$$(F_s(\pi_0(\alpha)), \pi_1(\alpha)) \; R \; (F_t(\pi_0(\beta)), \pi_1(\beta))$$

which in turn shows $(s, \alpha) \; R \; (t, \beta)$. This ends the proof of the theorem.

Lemma 4 *Let s and t be in S_{wf}.*
If $s \; E \; t$ and $\alpha \in I(s)_{\text{TOT}}$, then there is a $\beta \in I(t)_{\text{TOT}}$ with $(s, \alpha) \; R \; (t, \beta)$

Proof:
Let $\beta = [\alpha \cap |I(s \cap t)|]_t \in I(t)_{\text{TOT}}$.
Then

$$\beta \cap \alpha = \alpha \cap |I(s \cap t)| \in I(s \cap t)_{\text{TOT}}$$

so

$$(s, \alpha) \; R \; (t, \beta)$$

by Theorem 2.

Lemma 5 *If s and t are in S_{wf} and $s \; E \; t$, then s and t have the same rank in S_{wf}.*

Proof:
We use induction on s. Using Lemma 4, the proof is trivial.

Theorem 3 *The relations E and R are equivalence relations.*

Proof:
Symmetry is obvious from the definition.
Reflexivity follows from Theorem 2.
By Lemma 5, the relation E respects the rank in S_{wf}. Transitivity is proved by induction on this rank.
The induction start, when s represents a base type, is trivial because both relations are the identity relation on the domains in this case.
We prove transitivity of E in the $\prod$-case. Transitivity of R in the $\prod$-case and transitivity of E in the $\sum$-case will follow by the same argument. Transitivity of R in the $\sum$-case is trivial, since R is defined by coordinatewise satisfaction in this case.

Let $s = \prod(s_1, F_s)$, $t = \prod(t_1, F_t)$ and $r = \prod(r_1, F_r)$ be in S_{wf}, with $s \ E \ t$ and $t \ E \ r$.
Let $\alpha \in I(s_1)_{\mathrm{TOT}}$ and $\gamma \in I(r_1)_{\mathrm{TOT}}$ such that $(s_1, \alpha) \ R \ (r_1, \gamma)$.
By Lemma 4, let $\beta \in I(t_1)_{\mathrm{TOT}}$ such that $(s_1, \alpha) \ R \ (t_1, \gamma)$.
Then $(t_1, \beta) \ R \ (r_1, \gamma)$ by the induction hypothesis.
Then $F_s(\alpha) \ E \ F_t(\beta)$ and $F_t(\beta) \ E \ F_r(\gamma)$,
so $F_s(\alpha) \ E \ F_r(\gamma)$ by the induction hypothesis.
Then $s \ E \ t$ by definition. This ends the proof.

Remark
The relation E will represent extentional equality between types (or rather between the expressions for the types), and R will represent extentional equality between total elements of extentionally equal types. The relations are defined the way they are in order to make this explicit. In [7], Longo and Moggi showed that the continuous functionals of finite types will respect extentional equality without proving first that the total objects of a type is dense in the underlying domain. Their observation was that the intersection $F \cap G$ of two total objects F and G that are extentionally equivalent will be total, and thus for any Φ at the next level,

$$\Phi(F) = \Phi(F \cap G) = \Phi(G)$$

The proof of Theorem 2 is an elaboration of the idea from [7].
Reflexivity of the relations E and R simply means that all wellfounded parameterisations and all total functions will respect extentional equality, and Theorem 2 is essential in proving reflexivity.

Going back to the discussion in Section 2, we see that one of our reasonable axioms of totality is satisfied via Theorem 3. Another reasonable axiom, that a superobject of a total object is total, will not be satisfied, because a super-object might introduce multivaluednes on total input, and we do not want that.

We will see that each equivalence class will contain a unique, maximal element.

Theorem 4 *a) If $s \in S_{wf}$, then $\sqcup\{t \in S_{wf} \mid s \; E \; t\} \in S_{wf}$.*
b) If $s \in S_{wf}$, $t \in S_{wf}$, $s \; E \; t$ and $\alpha \in I(s)_{TOT}$, then

$$\sqcup\{\beta \in I(t)_{TOT} \mid (s, \alpha) \; R \; (t, \beta)\} \in I(t)_{TOT}.$$

Proof:
We use induction on the rank of s. Since E and R are relations on S_{wf} and the total objects, the restriction in the statement of the theorem is uneccessary. The base cases are trivial. We prove a) for the $\prod$-case, and leave the rest of the proof for the reader.

The intuitive idea is that we cannot break out of singlevaluednes by taking unions of equivalent objects. Corollary 2 makes it possible to prove that an object is well founded or total using estimates from both sides, and the estimates from above will be given by the induction hypothesis.

Let $s = \prod(s_1, F_s)$ and let $r = \sqcup\{t \mid s \; E \; t\}$.
Then

$$r = \prod(r_1, F_r)$$

and

$$s_1 \sqsubseteq r_1 = \sqcup\{t_1 \mid s \; E \; t\} \sqsubseteq \sqcup\{t_1 \mid s_1 \; E \; t_1\} \in S_{wf}$$

so $r_1 \in S_{wf}$ by Corollary 2.
If $\alpha \in I(r_1)_{TOT}$, then

$$F_s(\alpha \cap |I(s_1)|) \sqsubseteq F_r(\alpha) \sqsubseteq \sqcup\{F_t(\alpha) \mid s \; E \; t\} \sqsubseteq$$
$$\sqcup\{t' \mid F_s(\alpha \cap |I(s_1)|) \; E \; t'\} \in S_{wf},$$

so $F_r(\alpha) \in S_{wf}$.
Then $r \in S_{wf}$.
This completes the proof of the theorem.

Corollary 3 *Each equivalence-class of E, R and of the relations*

$$\alpha \; R_s \; \beta \; \text{if and only if} \; (s, \alpha) \; R \; (s, \beta)$$

will be lattices under $\sqcup$ and $\sqcap$ that are upwards complete.

The proof is trivial and is left for the reader.

6   Further discussion and results

In this section we will relate the construction of the hierarchy and its properties to previous constructions.

First we will see that the hierarchies of domains from Kristiansen and Normann [4] and Normann [10] can be seen as subhierarchies of the one constructed here. This is not entirely obvious since these hierarchies are based on singlevalued functions, and the hierarchy here is based on multivalued objects, and thus the underlying domains of even the lowest types in the hierarchies, like $\mathbf{N^N}$, are not the same. The underlying domains from [4] will in general not even be subdomains of the corresponding domains here. It will, however be the case that the spaces of equivalence classes essentially will be the same. We will, through stating the apropriate sequence of lemmas, indicate why. The proofs are tedious, but not difficult, and they are left for the reader.

Let D' and S' be the domains from [4] with $J(s)$ as the interpretation of $s \in S'$.

Lemma 6 *There is a map $\rho : |D'| \to |D|$ with the following three properties:*

i) $p \leq' q \Leftrightarrow \rho(p) \leq \rho(q)$.

ii) $\{p, q\}$ *is bounded in $|D'| \Rightarrow \{\rho(p), \rho(q)\}$ is bounded in D.*

iii) *If $q \leq \rho(p)$ then q is equivalent to some $\rho(p')$.*

ρ cannot be used to form a projection pair, because $\{p, q\}$ will not neccessarily be bounded in $|D'|$ when $\{\rho(p), \rho(q)\}$ is bounded in $|D|$.
We still can define the function $f^+ : D' \to D$ as the ρ-image, and the partial function $f^- : D \to D'$ as the inverse ρ-image.

Since we are working with domains that consist of sets of objects in a preordering, $f^-(\alpha)$ will be defined as a subset of $|D'|$ for all $\alpha \in D$, and this we will make use of below.
We will call a pair like f^+ and f^- a *weak projection pair*. There is a system of weak projection pairs that will connect the hierarchy from [4] and the hierarchy of this paper.

Lemma 7 *There is a continuous function*

$$g : S' \to S$$

and uniformly in $s \in S'$ there is a weak projection pair (f_s^+, f_s^-) from $J(s)$ to $I(g(s))$.

Theorem 5 *We use the notation from Lemma 7.*

a) $s \in S'_{\mathrm{wf}} \Rightarrow g(s) \in S_{\mathrm{wf}}$.

b) *If* $\alpha \in J(s)_{\mathrm{TOT}}$ *then* $f_s^+(\alpha) \in I(g(s))_{\mathrm{TOT}}$:

c) *If* $\beta \in I(g(s))_{\mathrm{TOT}}$ *then* $f_s^-(\beta) \in J(s)_{\mathrm{TOT}}$.

Corollary 4 *The hierarchy of topological spaces from* [4] *(when we have divided out by the consistency relation on* S'_{wf} *) is a subhierarchy of the corresponding hierarchy of this paper.*

Kleene [3] introduced recursion in higher types, and he also introduced the functional ^3E of quantification over Baire-space $\mathbf{N}^{\mathbf{N}}$. Recursion in ^3E is a natural transfinite extension of the *analytic* or *2nd. order definable* sets. Combining Theorem 5 with the main theorem from [10] we get

Corollary 5 *The rank of the hierarchy is the first ordinal not recursive in* ^3E *and any real.*

The first estimate of the complexity of a hierarchy like this can be found in the unpublished Normann [9]. We will give the construction from [9] here, because the construction is much more natural from a type-theorists point of view than the one in [10]. This proof makes use of the empty type.
We will assume that the reader is familiar with the basic definition of Kleene recursion and of ^3E. [10] contains sufficient background. For a deeper introduction, see the original paper [3], or Sacks [16].
When we write $\{e\}(\mathbf{f}) = n$, we actually mean that $\{e\}(\mathbf{f}_1, {}^3E, \mathbf{f}_2) = n$ where $\mathbf{f}$ is the concatenation of $\mathbf{f}_1$ and $\mathbf{f}_2$, and the position of ^3E is determined by e. When we write $\{e\}(\mathbf{f}) \downarrow$, we mean that for some n, $\{e\}(\mathbf{f}) = n$.

For the construction below, it will be an advantage to change and extend the notation a bit.
We introduce cartesian products $\times$ and disjoint sums $\oplus$ using products and sums of parameterisations over the type of boolean values.
We also introduce the function space constructor, $\rightarrow$, via constant parameterisations.
In formulating infinite products and sums, it will be an advantage to include the variable of the parameterisation in the notation. Thus, instead of $\prod(s, F)$, we may write $\prod(\alpha \in s)F(\alpha)$, ignoring in the notation the distinction between an $s \in S$ and $I(s)$.
When p is a compact, we let $[p]$ be the closure of p to an ideal.

Theorem 6 *Let e be a natural number, $\mathbf{f}$ a finite sequence of functions on the natural numbers.*
Uniformly in e and $\mathbf{f}$ there is an element

$$T(e, \mathbf{f}) \in S$$

and a continuous map

$$v(e, \mathbf{f}) : D \to \mathbf{N}$$

such that $\{e\}(\mathbf{f}) \downarrow$ *if and only if* $T(e, \mathbf{f}) \in S_{\mathrm{wf}}$, *and in this case*

- $I(T(e, \mathbf{f}))_{\mathrm{TOT}} \neq \emptyset$

- $v(e, \mathbf{f})$ *is constant* $\{e\}(\mathbf{f})$ *on* $I(T(e, \mathbf{f}))_{\mathrm{TOT}}$.

Proof:
T and v are defined as the least fixpoints of the following set of equations:

1. $\{e\}(x, \mathbf{f}) = x + 1$
   $T(e, x, \mathbf{f}) = [N]$
   $v(e, x, \mathbf{f})([n]) = x + 1$

2.,3.and 7. All other initial computations are treated the same way.

4. $\{e\}(\mathbf{f}) = \{e_1\}(\{e_2\}(\mathbf{f}), \mathbf{f})$
   $T(e, \mathbf{f}) = \Sigma(\alpha \in T(e_2, \mathbf{f}))T(e_1, v(e_2, \mathbf{f})(\alpha), \mathbf{f})$
   $v(e, \mathbf{f})(\gamma) = v(e_1, v(e_2, \mathbf{f})(\pi_0(\gamma)), \mathbf{f})(\pi_1(\gamma))$.

6. $\{e\}(\mathbf{f}) = \{e_1\}(\sigma(\mathbf{f}))$ where σ is a permutation.
   Let $T(e, \mathbf{f}) = [B] \times T(e_1, \sigma(\mathbf{f}))$.
   $v(e, \mathbf{f})(y) = v(e_1, \sigma(\mathbf{f}))(y([tt]))$.
   For technical reasons we want to increase the rank in this case.

8. $\{e\}(\mathbf{f}) = {}^3E(\lambda f\{e_1\}(f, \mathbf{f}))$.
   Let $T_1(f, y) = [N]$ and $T_2(f, y) = [O]$ if $v(e_1, f, \mathbf{f}) > 0$.
   Let $T_1(f, y) = [O]$ and $T_2(f, y) = [N]$ if $v(e_1, f, \mathbf{f}) = 0$.
   Let
   $T(e, \mathbf{f}) = \Pi(f \in \mathbf{N}^\mathbf{N})((T(e_1, f, \mathbf{f}) \to [O]) \to [O])$
   $\qquad \times [\Sigma(f \in \mathbf{N}^\mathbf{N})\Sigma(y \in T(e_1, f, \mathbf{f}))T_1(f, y)$
   $\qquad\qquad \oplus \Pi(f \in \mathbf{N}^\mathbf{N})\Pi(y \in T(e_1, f, \mathbf{f}))T_2(f, y)]$.
   Let $v(e, \mathbf{f})(x, r(y)) = 1$.
   Let $v(e, \mathbf{f})(x, l(y)) = 0$.

9. $\{e\}(e_1, \mathbf{f}) = \{e_1\}(\mathbf{f})$ is treated like 6.

Claim 1
If $\{e\}(\mathbf{f}) \downarrow$ then $T(e, \mathbf{f}) \in S_{\mathrm{wf}}$, $I(T(e, \mathbf{f}))_{\mathrm{TOT}} \neq \emptyset$ and $v(e, \mathbf{f})(\alpha) = \{e\}(\mathbf{f})$ for $\alpha \in I(T(e, \mathbf{f}))_{\mathrm{TOT}}$.

Proof:

We will use induction on the rank of the computation $\{e\}(\mathbf{f})$.
All cases but case 8 are left for the reader.
Let

$$\{e\}(\mathbf{f}) = {}^3E(\lambda f\{e_1\}(f,\mathbf{f}))\!\downarrow.$$

Then for all $f \in \mathbf{N}^{\mathbf{N}}$ we have that $\{e_1\}(f,\mathbf{f})\!\downarrow$, so $T(e_1,f,\mathbf{f}) \in S_{\mathrm{wf}}$ with a nonempty set of total elements.
This shows that

$$\Pi(f \in \mathbf{N}^{\mathbf{N}})((T(e_1,f,\mathbf{f}) \to [O]) \to [O]) \in S_{\mathrm{wf}}$$

with all its elements total.
Moreover, if $y \in I(T_1(e_1,f,\mathbf{f}))_{\mathrm{TOT}}$, then $T_1(f,y) \in S_{\mathrm{wf}}$ and $T_2(f,y) \in S_{\mathrm{wf}}$.
All together this shows that $T(e,\mathbf{f}) \in S_{\mathrm{wf}}$.
By the induction hypothesis there will be total elements in

$$I(\Pi(f \in \mathbf{N}^{\mathbf{N}})\,\Pi(y \in T(e_1,f,\mathbf{f}))T_2(f,y))$$

if and only if

$$\forall f \in \mathbf{N}^{\mathbf{N}}(\{e_1\}(f,\mathbf{f}) = 0),$$

and there will be total elements in

$$I(\Sigma(f \in \mathbf{N}^{\mathbf{N}})\,\Sigma(y \in T(e_1,f,\mathbf{f}))T_1(f,y))$$

if and only if

$$\exists f \in \mathbf{N}^{\mathbf{N}}(\{e_1\}(f,\mathbf{f}) > 0).$$

This shows that v acts as it should, and Claim 1 is proved.

Claim 2
If $T(e,\mathbf{f}) \in S_{\mathrm{wf}}$ then $\{e\}(\mathbf{f})\!\downarrow$.

Proof:
We use induction on the rank of $T(e,\mathbf{f})$. In cases S4 and S8 we will need Claim 1 as well.
It is in order to be able to prove this claim that we increase the rank in cases S6 and S9.
In case S8, observe that

$$\Pi(f \in \mathbf{N}^{\mathbf{N}})((T(e_1,f,\mathbf{f}) \to [O]) \to [O]) \in S_{\mathrm{wf}}$$

will imply that $T(e_1,f,\mathbf{f}) \in S_{\mathrm{wf}}$ for all $f \in \mathbf{N}^{\mathbf{N}}$, so $\{e_1\}(f,\mathbf{f})\!\downarrow$ for all $f \in \mathbf{N}^{\mathbf{N}}$.
The rest of the argument is trivial, and is left for the reader.
This ends the proof of the claim and of the theorem.

7 Concluding remarks

The main construction in this paper is based on two principles, the fact that equivalence between two total objects is the same as the intersection being total, and the use of multivaluedness to get all continuous functions defined on the total sets represented in the hierarchy even when the total sets are not dense. The idea to use the first principle evolved during a visit to Munich in spring 1993, when I discussed the possibilities of constructing semantics for type theory with Ulrich Berger and other members of the Munich group. It then seemed natural to rework the material from Normann [9] in this setting. In [9] types and total elements were represented by elements and subsets of $\mathbf{N^N}$ in a Kleene's associates kind of way. The representation based on domains seemed more natural. In addition, equivalence between representatives for types or total object was not an intergrated part of the hierarchies defined in [9].

One important concept in [9] is the concept of *typestream*. In the setting of this paper, a typestream will be a certain nonwellfounded element of S for which we still can define the total objects. Typestreams will generalise the class of types defined by strictly positive inductive definitions. The concept of a typestream is now developed for domains in Kristiansen and Normann [6].

In [9] we used recursion in ^3E and what we called E-structures to model a second order constructor. Continued work along these lines has indicated that this idea was not very fruitful, we do not obtain models of type theory with second order elements that are better, i.e. contains more information about the underlying theories, than the already known models.

In [9] we looked at an effective version of the hierarchy constructed there, i.e. a hierarchy where we only used recursive functions to code types and total elements. The result was that the hierarchy then corresponding to S_{wf} will have height ω_1^{CK}, the first non-recursive ordinal, while the corresponding hierarchy with typestreams will have the first recursively inaccessible ordinal as its height.This theorem is reproved in the setting of domains, and will appear in Normann [11]. This result is a semantical version of the proof-theoretic estimates of Martin-Löf type theory with induction, see Griffor and Rathjen [2] and Setzer [17].

If we take the set of equivalence classes of the well formed type expressions, S_{wf}/E or the equivalence classes of the total elements of one $I(s)$, $I(s)_{\mathrm{TOT}}/R_s$ we inherit natural topologies from the underlying domains. Waagbø[18] is investigating these topologies further. He shows a lifting theorem stating that a continuous map

$$\phi : I(s)_{\mathrm{TOT}}/R_s \to S_{\mathrm{wf}}/E$$

can be lifted to a total parameterisation in our sense, and he shows the analogue result for the continuous choice functions of the parameterisation.

He also extends our method to construct a model for some specific version of Per Martin-Löf type theory.

References

[1] Berger, U.: *Total sets and objects in domain theory*, Annals of pure and applied logic 60 (1993) 91-117

[2] Griffor, E. and Rathjen, M.: *The Strength of Some Martin-Löf Type Theories*, in manuscript.

[3] Kleene, S. C.: *Recursive Functionals and Quantifiers of Finite Types I*, T.A.M.S 91 (1959), 1-52

[4] Kristiansen, L. and Normann, D.: *Semantics for some constructors of type theory*, To appear in the proceedings of the Gauss symposium, der Ludwig-Maximilians-Universität München 1993

[5] Kristiansen, L. and Normann, D.: *Interpreting higher computations as types with totality*, to appear in Archives for Mathematical Logic

[6] Kristiansen, L. and Normann, D.: *Total objects in inductivly defined types*, in preparation

[7] Longo, G. and Moggi, E.: *The hereditarily partial effective functionals and recursion theory in higher types* JSL 49 (1984) 1319-1332

[8] Normann, D.: *Formalizing the notion of total information*, In P. P. Petkov (ed), Mathematical Logic, Plenum Press (1990) 67-94

[9] Normann, D.: *Wellfounded and non-wellfounded types of continuous functionals*, Oslo Preprint Series in Mathematics No 6 (1992)

[10] Normann, D.: *Closing the gap between the continuous functionals and recursion in ^3E*, Presented at the Sacks-symposium, MIT 1993

[11] Normann, D.: *Hereditarily Effective Typestreams*, in preparation

[12] Palmgren, E.: *An Information System Interpretation of Martin-Löf's Partial Type Theory with Universes*, Information and Computation 106 (1993) 26-60

[13] Palmgren, E. and Stoltenberg-Hansen, V.: *Domain interpretations of Martin-Löf's partial type theory*, Annals of pure and applied Logic 48 (1990) 135-196

[14] Palmgren, E. and Stoltenberg-Hansen, V.: *Remarks on Martin-Löf's partial type theory*, BIT 32 (1992) 70-83

[15] Stoltenberg-Hansen,V., Lindström, I. and Griffor,E.: *Mathematical Theory of Domains*, Cambridge Tracts in Theor. Comp. Science 22 (1994)

[16] Sacks, G. E.:*Higher recursion theory*, Springer-Verlag (1990)

[17] Setzer, A.: *Proof theoretical strength of Martin-Löf Type Theory with W-type and one universe*, Thesis, der Ludwig-Maximilians-Universität München (1993)

[18] Waagbø, G.:*Thesis*, in preparation

INDUCTIVE INFERENCE
OF TOTAL FUNCTIONS

Piergiorgio Odifreddi

University of Torino, Italy

In this paper we give an overview of an area of applied Recursion Theory that has attracted interest in recent years. The idea is to use recursion-theoretic notions to formalize the venerable problem of inductive inference, in the following way.

By taking time to be discrete and with a starting point, and events to be discrete and codifiable by natural numbers, a phenomenon to be inferred may be thought of as a function f on the natural numbers, given by the sequence of values

$$f(0), \ldots, f(n), \ldots$$

The function can be inferred if this is not just a sequence of accidents, but rather it has an intrinsic necessity. We can specify this internal structure of the sequence of values in at least two ways:

- If one is interested in tecnology, i.e. in the ability of reproducing effects, then one can require a method predicting the next value $f(n+1)$, once the values $f(0), \ldots, f(n)$ have been exhibited, for an arbitrary n.

- If one is interested in science, i.e. in the ability of understanding causes, then one can require a finite description compressing the infinite amount of information contained in the sequence of values.

Even in this vague formulation, one can identify the functions which are (in principle) inferrable w.r.t. any of the two methods with the recursive functions.

This completely disposes of the problem of which functions on natural numbers are *individually* inferrable, and one can thus turn the attention to *classes* of functions. The problem here takes the following form: for each class $\mathcal{C}$ of recursive functions, find a uniform method of inferring all members of $\mathcal{C}$.

Many possible formalizations of notions of inference for classes of total recursive functions have been considered in the literature. Here we will confine

ourselves to a few of them, and refer to the forthcoming volume II of our book
Classical Recursion Theory (Odifreddi [1997]) for a comprehensive treatment.
For background and notations we refer instead to volume I of the same book
(Odifreddi [1989]).

1   Identification by next value

Our first notion formalizes the idea of uniform method of prediction or ex-
trapolation.

Definition 1.1 (Barzdin [1972], Blum and Blum [1975]) *A class C of
total recursive functions is* **identifiable by next value** *($C \in \boldsymbol{NV}$) if there
is a total recursive function g (called a next-value function for C) such that,
for every $f \in C$ and almost every n,*

$$f(n+1) = g(\langle f(0), \ldots, f(n) \rangle).$$

Notice that we allow a finite number of wrong predictions for each element
of the class, i.e. g can take guesses and learn from its mistakes.

**Theorem 1.2 Number-Theoretic Characterization of $\boldsymbol{NV}$ (Barzdin
and Freivalds [1972])** *A class of total recursive function is in NV if and
only if it is a subclass of an r.e. class of total recursive functions.*

Proof. A next-value function g allows the computation of a recursive function
f, past the finitely many exceptions. Thus any function f for which g is a
next-value function is of the following form, for some sequence number a
(coding a list $\langle a_0, \ldots, a_n \rangle$ for some n):

$$f_a(x) = \begin{cases} a_x & \text{if } x \leq n \\ g(\langle f_a(0), \ldots, f_a(x-1) \rangle) & \text{otherwise.} \end{cases}$$

Any such f_a is recursive (by Course-of-Value Recursion) uniformly in a, and
by the S_n^m-Theorem there is then a recursive function h such that $\varphi_{h(a)} = f_a$.
Thus the class $\{f_a\}_{a \in \omega}$ is an r.e. class of total recursive functions. This shows
that any class of recursive functions identifiable by next value is a subclass
of an r.e. class of total recursive functions.

Conversely, for an r.e. class $\{\varphi_{h(e)}\}_{e \in \omega}$ of total recursive functions, we may
suppose it closed under finite variants (since the closure of an r.e. class under
finite variants is still r.e.). Let g be the recursive function defined as follows:

- on the empty list, g takes the value 0;

- on the list $\langle a_0, \ldots, a_n \rangle$, g takes the value $\varphi_{h(e)}(n+1)$, for the first e such that $\varphi_{h(e)}(x) = a_x$ for all $x \leq n$ (i.e. g takes the next value of the first function in the class that agrees with all values coded by the given list).

Since the class $\{\varphi_{h(e)}\}_{e \in \omega}$ is closed under finite variants, g is total. Since the class is r.e. (i.e. the functions $\varphi_{h(e)}$ are uniformly recursive), g is recursive. And g is a next-value function for every function in the given class (and hence in any subclass of it), by definition. $\square$

Theorem 1.3 Complexity-Theoretic Characterization of NV (Adleman) *A class of total recursive functions is in NV if and only if it is a subclass of a complexity class (w.r.t. some complexity measure).*

Proof. By 1.2, any class in NV is contained in an r.e. class of total recursive functions $\mathcal{C} = \{\varphi_{h(e)}\}_{e \in \omega}$. Consider the associated set $\{\Phi_{h(e)}\}_{e \in \omega}$ of step-counting functions w.r.t. any measure, and notice that if

$$t(x) = \max_{e \leq x} \Phi_{h(e)}(x)$$

then t is recursive because h is, and $\mathcal{C}$ is contained in the complexity class $\mathcal{C}_t$ named by t.

Conversely, given a recursive function t, notice that if φ_e is total and

$$(\forall_\infty x)[\Phi_e(x) \leq t(x)]$$

then there is a constant k such that

$$(\forall x)[\Phi_e(x) \leq t(x) + k].$$

Let g be the recursive function defined as follows:

- on the empty list, g takes the value 0

- on the list $\langle a_0, \ldots, a_n \rangle$, g takes the value $\varphi_e(n+1)$ for the smallest pair $\langle e, k \rangle$ defined as follows, if there is one, and 0 otherwise:

  - $\langle e, k \rangle \leq n$
  - for any $x \leq n+1$, $\Phi_e(x) \leq t(x) + k$
  - for any $x \leq n$, $\varphi_e(x) \simeq a_x$.

g is total recursive because all checks are recursive, and the second condition on $\langle e, k \rangle$ ensures that $\varphi_e(n+1)$ is defined.

It is easy to show that g identifies by next value the complexity class $\mathcal{C}_t$. $\square$

2 Identification by consistent explanation

We now turn to notions that formalize the idea of uniform method of explanation (via indices, that code descriptions of recursive functions). In a first attempt we require that the explanations agree with the available data.

Definition 2.1 (Gold [1967]) *A class C of total recursive functions is* **identifiable by consistent explanation** *($C \in \boldsymbol{EX_{cons}}$) if there is a total recursive function g (called a guessing function for C) such that, for every sequence number $\langle a_0, \ldots, a_n \rangle$:*

- $\varphi_{g(\langle a_0, \ldots, a_n \rangle)}(x) \simeq a_x$ *for all $x \leq n$,*

and for every $f \in C$:

- $\lim_{n \to \infty} g(\langle f(0), \ldots, f(n) \rangle)$ *exists, i.e.*

$$(\exists n_0)(\forall n \geq n_0)[g(\langle f(0), \ldots, f(n) \rangle) = g(\langle f(0), \ldots, f(n_0) \rangle)]$$

- $\varphi_{\lim_{n \to \infty} g(\langle f(0), \ldots, f(n) \rangle)} = f.$

In other words, $g(\langle f(0), \ldots, f(n) \rangle)$ provides a guess to an index of f consistent with the available information, and the guess stabilizes (from a certain point on) on an index of f.

The next result connects the two notions of identification introduced so far, and its two proofs suggest the two characterizations of EX_{cons} given in 2.5 and 2.7.

Proposition 2.2 (Gold [1967]) $NV \subseteq EX_{cons}$.

Proof. We can use the characterization of NV given in 1.2, and repeat almost verbatim the second half of its proof.

Alternatively, we can use the characterization of NV given in 1.3, and again repeat almost verbatim the second half of its proof. $\square$

The difficulty in proving the opposite inclusion is the following: given a guessing function g for a class of functions C, one might think of producing as a next-value function for C the one defined by

$$g_1(\langle a_0, \ldots, a_n \rangle) = \varphi_{g(\langle a_0, \ldots, a_n \rangle)}(n + 1),$$

i.e. to let the guessed program guess the next value. The problem with this is that $g(\langle a_0, \ldots, a_n \rangle)$ might be a program for a partial function (since we only know that the *final* guesses on initial segments of functions in C will be programs for total functions), and thus g_1 might not be total.

If one restricts the guesses to programs for total functions, then the problem disappears. One can argue that such a restriction is implicit in Popper Refutability Principle (Popper [1934]), according to which incorrect scientific explanations should be refutable: the unsolvability of the Halting Problem makes it in general impossible to decide whether a partial function is undefined at a given argument, and thus to refute an exlanation which is incorrect on such a basis. For more on this, see the discussion following 5.2.

We now show that if g is allowed to output indices for partial functions, then one is able to identify by consistent explanation more classes of functions. The result may be taken to show that technique has stronger requirements than science, and that being able to eventually explain a class of phenomena is not enough to be able to eventually predict them.

Proposition 2.3 (Blum and Blum [1975]) $EX_{cons} - NV \neq \emptyset$.

Proof. We want to find a class C of total recursive functions which is in EX_{cons} but not in NV. Let

$$C = \{\Phi_e : \Phi_e \text{ total}\},$$

i.e. the class of all total step-counting functions w.r.t. any measure.

To show $C \in EX_{cons}$ we can note that C is contained in the class $\{\Phi_e\}_{e \in \omega}$, which is a measured set of partial recursive functions, and prove the following result, of independent interest: *any class of total recursive functions contained in a measured set of partial recursive functions is in EX_{cons}.* The proof is a trivial extension of the first proof of 2.2.

Alternatively, to show $C \in EX_{cons}$ we can note that C is contained in the class $\{\Phi_e\}_{e \in \omega}$, which is a honest set of partial recursive functions, and prove the following result, of independent interest: *any class of total recursive functions contained in a honest set of partial recursive functions is in EX_{cons}.* The proof is a trivial extension of the second proof of 2.2. $\square$

We turn now to a characterization of EX_{cons}. The characterization of NV given in 1.2 already used up all r.e. classes of total recursive functions, and thus we will look at r.e. classes of *partial* recursive functions.

We consider a notion that isolates just what is needed to make the proofs of 2.2 and 2.3 work.

Definition 2.4 *An r.e. class $\{\varphi_{h(e)}\}_{e \in \omega}$ is called* **quasi-measured** *if there is a uniform recursive procedure to decide, given any e and any finite initial segment σ, whether $\varphi_{h(e)}$ extends σ.*

Obviously, a measured set is quasi-measured: to check whether $\varphi_{h(e)}$ extends σ, one simply checks whether $\varphi_{h(e)}(x) \simeq \sigma(x)$ for every x in the domain of σ.

But the converse does not hold: quasi-measuredness allows us to check whether $\varphi_{h(e)}$ extends or not a given finite initial segment, but not whether it has a certain value on an isolated argument (unless we already know the values on all previous arguments).

Theorem 2.5 Number-Theoretic Characterization of EX_{cons} (Viviani) *A class of total recursive functions is in EX_{cons} if and only if it is a subclass of a quasi-measured set of partial recursive functions.*

Proof. The first proof of 2.3, showing that any class of total recursive functions contained in a measured set of partial recursive functions is in EX_{cons}, can easily be adapted to quasi-measured sets.

Conversely, let $\mathcal{C}$ be identifiable by consistent explanation via g. Then every function f in $\mathcal{C}$ is of the following form, for some sequence number a (coding a list $\langle a_0, \ldots, a_n \rangle$ for some n):

$$f_a(x) \simeq \begin{cases} a_x & \text{if } x \leq n \\ \varphi_{g(\langle a_0, \ldots, a_n \rangle)}(x) & \text{if } x > n \text{ and } g \text{ is not forced to change} \\ \text{undefined} & \text{otherwise} \end{cases}$$

(more precisely, $f \in \mathcal{C}$ is equal to f_a for any sequence number a coding the first n values of f, for any n such that $g(\langle f(0), \ldots, f(n) \rangle)$ has reached its limit). Any such f_a is partial recursive uniformly in a, and by the S_n^m-Theorem there is then a recursive function h such that $\varphi_{h(a)} = f_a$. Thus the class $\{f_a\}_{a \in \omega}$ is an r.e. class of partial recursive functions containing $\mathcal{C}$.

It remains to show that $\{f_a\}_{a \in \omega}$ is quasi-measured. Given a sequence number $a = \langle a_0, \ldots, a_n \rangle$ and an initial segment σ of length $m + 1$, f_a extends σ if and only if:

1. either σ is contained in a (as a partial function), i.e. $\sigma(x) \simeq a_x$ for all $x \leq m$;

2. or σ extends a and $g(\langle a_0, \ldots, a_n \rangle)$ does not change on σ, i.e.

$$\begin{aligned} g(\langle a_0, \ldots, a_n \rangle) &= g(\langle a_0, \ldots, a_n, \sigma(n+1) \rangle) \\ &\cdots \\ &= g(\langle a_0, \ldots, a_n, \sigma(n+1), \ldots, \sigma(m) \rangle). \end{aligned}$$

The condition is obviously necessary, by definition of f_a, and we now show that it is also sufficient. In case 1, f_a agrees with a up to n, and hence with

σ up to m. In case 2, f_a agrees with a, and hence with σ, up to n. For $n+1$, notice that

$$\varphi_{g(\langle a_0,\ldots,a_n\rangle)}(n+1) = \varphi_{g(\langle a_0,\ldots,a_n,\sigma(n+1)\rangle)}(n+1) = \sigma(n+1),$$

where the first equality holds because $g(\langle a_0,\ldots,a_n\rangle)$ does not change on σ, and the second does by consistency of g; this shows in particular that $f_a(n+1)$ is defined, because g is not forced to change, and that

$$f_a(n+1) = \sigma(n+1).$$

The proof for $n+2$ is similar, using the fact just proved that $f_a(n+1) = \sigma(n+1)$, and that g does not change on σ by hypothesis. In the same way one can proceed all the way to m. $\quad\square$

As we have already done for the notion of measuredness in 2.4, we now consider a weakening of the notion of honesty. Recall that f is h-honest if

$$(\exists e)[f \simeq \varphi_e \;\wedge\; (\forall_\infty x)(\Phi_e(x) \le h(x,\varphi_e(x)))].$$

Definition 2.6 *If f and h are recursive functions, then f is* **quasi-h-honest** *if*

$$(\exists e)[f \simeq \varphi_e \;\wedge\; (\forall_\infty x)(\Phi_e(x) \le h(x, \max_{y \le x} \varphi_e(y)))].$$

Obviously, if f is h-honest and h is monotone then f is quasi-h-honest:

$$\Phi_e(x) \le h(x,\varphi_e(x)) \le h(x, \max_{y \le x} \varphi_e(y)).$$

But the converse does not hold: quasi-h-honesty provides a bound to $\Phi_e(x)$ in terms only of $\max_{y \le x} \varphi_e(y)$, not just of $\varphi_e(x)$.

Theorem 2.7 Complexity-Theoretic Characterization of EX_{cons}. *A class of total recursive functions is in EX_{cons} if and only if it is a class of quasi-h-honest functions, for some recursive function h.*

Proof. The second proof of 2.3, showing that any class of total recursive functions contained in a set of honest functions is in EX_{cons}, can easily be adapted to quasi-honest sets, by substituting $h(x, \max_{y \le x} a_y)$ for $h(x, a_x)$.

Conversely, let $\mathcal{C}$ be identifiable by consistent explanation via g. Then one can define the following function:

$$h(x,z) = \max\{\Phi_{g(\langle a_0,\ldots,a_x\rangle)}(x) : z = \max_{y \le x} a_y\}.$$

Since g is consistent, $\varphi_{g(\langle a_0,\ldots,a_x\rangle)}(x) \simeq a_x$, and so $\Phi_{g(\langle a_0,\ldots,a_x\rangle)}(x)$ is defined; moreover, there are only finitely many sequence numbers $\langle a_0,\ldots,a_x\rangle$ such that $z = \max_{y\leq x} a_y$, and hence h is total recursive.

If $f \in C$ then g has a limit e on f, and $\varphi_e \simeq f$, because g identifies C. If x_0 is a point after which g does not change anymore on f, then

$$\Phi_e(x) \leq h(x, \max_{y\leq x} f(y))$$

for all $x \geq x_0$. Thus e is a witness of the fact that f is quasi-h-honest. $\qquad\square$

3 Identification by reliable explanation

The next notion formalizes the idea of a method of explanation that never permanently settles on a false hypothesis, and thus gives indirect information about its mistakes.

Definition 3.1 (Blum and Blum [1975]) *A class C of total recursive functions is **identifiable by reliable explanation** ($C \in \mathbf{EX_{rel}}$) if there is a total recursive function g such that, for every $f \in C$:*

- $\lim_{n\to\infty} g(\langle f(0),\ldots,f(n)\rangle)$ *exists, i.e.*

$$(\exists n_0)(\forall n \geq n_0)[g(\langle f(0),\ldots,f(n)\rangle) = g(\langle f(0),\ldots,f(n_0)\rangle)]$$

and for every total recursive f:[1]

- $\lim_{n\to\infty} g(\langle f(0),\ldots,f(n)\rangle)$ *exists* $\Rightarrow \varphi_{\lim_{n\to\infty} g(\langle f(0),\ldots,f(n)\rangle)} = f.$

In other words, $g(\langle f(0),\ldots,f(n)\rangle)$ provides a guess to an index of f, which stabilizes (from a certain point on) on an index of f whenever it stabilizes, and it does stabilize if $f \in C$.

Notice that, despite the fact that we do not require from our guesses that they be consistent with the available information, such information cannot be disregarded, lest we proceed independently of f and never be able to stabilize our guess.

Proposition 3.2 $EX_{cons} \subsetneq EX_{rel}.$

---

[1]This is the most we can ask in our setting, since $g(\langle f(0),\ldots,f(n)\rangle)$ can be defined for all n only if f is total, and its limit can be an index of f only if f is recursive.

Proof. Let g identify C by consistent explanation. It is enough to show that

$$\lim_{n\to\infty} g(\langle f(0),\dots,f(n)\rangle) \text{ exists} \Rightarrow \varphi_{\lim_{n\to\infty} g(\langle f(0),\dots,f(n)\rangle)} = f.$$

If $\lim_{n\to\infty} g(\langle f(0),\dots,f(n)\rangle)$ exists, let n_0 be such that

$$(\forall n \geq n_0)[g(\langle f(0),\dots,f(n)\rangle) = g(\langle f(0),\dots,f(n_0)\rangle)].$$

For every x, if n is greater than both n_0 and x then

$$\varphi_{g(\langle f(0),\dots,f(n_0)\rangle)}(x) \simeq \varphi_{g(\langle f(0),\dots,f(n)\rangle)}(x) \simeq f(x),$$

where the first equality holds because $n \geq n_0$, and the second one because $n \geq x$ and g is consistent. $\square$

Having showed that reliable identification is at least as powerful as consistent identification, we now show that it is strictly more powerful.

Proposition 3.3 (Blum and Blum [1975], Fulk [1988]) $EX_{rel} - EX_{cons} \neq \emptyset$.

Proof. We want to find a class C of total recursive functions which is in EX_{rel} but not in EX_{cons}. Let

$$C = \{f : (\exists e)[f \simeq \varphi_e \wedge (\forall x)(\Phi_e(x) \leq f(x+1))]\}.$$

To show that $C \in EX_{rel}$, define g as follows: $g(\langle \rangle) = 0$; given $\langle a_0,\dots,a_n\rangle$, look for the smallest $e \leq n$ such that $\Phi_e(x) \leq a_{x+1}$ and $\varphi_e(x) \simeq a_x$ for all $x < n$, and let $g(\langle a_0,\dots,a_n\rangle)$ be such an e if one exists, and n otherwise.[2]

To show that $C \notin EX_{cons}$, suppose g is a consistent recursive guessing function such that

$$f \in C \Rightarrow f = \varphi_{\lim_{n\to\infty} g(\langle f(0),\dots,f(n)\rangle)}.$$

The idea is to construct $f \in C$ such that either $\lim_{n\to\infty} g(\langle f(0),\dots,f(n)\rangle)$ does not exist (because g changes its guess infinitely often), or it exists but it is not an index of f (because f differs from it on some argument).

By the S^m_n-Theorem, there is a recursive function t such that $\varphi_{t(e)}$ is defined as follows. If $x = 0$, then $\varphi_{t(e)}(0) \simeq 0$. If $x > 0$, wait until all of $\varphi_e(0),\dots,\varphi_e(x-1)$, and hence also $\Phi_e(x-1)$, are defined. There are two possible cases:

---

[2]Notice that this does not even imply that $\varphi_e(n)$ is defined, let alone that it is equal to a_n; thus the proposed identification procedure is not necessarily consistent (by the second part of the proof, it is provably not consistent).

1. If $g(\langle\varphi_e(0),\ldots,\varphi_e(x-1)\rangle) \neq g(\langle\varphi_e(0),\ldots,\varphi_e(x-1),\Phi_e(x-1)+1\rangle)$
   then we define $\varphi_{t(e)}(x) \simeq \Phi_e(x-1)+1$, thus ensuring that g has changed
   its value once.

2. Otherwise, we define $\varphi_{t(e)}(x) \simeq \Phi_e(x-1)$, thus ensuring that $\varphi_{t(e)}$ is
   different from the function guessed by g on the initial segment of $\varphi_{t(e)}$
   of length x.

By the Fixed-Point Theorem, there is e such that $\varphi_e \simeq \varphi_{t(e)}$. Let $f \simeq \varphi_e$:
then f is total by induction, and in $\mathcal{C}$ because $\Phi_e(x) \leq f(x+1)$ for every x.
There are now two possible cases:

- $\lim_{n\to\infty} g(\langle f(0),\ldots,f(n)\rangle)$ *does not exist*
  Then g does not identify f by consistent explanation.

- $\lim_{n\to\infty} g(\langle f(0),\ldots,f(n)\rangle)$ *exists*
  If $g(\langle f(0),\ldots,f(x)\rangle)$ is this limit, then case 1 cannot take place at x,
  otherwise g would have changed its value at least once more, and so

$$\varphi_{g(\langle f(0),\ldots,f(x-1)\rangle)}(x) \simeq \varphi_{g(\langle f(0),\ldots,f(x-1),\Phi_e(x-1)+1\rangle)}(x) \simeq \Phi_e(x-1)+1,$$

  where the second equality holds by consistency of g. But since case 1
  does not take place, $f(x) = \Phi_e(x-1)$; thus f differs from
  $\varphi_{\lim_{n\to\infty} g(\langle f(0),\ldots,f(n)\rangle)}$ at x. □

We do not know of any number-theoretic characterization of EX_{rel}. For a
complexity-theoretic characterization, a hint comes from the notion of quasi-
honesty used in 2.7, and the observation that the expression

$$h(x, \max_{y\leq x} \alpha(y)),$$

used in it defines a partial recursive functional $H(\alpha, x)$.

Definition 3.4 *If f is a recursive function and H is a partial recursive functional, then f is **H**-honest if*

$$(\exists e)[f \simeq \varphi_e \wedge (\forall_\infty x)(H(\varphi_e, x)\downarrow \wedge \Phi_e(x) \leq H(\varphi_e, x))].$$

**Theorem 3.5 Complexity-Theoretic Characterization of EX_{rel} (Blum
and Blum [1975])** *A class $\mathcal{C}$ of total recursive functions is in EX_{rel} if and
only if it is a class of H-honest functions, for some partial recursive functional
H which is total on all total recursive functions.*

Proof. Given a partial recursive functional H total on all total recursive functions, notice that if φ_e is total and H-honest then there is a constant k such that

$$(\forall x)[\Phi_e(x) \leq H(\varphi_e, x) + k].$$

One can define a recursive function g that reliably identifies the class of all total H-honest functions, as follows:

- on the empty list, g takes the value 0

- on the list $\langle a_0, \ldots, a_n \rangle$, g takes the value e for the smallest pair $\langle e, k \rangle$ defined as follows, if there is one, and n otherwise:

  - $\langle e, k \rangle \leq n$
  - for all $x \leq n$, if $H(\langle a_0, \ldots, a_n \rangle, x)$ converges in at most n steps (where $\langle a_0, \ldots, a_n \rangle$ is considered as the partial function giving value a_x to $x \leq n$, and undefined otherwise) then

    $$\Phi_e(x) \leq H(\langle a_0, \ldots, a_n \rangle, x) + k \quad \text{and} \quad \varphi_e(x) \simeq a_x.$$

It is easy to show that g identifies by reliable explanation the class of all total H-honest functions.

In the opposite direction, let $\mathcal{C}$ be identifiable by reliable explanation via g, and define

$$H(\alpha, x) \simeq \begin{cases} \Phi_{g(\langle \alpha(0), \ldots, \alpha(x) \rangle)}(x) & \text{if } \varphi_{g(\langle \alpha(0), \ldots, \alpha(x) \rangle)}(x) \text{ converges first} \\ 0 & \text{if } g \text{ changes its mind on } \alpha \text{ after } x \text{ first} \\ \text{undefined} & \text{otherwise.} \end{cases}$$

H is a partial recursive functional by definition.

If f is a total recursive function, we show that $H(f, x)$ is defined for any x, so that H is total on the total recursive functions. Either $g(\langle f(0), \ldots, f(x) \rangle)$ changes after x, and then $H(f, x)$ is defined by the second clause if not otherwise; or $g(\langle f(0), \ldots, f(x) \rangle)$ is the limit of g on f and hence, by reliability, an index of the total function f, so that $H(f, x)$ is defined by the first clause.

Finally, if $f \in \mathcal{C}$ then g has a limit e on f, and $\varphi_e \simeq f$, because g identifies $\mathcal{C}$. If x_0 is a point after which g does not change anymore on f, then $H(f, x) \simeq \Phi_e(x)$ for all $x \geq x_0$, because $\varphi_e(x)$ converges (since f is total). Thus e is a witness of the fact that f is H-honest. $\square$

4 Identification by explanation

There is an obvious way of relaxing the definitions of EX_{cons} or EX_{rel}: just drop any consistency or reliability requirement.

Definition 4.1 (Gold [1967]) *A class C of total recursive functions is* **identifiable by explanation** *($C \in \mathbf{EX}$) if there is a total recursive function g such that, for every $f \in C$:*

- $\lim_{n \to \infty} g(\langle f(0), \ldots, f(n) \rangle)$ *exists, i.e.*

$$(\exists n_0)(\forall n \geq n_0)[g(\langle f(0), \ldots, f(n) \rangle) = g(\langle f(0), \ldots, f(n_0) \rangle)]$$

- $\varphi_{\lim_{n \to \infty} g(\langle f(0), \ldots, f(n) \rangle)} = f.$

In other words, $g(\langle f(0), \ldots, f(n) \rangle)$ provides a guess to an index of f, and the guess stabilizes (from a certain point on) on an index of f.

Proposition 4.2 (Barzdin [1974], Blum and Blum [1975]) $EX_{rel} \subset EX$.

Proof. The inclusion is obvious by definition. To show that it is proper, we want to find a class C of total recursive functions which is in EX but not in EX_{rel}. Let

$$C = \{f : f = \varphi_{f(0)}\},$$

i.e. the class of total recursive functions such that $f(0)$ is an index of f.

$C \in EX$, by letting $g(\langle \; \rangle) = 0$ and $g(\langle a_0, \ldots, a_n \rangle) = a_0$. In other words, every function in C gives away a program for itself as its first value, thus making identification by explanation trivial.

To show that $C \notin EX_{rel}$, suppose g is a reliable recursive guessing function such that

$$f = \varphi_{f(0)} \Rightarrow f = \varphi_{\lim_{n \to \infty} g(\langle f(0), \ldots, f(n) \rangle)}.$$

By reliability, for any f

$$\lim_{n \to \infty} g(\langle f(0), \ldots, f(n) \rangle) \text{ exists} \Rightarrow f = \varphi_{\lim_{n \to \infty} g(\langle f(0), \ldots, f(n) \rangle)}.$$

Thus we cannot construct an $f \in C$ such that if $\lim_{n \to \infty} g(\langle f(0), \ldots, f(n) \rangle)$ exists then it is not an index of f (as was one option in the proof of 3.3), and we are forced to find a function f such that

$$f = \varphi_{f(0)} \quad \text{and} \quad \lim_{n \to \infty} g(\langle f(0), \ldots, f(n) \rangle) \text{ does not exist.}$$

By the S_n^m-Theorem, there is a recursive function t such that $\varphi_{t(e)}$ is defined as follows. Start with $\varphi_{t(e)}(0) \simeq e$. Consider the functions f_i ($i = 0, 1$) equal to e for $x = 0$, and identically equal to i for $x > 0$: of the two limits

$$\lim_{n \to \infty} g(\langle f_1(0), \ldots, f_1(n) \rangle) \quad \text{and} \quad \lim_{n \to \infty} g(\langle f_2(0), \ldots, f_2(n) \rangle)$$

either at least one does not exist, or they both exist and are different (by reliability of g, since the two functions f_0 and f_1 are different). Thus there must exist n and i ($i = 0$ or $i = 1$) such that

$$g(\langle \varphi_{t(e)}(0) \rangle) \neq g(\langle f_i(0), \ldots, f_i(n) \rangle),$$

and we can effectively find them. Let $\varphi_{t(e)}(x)$ be equal to $f_i(x)$ for every $x \leq n$, and iterate the procedure (by extending, at each step, the values already obtained by a sequence of either 0's or 1's).

By the Fixed-Point Theorem, there is e such that $\varphi_e \simeq \varphi_{t(e)}$. Let $f \simeq \varphi_e$: then f is in $\mathcal{C}$ because its value on 0 is an index for it, and g does not identify f because, by construction, $\lim_{n \to \infty} g(\langle f(0), \ldots, f(n) \rangle)$ does not exist. $\square$

Notice that the class $\{f : f = \varphi_{f(0)}\}$ used in the previous proof is a subclass of the r.e. class of partial recursive functions $\{\varphi_{h(e)}\}_{e \in \omega}$ defined as follows:

$$\varphi_{h(e)}(x) \simeq \begin{cases} e & \text{if } x = 0 \\ \varphi_e(x) & \text{otherwise.} \end{cases}$$

We can effectively tell apart all pairs of members of such a class, since they all differ on their first arguments.

We turn now to a characterization of EX, in terms of the following generalization of the property just noticed.

Definition 4.3 *An r.e. class $\{\varphi_{h(e)}\}_{e \in \omega}$ is called* **effectively separable** *if there is a uniform recursive procedure to determine, for every different e and i, an upper bound to an argument on which $\varphi_{h(e)}$ and $\varphi_{h(i)}$ disagree.*

Theorem 4.4 First Number-Theoretic Characterization of EX (Wiehagen and Jung [1977], Wiehagen [1978]) *A class $\mathcal{C}$ of total recursive functions is in EX if and only if it is a subclass of an effectively separable r.e. class of partial recursive functions.*

Proof. Given an effectively separable r.e. class $\{\varphi_{h(e)}\}_{e \in \omega}$, there is a recursive function d (for 'disagreement') such that, for every different e and i, there is $x \leq d(e, i)$ such that $\varphi_{h(e)}(x) \not\simeq \varphi_{h(i)}(x)$. One can define a recursive function g that identifies by explanation any subclass $\mathcal{C}$ of the given class, as follows:

- on the empty list, g takes the value $h(0)$

- suppose the value of g on the list $\langle a_0, \ldots, a_{n-1} \rangle$ was $h(e)$; then on the list $\langle a_0, \ldots, a_n \rangle$ the value of g is still $h(e)$, unless there is an i as follows, in which case g takes the value $h(e+1)$:

  - i is different from e

- $i \leq n$
- $d(e, i) \leq n$
- for all $x \leq d(e, i)$, $\varphi_{h(i)}(x) \simeq a_x$, with all computations convergent in at most n steps.

It is easy to show that C is identified by explanation via g.

In the opposite direction, let C be identifiable by explanation via g. Then every function in C is of the following form, for some sequence number a (coding a list $\langle a_0, \ldots, a_n \rangle$ for some n):

$$f_a(x) \simeq \begin{cases} a_x & \text{if } x \leq n \\ \varphi_{g(\langle a_0, \ldots, a_n \rangle)}(x) & \text{otherwise} \end{cases}$$

(more precisely, $f \in C$ is equal to f_a for any sequence number a coding the first n values of f, for any n such that $g(\langle f(0), \ldots, f(n) \rangle)$ has reached its limit). Any such f_a is partial recursive uniformly in a, and by the S_n^m-Theorem there is then a recursive function h such that $\varphi_{h(a)} = f_a$. Thus the class $\{f_a\}_{a \in \omega}$ is an r.e. class of partial recursive functions containing C. But not all members of it are different (since, by the parenthetical remark above, each function in C appears infinitely often in the class).

Incompatible sequence numbers are not problematic, since the associated functions agree with them, and hence they differ among each other. Among the compatible sequence numbers, we can certainly cut down a number of repetitions, by considering only those for which g provides new guesses: they form an r.e. class, and hence so does the set of associated functions. But we can still have two compatible sequence numbers a and b such that $f_a = f_b$: for example, both $g(a)$ and $g(b)$ might be the final guess of an index of a function in C, but in between them g could have changed its mind.

We thus take two complementary actions: on the one hand, we start defining a new function only when we hit a sequence number that provides a new guess of g, and on the other hand we stop defining the new function as soon as we discover that the guess of g changes. In other words, for any sequence number $a = \langle a_0, \ldots, a_n \rangle$ such that

$$g(\langle a_0, \ldots, a_{n-1} \rangle) \neq g(\langle a_0, \ldots, a_n \rangle)$$

we let:

$$f_a(x) \simeq \begin{cases} a_x & \text{if } x \leq n \\ \varphi_{g(\langle a_0, \ldots, a_n \rangle)}(x) & \text{if } x > n \text{ and } g \text{ is not forced to change} \\ \text{undefined} & \text{otherwise.} \end{cases}$$

As argued above, we thus obtain an r.e. class of partial recursive functions containing C. It only remains to show that there is a recursive function d

such that, if a and b are sequence numbers among the ones considered, then $d(a, b)$ is an upper bound to an argument on which f_a and f_b disagree.

It is enough to let $d(a, b)$ be the maximum of the lengths of a and b. There are two cases to consider:

- If a and b are incompatible, then they differ on some component below $d(a, b)$. But f_a and f_b agree with a and b, respectively, and so they differ on some argument below $d(a, b)$.

- If a and b are compatible, suppose e.g. that a is contained in b, and thus $d(a, b)$ is just the length of b.

  Since both a and b have been considered, g must have changed its guess on some intermediate sequence number, i.e. on some initial segment of b of length $n < d(a, b)$. Then either f_a did not agree with b up to n, and then it differs from f_b for one reason (because by definition f_b does agree with b up to its length), or it did agree, and then it differs from f_b for another reason (because by definition f_a stops being defined at n if not before, while f_b is defined there). $\square$

We now introduce a variation of 4.3.

Definition 4.5 *An r.e. class* $\{\varphi_{h(e)}\}_{e \in \omega}$ *of partial recursive functions is* **effectively discrete** *if there is a uniform procedure to determine, for every e and all almost all i, an upper bound to arguments on which $\varphi_{h(e)}$ disagrees with $\varphi_{h(i)}$.*

Notice how effective discreteness relates to effective separability: it is weaker because it does not require the existence of a recursive enumeration of the given class without repetitions, and it allows instead for finitely many repetions of each function; and it is stronger because it provides a bound that depends only on e, and not on both e and i.

The next result shows that weakening and strengthening compensate, and that the new notion still characterizes the same classes as the old one.

Theorem 4.6 Second Number-Theoretic Characterization of *EX* (Freivalds, Kinber and Wiehagen [1984]) *A class C of total recursive functions is in EX if and only if it is a subclass of an effectively discrete r.e. class of partial recursive functions.*

Proof. Given an effectively discrete r.e. class $\{\varphi_{h(e)}\}_{e \in \omega}$, there is a recursive function d (for 'discreteness') such that, for every e, there are at most finitely many i such that $\varphi_{h(e)}(x) \simeq \varphi_{h(i)}(x)$ for all $x \leq d(e)$. One can define a recursive function g that identifies by explanation any subclass C of the given class, as follows:

- on the empty list, g takes value 0

- on the list $\langle a_0, \ldots, a_n \rangle$, g takes value 0, unless there are indices e such that:

    - $e \leq n$
    - $d(e) \leq n$
    - for all $x \leq d(e)$, $\varphi_{h(e)}(x) \simeq a_x$ in at most n steps
    - for all x such that $d(e) < x \leq n$, if $\varphi_{h(e)}(x)$ converges in at most n steps, then $\varphi_{h(e)}(x) \simeq a_x$.

In this case one lets $I_{\langle a_0, \ldots, a_n \rangle}$ be the set of the $h(e)$'s corresponding to all such e's, and defines $g(\langle a_0, \ldots, a_n \rangle)$ as an index of the function that, on any input x, dovetails computations of $\varphi_{h(e)}(x)$ for all $h(e) \in I_{\langle a_0, \ldots, a_n \rangle}$, and outputs the first convergent value.

It is easy to show that $\mathcal{C}$ is identified by explanation via g.

In the opposite direction, let $\mathcal{C}$ be identifiable by explanation via g. By letting $d(a)$ be the length of a, one sees that the r.e. class $\{f_a\}_{a \in \omega}$ defined in the proof of 4.4 is effectively discrete. $\square$

Turning now to complexity-theoretic characterizations, the most natural weakening of 3.5 is obtained by dropping the restriction that the functional H be total on all total recursive functions: this is however too weak, and will be used in 5.3 to characterize EX^*. For a characterization of EX, the following stronger notion is appropriate.

Definition 4.7 *If f is a recursive function and H is a partial recursive functional, then f is **very H-honest** if*

$$(\exists e)[f \simeq \varphi_e \ \wedge \ (\forall_\infty x)(H(\varphi_e, x)\downarrow \wedge (\max_{y \leq x} \Phi_e(y)) \leq H(\varphi_e, x))].$$

Thus, while the value $H(\varphi_e, x)$ almost always bounds $\Phi_e(x)$ for H-honest functions, it actually bounds $\max_{y \leq x} \Phi_e(y)$ for very H-honest functions.

Theorem 4.8 Complexity-Theoretic Characterization of EX (Wiehagen and Liepe [1976], Wiehagen [1978]) *A class $\mathcal{C}$ of total recursive functions is in EX if and only if it is a class of very H-honest functions, for some partial recursive functional H.*

Proof. Given a partial recursive functional H, notice that if φ_e is total and very H-honest then there is a constant k such that

$$(\forall x \geq k)[H(\varphi_e, x)\downarrow \ \Rightarrow \ (\max_{y \leq x} \Phi_e(y)) \leq H(\varphi_e, x)].$$

One can define a recursive function g that that identifies by explanation the class of all total very H-honest functions, as follows:

- on the empty list, g takes the value 0

- on the list $\langle a_0, \ldots, a_n \rangle$, g takes the value e for the smallest pair $\langle e, k \rangle$ defined as follows, if there is one, and n otherwise:

    - $\langle e, k \rangle \leq n$
    - for all x such that $k \leq x \leq n$, if $H(\langle a_0, \ldots, a_n \rangle, x)$ converges in at most n steps (where $\langle a_0, \ldots, a_n \rangle$ is considered as the partial function giving value a_x to $x \leq n$ and undefined otherwise) then, for all $y \leq x$,

$$\Phi_e(y) \leq H(\langle a_0, \ldots, a_n \rangle, x) \quad \text{and} \quad \varphi_e(y) \simeq a_y.$$

It is easy to show that g identifies by explanation the class of all total very H-honest functions.

In the opposite direction, let $\mathcal{C}$ be identifiable by explanation via g, and define

$$H(\alpha, x) \simeq \max\{\Phi_{g(\langle \alpha(0), \ldots, \alpha(x) \rangle)}(y) : y \leq x\}.$$

H is a partial recursive functional by definition.

Since g identifies $\mathcal{C}$, if $f \in \mathcal{C}$ then g has a limit e on f, and $\varphi_e \simeq f$. If x_0 is a point after which g does not change anymore on f then, for all $x \geq x_0$,

$$H(f, x) \simeq \max\{\Phi_{g(\langle f(0), \ldots, f(x) \rangle)}(y) : y \leq x\} = \max_{y \leq x} \Phi_e(y).$$

Then e is a witness of the fact that f is very H-honest. $\qquad \square$

5 Identification by explanation with finite errors

There are two opposite ways of relaxing the requirements imposed in the definition of EX: we can be less restrictive on the function we identify in the limit, and ask not for the real f, but only for a finite approximation to it; or we can be less restrictive on the convergence of our guesses. We investigate here the first option, and in next sections the second one.

Finitely many exceptions to the range of an explanation are readily accepted in science: for example, Newtonian mechanics was accepted even if it did not account correctly for the motion of Mercury's perielion. The next notion is thus not without interest.

Definition 5.1 (Blum and Blum [1975]) *A class C of total recursive functions is* **identifiable by explanation with finitely many errors** *($C \in EX^*$) if there is a total recursive function g such that, for every $f \in C$:*

- $\lim_{n \to \infty} g(\langle f(0), \dots, f(n) \rangle)$ *exists*

- $\varphi_{\lim_{n \to \infty} g(\langle f(0), \dots, f(n) \rangle)} \simeq^* f$.

In other words, $g(\langle f(0), \dots, f(n) \rangle)$ provides a guess to an index of f, and the guess stabilizes (from a certain point on) on an index of a finite variant of f.

Notice how, since we are now interested in guessing not the real function f but only a finite variant of it, we cannot request that the current guess agrees with the available information, and thus no analogue of EX_{cons} makes sense in this context.

Also, while in the definitions of EX_{cons} and EX the temporary guesses could be programs for partial functions but the final guesses had to be programs for a total function, here *all* guesses may actually be programs for partial functions (although the final guess must compute a function that can be undefined only on finitely many arguments, since it has to be a finite variant of a total functions).

Proposition 5.2 (Blum and Blum [1975]) $EX \subset EX^*$.

Proof. The inclusion is obvious by definition. To show that it is proper, we want to find a class C of total recursive functions which is in EX^* but not in EX. Let

$$C = \{f : f \simeq^* \varphi_{f(0)}\},$$

i.e. the class of total recursive functions such that $f(0)$ is an index of a finite variant of f.

$C \in EX^*$, by letting $g(\langle \ \rangle) = 0$ and $g(\langle a_0, \dots, a_n \rangle) = a_0$. In other words, every function in C gives away a program for a finite variant of itself as its first value, thus making identification by explanation with finite errors trivial.

To show that $C \notin EX$, suppose g is a recursive function such that

$$f \simeq^* \varphi_{f(0)} \ \Rightarrow \ f = \varphi_{\lim_{n \to \infty} g(\langle f(0), \dots, f(n) \rangle)}.$$

The idea is to construct $f \in C$ such that either $\lim_{n \to \infty} g(\langle f(0), \dots, f(n) \rangle)$ does not exist (because g changes its guess infinitely often), or it exists but it is not an index of f (being an index of a partial function).

There is no problem in forcing $f(0)$ to be an index of the function f to be defined, using the Fixed-Point Theorem as in the proof of 4.2. We thus concentrate on the definition of the remaining values of f. At each stage $n+1$ we have already defined all values of f up to n with one exception a_n, which is used to satisfy the idea above. There are three possible cases:

1. If $g(\langle f(0), \ldots, f(a_n-1) \rangle) \neq g(\langle f(0), \ldots, f(a_n-1), 0, f(a_n+1), \ldots, f(n) \rangle)$ then we define $f(a_n) = 0$, thus ensuring that g has changed its value once, and let $a_{n+1} = n + 1$ (since now every value up to n has been defined).

2. If case 1 does not hold, and $\varphi_{g(\langle f(0), \ldots, f(a_n-1) \rangle)}(a_n)$ converges in at most n steps, then we define $f(a_n) = 1 - \varphi_{g(\langle f(0), \ldots, f(a_n-1) \rangle)}(a_n)$, thus ensuring that f is different from the function guessed on the initial segment of f of length a_n, and let $a_{n+1} = n + 1$ as above.

3. If cases 1 and 2 do not hold, we let $f(n+1) = 0$, thus defining f on one more value, and $a_{n+1} = a_n$ (to have a new shot at it at the next stage).

At the end, there are two possible cases:

- $\lim_{n \to \infty} a_n$ *does not exist*
  This means that we move a_n infinitely often, in particular the function f is total. Moreover, $f = \varphi_{f(0)}$ by construction, in particular $f \in C$. By hypothesis then $\lim_{n \to \infty} g(\langle f(0), \ldots, f(n) \rangle)$ should exist, and f should be $\varphi_{\lim_{n \to \infty} g(\langle f(0), \ldots, f(n) \rangle)}$.

  But if $\lim_{n \to \infty} g(\langle f(0), \ldots, f(n) \rangle)$ exists, then case 1 holds only finitely many times. Since a_n moves infinitely often, after a certain stage it must do so because of case 2. But then f is different from $\varphi_{\lim_{n \to \infty} g(\langle f(0), \ldots, f(n) \rangle)}$ (because it disagrees with infinitely many guesses, and hence it must disagree with the final guess).

- $\lim_{n \to \infty} a_n$ *exists*
  If a is this limit, then f is defined everywhere except in a, since otherwise a would have moved as soon as $f(a)$ had been defined. Let f_1 be the extension of f obtained by letting $f_1(a) = 0$, and $f_1(x) = f(x)$ if $x \neq a$.

  Then $g(\langle f_1(0), \ldots, f_1(a-1) \rangle) = \lim_{n \to \infty} g(\langle f_1(0), \ldots, f_1(n) \rangle)$, otherwise case 1 would have taken place, and a would have moved.

  Moreover, $\varphi_{\lim_{n \to \infty} g(\langle f_1(0), \ldots, f_1(n) \rangle)}(a)$ must be undefined, otherwise case 2 would have taken place, and a would have moved.

  $f_1(0) = f(0)$ is by definition an index of f, and hence of a finite variant of f_1. So $f_1 \simeq^* \varphi_{f_1(0)}$, but $f_1 \not\simeq \varphi_{\lim_{n \to \infty} g(\langle f_1(0), \ldots, f_1(n) \rangle)}$, because the former is defined on a but the latter is not. $\square$

The proof of the previous result shows that actually

$$C = \{f : f \simeq^* \varphi_{f(0)}, \text{ with at most one disagreement point}\}$$

is in $EX^* - EX$. In particular, even allowing for a *single* exception in the explanation of a class of phenomena already gives more power (in the sense of being able to explain more classes of phenomena) than requiring an explanation to be always correct.

Moreover, the proof also shows how the exception might occur in a place in which the explanation does not give any answer, being undefined. Since the Halting Problem is unsolvable, such an explanation (being correct when it does provide an answer, and incorrect in a divergent point) cannot be refuted (in a recursive way). In other words, *there are incomplete and irrefutable explanations*, and they do not satisfy a form of Popper Refutability Principle (Popper [1934]), according to which incorrect scientific explanations should be refutable.

We do not know of any number-theoretic characterization of EX^*. For a complexity-theoretic characterization, we already have at hand the appropriate notion.

Theorem 5.3 Complexity-Theoretic Characterization of EX^* (Wiehagen [1978], Kinber) *A class $\mathcal{C}$ of total recursive functions is in EX^* if and only if it is a class of H-honest functions, for some partial recursive functional H and w.r.t. space complexity measure.*

Proof. Given a partial recursive functional H, notice that if φ_e is H-honest then there is a constant k such that

$$(\forall x \geq k)[H(\varphi_e, x)\!\downarrow \,\Rightarrow\, \Phi_e(x) \leq H(\varphi_e, x)].$$

One can define a recursive function g that identifies by explanation with finite errors the class of all total H-honest functions, as follows:

- on the empty list, g takes the value 0

- on the list $\langle a_0, \ldots, a_n \rangle$, g takes the value e for the smallest pair $\langle e, k \rangle$ defined as follows, if there is one, and n otherwise:

    - $\langle e, k \rangle \leq n$
    - for all x such that $k \leq x \leq n$, if $H(\langle a_0, \ldots, a_n \rangle, x)$ converges in at most n steps (where $\langle a_0, \ldots, a_n \rangle$ is considered as the partial function giving value a_x to $x \leq n$ and undefined otherwise) then

$$\Phi_e(x) \leq H(\langle a_0, \ldots, a_n \rangle, x) \quad \text{and} \quad \varphi_e(x) \simeq a_x.$$

It is easy to show that g identifies by explanation with finite errors the class of all total H-honest functions.

In the opposite direction, let $\mathcal{C}$ be identifiable by explanation with finite errors via g, and define

$$H(\alpha, x) \simeq \Phi_{g(\langle \alpha(0), \ldots, \alpha(x) \rangle)}(x).$$

H is a partial recursive functional by definition.

Since g identifies C with finite errors, if $f \in C$ then g has a limit e on f, and $\varphi_e \simeq^* f$. If x_0 is a point after which g does not change anymore on f then, for all $x \geq x_0$,

$$H(f,x) \simeq \Phi_{g(\langle f(0),\ldots,f(x)\rangle)}(x) \simeq \Phi_e(x).$$

If moreover $x_1 \geq x_0$ is a point after which φ_e and f agree, then $\varphi_e(x)$ converges for all $x \geq x_1$, and hence

$$(\forall_\infty x)[H(f,x)\downarrow \wedge \Phi_e(x) \leq H(f,x)].$$

We cannot claim that f is H-honest yet, since e is not an index of f (as required by Definition 3.4), but only of a finite variant of it. However, if we consider a complexity measure such as space, whose complexity classes are closed under finite variants, the result then follows. □

6 Behaviorally correct (consistent) identification

In the definitions of EX_{cons} and EX we required the final explanation of each phenomenon in a given class to be *intensionally* unique. The next definition only asks for *extensional* uniqueness, and allows for the possibility of not having a final intensional explanation (i.e. it allows infinitely many changes in the program, although still only finitely many in the function defined by it).

Definition 6.1 (Feldman [1972], Barzdin [1974]) *A class C of total recursive functions is* **behaviorally correctly and consistently identifiable** *($C \in BC_{cons}$) if there is a total recursive function g such that, for every sequence number $\langle a_0,\ldots,a_n\rangle$:*

- $\varphi_{g(\langle a_0,\ldots,a_n\rangle)}(x) \simeq a_x$ *for all $x \leq n$,*

and for every $f \in C$ and almost every n,

- $\varphi_{g(\langle f(0),\ldots,f(n)\rangle)} = f$.

In other words, $g(\langle f(0),\ldots,f(n)\rangle)$ provides a guess to an index of f consistent with the available information, and the guess stabilizes (from a certain point on) on indices of f.

C is **behaviorally correctly identifiable** *($C \in BC$) if the first condition on g is dropped.*

First of all we see that the new notions of inference just introduced coincide, so that no analogue of 4.2 holds.

Proposition 6.2 $BC_{cons} = BC$.

Proof. $BC_{cons} \subseteq BC$ by definition. Conversely, if a procedure g outputs the guesses for functions in a class $C \in BC$, then these guesses on $f \in C$ are correct, and hence consistent in the limit. One can modify such a procedure, and output at stage n an index of the function that agrees with f on the arguments $\leq n$, and with $\varphi_{g(\langle f(0),...,f(n)\rangle)}$ otherwise. I.e. one outputs $g_1(\langle f(0), \ldots, f(n)\rangle)$, for any g_1 such that

$$\varphi_{g_1(\langle a_0,...,a_n\rangle)}(x) \simeq \begin{cases} a_x & \text{if } x \leq n \\ \varphi_{g(\langle a_0,...,a_n\rangle)}(x) & \text{otherwise.} \end{cases}$$

Since if $f \in C$ then $g(\langle f(0), \ldots, f(n)\rangle)$ stabilizes on indices of f, g_1 still identifies C in a behaviorally correct way, and is consistent by definition. Thus C is in BC_{cons}. $\square$

We now show that the new notion of inference is related to (and weaker than) the ones studied so far.

Proposition 6.3 (Steel) $EX^* \subseteq BC$.

Proof. If a procedure g outputs the guesses for functions in a class $C \in EX^*$, then these guesses on $f \in C$ are correct in the limit, but only for finite variants of f. One can modifiy such a procedure as in 6.2, by considering g_1 such that

$$\varphi_{g_1(\langle a_0,...,a_n\rangle)}(x) \simeq \begin{cases} a_x & \text{if } x \leq n \\ \varphi_{g(\langle a_0,...,a_n\rangle)}(x) & \text{otherwise.} \end{cases}$$

Since if $f \in C$ then $g(\langle f(0), \ldots, f(n)\rangle)$ stabilizes on an index of a finite variant of f, after a finite number of stages all modifications will compute the same function, and will be correct on all values. $\square$

Notice that in the previous proof we cannot conclude $C \in EX$, because if $f \in C$ then $g_1(\langle f(0), \ldots, f(n)\rangle)$ codes a different program for every n, although a program obtained by patching up an eventually fixed program (for a finite variant of f) on an increasing number of arguments.

Proposition 6.4 (Barzdin [1974], Case and Smith [1983], Harrington) $BC - EX^* \neq \emptyset$.

Proof. We want to find a class of $\mathcal{C}$ of total recursive functions which is in BC but not in EX^*. Let

$$\mathcal{C} = \{f : (\forall_\infty x)(f = \varphi_{f(x)})\},$$

i.e. the class of total recursive functions such that $f(x)$ is an index of f, for almost every x.

$\mathcal{C} \in BC$, by letting $g(\langle\ \rangle) = 0$ and $g(\langle a_0, \ldots, a_n \rangle) = a_n$. In other words, every function in $\mathcal{C}$ almost always gives away programs for itself as values, thus making behaviorally correct identification trivial.

To show that $\mathcal{C} \notin EX^*$, suppose g is a recursive function such that

$$(\forall_\infty x)(f = \varphi_{f(x)}) \;\Rightarrow\; f \simeq^* \varphi_{\lim_{n\to\infty} g(\langle f(0),\ldots,f(n)\rangle)}.$$

The idea is to construct $f \in \mathcal{C}$ such that either $\lim_{n\to\infty} g(\langle f(0), \ldots, f(n)\rangle)$ does not exist (because g changes its guess infinitely often), or it exists but it is not an index of f (being an index of a function which is not a finite variant of f).

We construct f by initial segments σ_s, starting from $\sigma_0 = \emptyset$, and let n_s be the greatest argument on which σ_s is defined. A natural strategy to satisfy the condition

$$f \not\simeq^* \varphi_{\lim_{n\to\infty} g(\langle f(0),\ldots,f(n)\rangle)},$$

if the limit exists, is to make

$$f(x) \neq \varphi_{g(\langle \sigma_s(0),\ldots,\sigma_s(n_s)\rangle)}(x),$$

for a new x at every stage.

At stage $s + 1$ we thus wait until $\varphi_{g(\langle \sigma_s(0),\ldots,\sigma_s(n_s)\rangle)}(x)$ converges for some $x > n_s$, and then diagonalize. But since at the same time we want $f \in \mathcal{C}$, we will have to give f its own indices as values; we then choose two distinct indices e_0 and e_1 of f (by the Fixed-Point Theorem), and when $\varphi_{g(\langle \sigma_s(0),\ldots,\sigma_s(n_s)\rangle)}(x)$ converges for some $x > n_s$ we extend σ_s by letting $\sigma_{s+1}(x) = e_i$, for an i (equal to 0 or 1) such that

$$e_i \neq \varphi_{g(\langle \sigma_s(0),\ldots,\sigma_s(n_s)\rangle)}(x).$$

To have an initial segment, we also give all the arguments less than x and not in the domain of σ_s one of the values e_0 or e_1 (say, e_0).

Obviously, there is no guarantee that $\varphi_{g(\langle \sigma_s(0),\ldots,\sigma_s(n_s)\rangle)}(x)$ will ever converge, for any $x > n_s$; and if it does not, then the construction would stall. We thus need an alternative strategy, and at stage $s + 1$ we also build an additional function f_s, that will work if the construction of f stalls at that step.

Again, to ensure $f_s \in C$ we will have to give f_s its own indices as values; we thus choose an index i_s of f_s (by the Fixed-Point Theorem), and at every step in the dovetailed computations of $\varphi_{g(\langle \sigma_s(0),\ldots,\sigma_s(n_s)\rangle)}(x)$ for $x > n_s$ we define f_s as i_s on a new argument, with the intent that if $\varphi_{g(\langle \sigma_s(0),\ldots,\sigma_s(n_s)\rangle)}(x)$ does not converge for any $x > n_s$, then f_s will be equal to i_s from a certain point on, and then automatically in C.

This would not be of great help, unless we also ensured that

$$f_s \not\simeq^* \varphi_{\lim_{n\to\infty} g(\langle f_s(0),\ldots,f_s(n)\rangle)}.$$

We are working under the hypothesis that $\varphi_{g(\langle \sigma_s(0),\ldots,\sigma_s(n_s)\rangle)}(x)$ does not converge for any $x > n_s$, and hence that $\varphi_{g(\langle \sigma_s(0),\ldots,\sigma_s(n_s)\rangle)}$ is a finite function. It is thus natural to use this, by letting f_s extend σ_s; thus we will define f_s as equal to i_s only for $x > n_s$, and equal to $\sigma_s(x)$ if $x \leq n_s$. If

$$g(\langle \sigma_s(0),\ldots,\sigma_s(n_s)\rangle) = \lim_{n\to\infty} g(\langle f_s(0),\ldots,f_s(n)\rangle),$$

then

$$f_s \not\simeq^* \varphi_{\lim_{n\to\infty} g(\langle f_s(0),\ldots,f_s(n)\rangle)}$$

because f_s is total, while the right hand side is a finite function.

Obviously, there is no guarantee that

$$g(\langle \sigma_s(0),\ldots,\sigma_s(n_s)\rangle) = \lim_{n\to\infty} g(\langle f_s(0),\ldots,f_s(n)\rangle).$$

But if this does not hold, then g changes value on f_s sooner or later, and we can take advantage of this to go back to the definition of f. Indeed, by the first strategy we tried to make $\lim_{n\to\infty} g(\langle f(0),\ldots,f(n)\rangle)$ not an index of f, while with the present back up strategy we are trying to ensure that the limit does not exist.

In other words, we continue to define f_s as i_s for more and more arguments, until we discover that for an initial segment τ of f_s,

$$g(\tau) \neq g(\langle \sigma_s(0),\ldots,\sigma_s(n_s)\rangle).$$

If this happens, then we let $\sigma_{s+1} = \tau$. This is consistent with what was previously done, i.e. $\sigma_{s+1} \supseteq \sigma_s$, because f_s extends σ_s.

Notice that defining $\sigma_{s+1} = \tau$ gives f value i_s for a number of arguments; to avoid ruining the strategy for $f \in C$, we better make i_s an index of f. Thus, we stop defining new values of f_s as i_s (since, in any case, f_s has lost its role as possible witness of the fact that g does not identify C), and from now on the definition of f_s will just copy the definition of f.

We now have to argue that the proposed construction works. There are two possible cases:

- *all stages terminate*

  Each value of f is either one of e_0 and e_1, or some i_s; hence an index of f, either by the initial choice of e_0 and e_1, or by construction of f_s. In particular, $(\forall x)(f = \varphi_{f(x)})$, and $f \in C$.

  If g has no limit on f, there is nothing to prove. If g does have a limit on f, then we look at the construction of f. Every time the second part of the construction is applied at some stage, g changes value on f; since g has a limit on f by hypothesis, the second part of the construction can thus by applied only finitely many times. Then the first part of the construction is applied almost always; but every time it is applied, a disagreement between f and the current guess is enforced, and once the guess has stabilized, a disagreement with the final guess is enforced. In particular, $f \not\simeq^* \varphi_{\lim_{n \to \infty} g(\langle f(0),...,f(n)\rangle)}$.

- *some stage $s + 1$ does not terminate*

  By construction, f_s is then total. Moreover, almost all of its values are i_s, hence an index of f_s by the choice of i_s. In particular, $(\forall_\infty x)(f_s = \varphi_{f_s(x)})$, and $f_s \in C$.

  By construction

  $$g(\langle \sigma_s(0), \ldots, \sigma_s(n_s)\rangle) = \lim_{n \to \infty} g(\langle f_s(0), \ldots, f_s(n)\rangle),$$

  otherwise stage $s+1$ would terminate by the second part of the construction. Moreover, $\varphi_{g(\langle \sigma_s(0),...,\sigma_s(n_s)\rangle)}$ is a finite function, because if it were defined on some $x > n_s$ then stage $s + 1$ would terminate by the first part of the construction. In particular, $f_s \not\simeq^* \varphi_{\lim_{n \to \infty} g(\langle f_s(0),...,f_s(n)\rangle)}$, because f_s is total.

We have made infinitely many appeals to the Fixed-Point Theorem in the constructions of f and of the f_s (which is not surprising, since f had to be self-referential infinitely often, to be in C). There would be no problem if these constructions were independent, but they are instead related one to the others (f may use the index i_s, and f_s may mimic f). We thus still have to make sure that it is really possible to construct an infinite sequence of functions, simultaneously using indices for all of them.

Since a sequence of indices can be thought of as the range of a function f, and the previous construction actually produces a recursive functional F, the next result provides a formal justification.

- **Functional Recursion Theorem (Case [1974])** *Given a partial recursive functional F, there is a total recursive function f such that*

  $$(\forall e)(\forall x)[\varphi_{f(e)}(x) \simeq F(f, e, x)].$$

Proof. It is enough to find a recursive function g such that

$$\varphi_{\varphi_{g(i)}(e)}(x) \simeq F(\varphi_i, e, x).$$

Then the usual Fixed-Point Theorem gives an i such that $\varphi_i \simeq \varphi_{g(i)}$, and

$$\varphi_{\varphi_i(e)}(x) \simeq \varphi_{\varphi_{g(i)}(e)}(x) \simeq F(\varphi_i, e, x).$$

By taking $f \simeq \varphi_i$, we obtain the result.

To get g, consider the partial recursive function

$$\psi(i, e, x) \simeq F(\varphi_i, e, x).$$

By the S_n^m-Theorem, there is a recursive function h such that

$$\varphi_{h(i,e)}(x) \simeq \psi(i, e, x).$$

By the S_n^m-Theorem again, there is a recursive function g such that

$$\varphi_{g(i)}(e) \simeq h(i, e).$$

Then

$$\varphi_{\varphi_{g(i)}(e)}(x) \simeq \varphi_{h(i,e)}(x) \simeq \psi(i, e, x) \simeq F(\varphi_i, e, x).$$

Notice that h is total because defined by the S_n^m-Theorem, and hence so are $\varphi_{g(i)}$ and φ_i, i.e. f. □

We do not know of any complexity-theoretic characterization of BC. The next result provides a number-theoretic one in the style of the characterization 4.6 of EX.

Definition 6.5 *Given an r.e. class $\{\varphi_{h(e)}\}_{e \in \omega}$ of partial recursive functions, a subclass C of total recursive functions is called* **effectively weakly discrete** *in it if there is a uniform procedure to determine, for every $f \in C$ and almost every i, an upper bound to arguments on which f disagrees with $\varphi_{h(i)}$, if they disagree at all.*

The notion just introduced is a double weakening of effective discreteness: first of all, it allows for infinitely many repetitions in the given class, while only finitely many were allowed in 4.5; secondly, it is not a global condition on the given class, but rather a local one on the given subclass.

Theorem 6.6 Characterization of BC (Wiehagen) *A class C of total recursive functions is in BC if and only if it is an effectively weakly discrete subclass of an r.e. class of partial recursive functions.*

Proof. Given a class $\mathcal{C}$ effectively weakly discrete in an r.e. class $\{\varphi_{h(e)}\}_{e \in \omega}$, there is a recursive function d such that, for every $\varphi_{h(e)} \in \mathcal{C}$, there are at most finitely many i such that, if $\varphi_{h(e)}(x) \simeq \varphi_{h(i)}(x)$ for all $x \leq d(e)$, then $\varphi_{h(e)}$ and $\varphi_{h(i)}$ are different. One can define a recursive function g that behaviorally correctly identifies $\mathcal{C}$ as in the proof of 4.6.

In the opposite direction, let $\mathcal{C}$ be behaviorally correctly identifiable via g. Then every function in $\mathcal{C}$ is of the following form, for some sequence number a (coding a list $\langle a_0, \ldots, a_n \rangle$ for some n):

$$f_a(x) \simeq \begin{cases} a_x & \text{if } x \leq n \\ \varphi_{g(\langle a_0, \ldots, a_n \rangle)}(x) & \text{otherwise} \end{cases}$$

(more precisely, $f \in \mathcal{C}$ is equal to f_a for any sequence number a coding the first n values of f, for any n greater than the point after which g only outputs indices of f). Any such f_a is partial recursive uniformly in a, and by the S_n^m-Theorem there is then a recursive function h such that $\varphi_{h(a)} = f_a$. Thus the class $\{f_a\}_{a \in \omega}$ is an r.e. class of partial recursive functions containing $\mathcal{C}$.

Let $d(a)$ be the length of a. Suppose $f \in \mathcal{C}$, and f and f_a agree up to $d(a)$, i.e. a is an initial segment of f. There are two cases: either $d(a)$ is greater than the point after which g only outputs indices of f, and then $f_a = f$ by definition; or $d(a)$ is not greater than such a point, and this can happen only for finitely many a's. $\square$

7 Behaviorally correct identification with finite errors

We next relax BC in a way similar to what we did for EX in 5.1.

Definition 7.1 (Osherson and Weinstein [1982], Case and Smith [1983]) *A class $\mathcal{C}$ of total recursive functions is **behaviorally correctly identifiable with finitely many errors** ($\mathcal{C} \in \boldsymbol{BC^*}$) if there is a recursive function g such that, for every $f \in \mathcal{C}$ and almost every n,*

$$\varphi_{g(\langle f(0), \ldots, f(n) \rangle)} \simeq^* f.$$

In other words, $g(\langle f(0), \ldots, f(n) \rangle)$ provides a guess to an index of f, and the guess stabilizes (from a certain point on) on indices of finite variants of f.

Notice that BC^* is a very weak notion: not only the guesses can change infinitely often, as they did for BC, but even the functions that they compute can do so. The only requirement is that each such function eventually be a finite variant of f.

The next result shows that we have finally reached the possible limits of inductive inference.

Theorem 7.2 Characterization of BC^* (Harrington) *The class of all (and hence any class of) total recursive functions is in BC^*.*

Proof. Let $\mathcal{C}$ be the class of all total recursive functions. To show that $\mathcal{C} \in BC^*$, the idea is to take advantage of the fact that if f is a total recursive function and e is its least index, then

$$(\forall i < e)(\exists x)[\varphi_i(x) \not\simeq f(x)],$$

i.e. for every smaller index there is a disagreement with f. Notice that, however, φ_i might disagree with f on x either because it converges to a different value, or because it does not converge (while f does, being total).

Let $g(\langle\ \rangle) = 0$. Given $a_0, \ldots, a_n$, we let $g(\langle a_0, \ldots, a_n \rangle)$ be the program of the total recursive function defined as follows. On input x, see if there is e_x such that:

- $e_x \leq n$

- for every $m \leq n$, $\varphi_{e_x}(m)$ converges in at most x steps to a_m.

Then

$$\varphi_{g(\langle a_0, \ldots, a_n \rangle)}(x) \simeq \begin{cases} \varphi_{e_x}(x) & \text{for the least such } e_x, \text{ if one exists} \\ 0 & \text{otherwise.} \end{cases}$$

Let now f be a total recursive function. We show that, for any sufficiently big n, $g(\langle f(0), \ldots, f(n) \rangle)$ is an index of a finite variant of f. Indeed, let e be the least index of f, and n_0 be big enough so that:

- $e \leq n_0$

- for every $i < e$ there is an $m \leq n_0$ such that $\varphi_i(m) \not\simeq f(m)$.

For any $n \geq n_0$, let x_n be big enough so that:

- for every $m \leq n$, $\varphi_e(m)$ converges in at most x_n steps.

Then, for every $x \geq x_n$,

$$\varphi_{g(\langle f(0), \ldots, f(n) \rangle)}(x) \simeq \varphi_e(x) \simeq f(x),$$

and thus $g(\langle f(0), \ldots, f(n) \rangle)$ is an index of a finite variant of f. $\square$

Proposition 7.3 (Bardzin [1974], Case and Smith [1983]) $BC \subset BC^*$.

Proof. The inclusion is obvious by definition. To show that it is proper, it is enough to show that the class of total recursive functions is not in BC. Let g be a recursive function such that

$$f \text{ total recursive } \Rightarrow \lim_{n \to \infty} \varphi_{g(\langle f(0),...,f(n) \rangle)} = f.$$

We define a total recursive function f such that $\lim_{n \to \infty} \varphi_{g(\langle f(0),...,f(n) \rangle)}$ does not exist, contradicting the hypothesis on g. The idea is that f will repeat the same value a number of times sufficient to make g converge, and then will diagonalize against the program thus obtained: since we do this infinitely often, then the function defined by g changes infinitely often on f, and thus it has no limit.

Start with $f(0) = 0$. The function f_0 identically equal to 0 is recursive and, by hypothesis on g, $\lim_{n \to \infty} \varphi_{g(\langle f_0(0),...,f_0(n) \rangle)}$ exists. Thus there must exist n and $m > n$ such that $\varphi_{g(\langle f_0(0),...,f_0(n) \rangle)}(m)\downarrow$, and we can effectively find them (by dovetailing computations). Let

$$f(x) = \begin{cases} f_0(x) & \text{if } x \leq n \\ 0 & \text{if } n \leq x < m \\ \varphi_{g(\langle f_0(0),...,f_0(n) \rangle)}(x) + 1 & \text{if } x = m, \end{cases}$$

and iterate the procedure (by considering, at each step s, the function f_s extending the values already obtained by a sequence of 0's). $\square$

In conclusion, the sequence of results proved in 2.2, 2.3, 3.2, 3.3, 4.2, 5.2, 6.2, 6.3, 6.4, and 7.3 provide the following picture of inductive inference classes:

$$NV \subset EX_{cons} \subset EX_{rel} \subset EX \subset EX^* \subset BC_{cons} = BC \subset BC^*.$$

Bibliography

Barzdin, J.M.
[1972] Prognostication of automata and functions, *Information Processing '71*, Freiman ed., North Holland 1972, 1972, pp. 81–84.
[1974] Two theorems on the limiting synthesis of functions, *Latv. Gos. Univ. Uce. Zap.* 210 (1974) 82–88.

Barzdin, J.M., and Freivalds, R.V.
[1972] On the prediction of general recursive functions, *Sov. Math. Dokl.* 13 (1972) 1224–1228.

Blum, L., and Blum, M.
[1975] Toward a mathematical theory of inductive inference, *Inf. Contr.* 28 (1975) 125–155.

Case, J.

[1974] Periodicity in generations of automata, *Math. Syst. Th.* 8 (1974) 15–32.

Case, J., and Smith, C.

[1983] Comparison of identification criteria for machine inductive inference, *Theor. Comp. Sci.* 25 (1983) 193–220.

Feldman, J.

[1972] Some decidability results on grammatical inference and complexity, *Inf. Contr.* 20 (1972) 244–262.

Freivalds, R.V., Kinber, E.B., and Wiehagen, R.

[1984] Conncetions between identifying functionals, standardizing operations, and computable numberings, *Zeit. Math. Log. Grund. Math.* 30 (1984) 145–164.

Fulk, M.A.

[1988] Saving the phenomena: requirements that inductive inference machines not contradict known data, *Inf. Comp.* 79 (1988) 193–209.

Gold, E.M.

[1967] Language identification in the limit, *Inf. Contr.* 10 (1967) 447–474.

Odifreddi, P.

[1989] *Classical Recursion Theory*, North Holland, 1989.

[1997] *Classical Recursion Theory*, volume II, North Holland, 1997.

Osherson, D.N., and Weinstein, S.

[1982] Criteria of language learning, *Inf. Contr.* 52 (1982) 123–138.

Popper, K.

[1934] *The logic of scientific discovery*, 1934.

Wiehagen, R.

[1978] Characterization problems in the theory of inductive inference, *Springer Lect. Not. Comp. Sci.* 62 (1978) 494–508.

Wiehagen, R., and Jung, H.

[1977] Rekursionstheoretische characterisierung von erkennbaren klassen rekursiver funktionen, *J. Inf. Proc. Cyb.* 13 (1977) 385–397.

Wiehagen, R., and Liepe, W.

[1976] Charakteristche Eigenschaften von erkennbaren Klassen rekursiver Funktionen, *J. Inf. Proc. Cyb.* 12 (1976) 421–438.

The Medvedev Lattice of Degrees of Difficulty

Andrea Sorbi *
Department of Mathematics
Via del Capitano 15
53100 Siena, Italy

1 Introduction

The Medvedev lattice was introduced in [5] as an attempt to make precise the idea, due to Kolmogorov, of identifying true propositional formulas with identically "solvable" problems. A *mass problem* is any set of functions (throughout this paper "function" means total function from ω to ω; the small Latin letters $f, g, h, \ldots$ will be used as variables for functions). Mass problems correspond to informal problems in the following sense: given any "informal problem", a mass problem corresponding to it is a set of functions which "solve" the problem, and at least one such function can be "obtained" by any "solution" to the problem (see [10]).

Example 1.1 If $A, B \subseteq \omega$ are sets, and ϕ is a partial function, then the following are mass problems:

1. $\{c_A\}$ (where c_A is the characteristic function of A): this is called the *problem of solvability of A*; this mass problem will be denoted by the symbol $\mathcal{S}_A$;

2. $\{f : \text{range}(f) = A\}$: the *problem of enumerability of A*; this mass problem will be denoted by the symbol $\mathcal{E}_A$;

3. (Other examples) The *problem of separability of A and B*, i.e. $\{f : f^{-1}(0) = A \ \& \ f^{-1}(1) = B\}$; of course, this mass problem is empty if $A \cap B \neq \emptyset$: it is absolutely impossible to "solve" the problem in this case. The *problem of many-one reducibility of A to B*: $\{f : f^{-1}(B) = A\}$. The *problem of extendibility of ϕ*: $\{f : f \supseteq \phi\}$.

*Research partially supported by Human Capital and Mobility network *Complexity, Logic and Recursion Theory*

A mass problem is *solvable* if it contains a recursive function, i.e., informally, if we have some ways of effectively computing a solution to the corresponding informal problem. Corresponding to this definition of solvable mass problem, there is an intuitive notion of reducibility between mass problems. If $\mathcal{A}$ and $\mathcal{B}$ are mass problems, we say that $\mathcal{A}$ is (informally) *reducible* to $\mathcal{B}$ if there is an effective procedure which, given any member of $\mathcal{B}$, yields a member of $\mathcal{A}$, i.e. we have an effective procedure for solving the problem corresponding to $\mathcal{A}$ given any solution to the problem corresponding to $\mathcal{B}$. The degree structure originated by this notion of reducibility turns out to be a distributive lattice (in fact a Brouwer algebra), called the *Medvedev lattice*. Despite its richness and, at the same time, the pleasant regularity of its properties, the Medvedev lattice has not been extensively studied. Rogers ([10]) raised several questions about this lattice: many of these questions are still open. The Medvedev lattice provides a more extensive context for both Turing reducibility and enumeration reducibility. It is to be expected that many of the global questions concerning these reducibilities can be profitably investigated in this wider context. There are several problems concerning automorphisms, automorphism bases and definability. Very little is known about filters, ideals and congruences. There is a variety of questions related to the Brouwer algebra structure of the Medvedev lattice; some of these questions deal with the problem of embedding Brouwer algebras in the Medvedev lattice and its initial segments, with consequent applications to intermediate propositional logics.

In this paper we review some of the existing literature on the Medvedev lattice. During the exposition, we list several open problems.

1.1   The formal definition of reducibility

The formal definition of reducibility between mass problems is given through the notion of a recursive operator. We shall refer to a listing $\{\Psi_z : z \in \omega\}$ of a large enough set of recursive operators, i.e. such that

1. $\Psi(\phi)$ is defined (i.e. a partial function), for all partial functions ϕ;

2. $(\forall$ recursive operator $\Omega)(\exists z)(\forall f)[\Psi_z(f) = \Omega(f)]$:

see [10] for a proof that such a listing exists.

Definition 1.2 We say that the mass problem $\mathcal{A}$ is *reducible* to the mass problem $\mathcal{B}$ (notation: $\mathcal{A} \leq \mathcal{B}$) if $(\exists z)[\Psi_z(\mathcal{B}) \subseteq \mathcal{A}]$. (Notice that $\Psi_z(\mathcal{B}) \subseteq \mathcal{A}$ implies that $\Psi_z(f)$ is total, for every $f \in \mathcal{B}$.)

The equivalence class $[\mathcal{A}]$ of $\mathcal{A}$ under the equivalence relation $\equiv$ generated by $\leq$ is called the *degree of difficulty* of $\mathcal{A}$. Let M be the set of degrees of difficulty: M is partially ordered by the relation $\leq$ defined by $[\mathcal{A}] \leq [\mathcal{B}]$ if and only if $\mathcal{A} \leq \mathcal{B}$. In fact,

Theorem 1.3 ([5]) *The structure* $\mathfrak{M} = \langle M, \leq \rangle$ *is a distributive lattice with* $0, 1$.

Proof. We just recall how the lattice–theoretic operations are defined. Given $n \in \omega$, functions f, g, and mass problems $\mathcal{A}$, $\mathcal{B}$, let $n * f$ denote the function defined by the clauses: $n * f(0) = n$, and $n * f(x) = f(x-1)$, if $x > 0$; let $f \vee g$ denote the function defined by the following clauses: $f \vee g(2x) = f(x)$, and $f \vee g(2x+1) = g(x)$; finally let $n * \mathcal{A} = \{n * f : f \in \mathcal{A}\}$, and $\mathcal{A} \vee \mathcal{B} = \{f \vee g : f \in \mathcal{A} \& g \in \mathcal{B}\}$. Let now $\mathcal{A}$, $\mathcal{B}$ be mass problems: define $[\mathcal{A}] \wedge [\mathcal{B}] = [0 * \mathcal{A} \cup 1 * \mathcal{B}]$; $[\mathcal{A}] \vee [\mathcal{B}] = [\mathcal{A} \vee \mathcal{B}]$. Finally, let $\mathbf{0} = [\mathcal{A}]$, where $\mathcal{A}$ contains at least a recursive function, and $\mathbf{1} = [\emptyset]$.

It is easy to see that the above are well given definitions and make $\mathfrak{M}$ a distributive lattice, $\mathbf{0}$ being the least element, and $\mathbf{1}$ being the greatest element. $\square$

Notice that $\mathbf{0}$ is the "easiest" degree of difficulty, containing only solvable problems; $\mathbf{1}$ is the "most difficult" degree of difficulty: it is absolutely impossible to solve the empty mass problem!

In the following, if P is any property of mass problems, we say that a degree of difficulty $\mathbf{A}$ has the property P if $\mathbf{A}$ contains some mass problem $\mathcal{A}$ having the property P. Thus, for instance $\mathbf{A}$ is *discrete*, if $\mathbf{A}$ contains some discrete mass problem in the Baire topology of ω^ω, etc.

2 A common framework for Turing degrees and enumeration degrees

Definition 2.1 For every set A, let $\mathbf{S}_A = [\mathcal{S}_A]$, and $\mathbf{E}_A = [\mathcal{E}_A]$. The degree $\mathbf{S}_A$ is called the *degree of solvability* of A; $\mathbf{E}_A$ is called the *degree of enumerability* of A.

Theorem 2.2 ([5]) *1. The mapping* $[A]_T \mapsto \mathbf{S}_A$ *is an embedding of the structure* $\mathfrak{D}_T$ *of the Turing degrees onto the degrees of solvability, preserving* $0, \vee$.

2. The mapping $[A]_e \mapsto \mathbf{E}_A$ *is an embedding of the structure* $\mathfrak{D}_e$ *of the enumeration degrees onto the degrees of enumerability, preserving* $0, \vee$.

Proof. The proof of 1. is immediate. As to 2, it is enough to observe that if $A \leq_e B$ then there is an effective procedure for enumerating A given any enumeration of B; use this procedure to define a recursive operator Ψ such that, for every f, if $\text{range}(f) = B$, then $\text{range}(\Psi(f)) = A$. On the other hand if $A \not\leq_e B$, then it is not difficult to construct a function g such that range$(g) = B$, but range$(\Psi_z(g)) \neq A$, all z, giving $\mathcal{E}_A \not\leq \mathcal{E}_B$. □

Remarkably, the degrees of solvability are definable in $\mathfrak{M}$, as is shown in the following theorem, which solves a problem raised in [10].

Theorem 2.3 ([2], [5]) *Let* $p(x) =^{df} (\exists y)[x < y \;\&\; (\forall z)[x < z \Rightarrow y \leq z]]$. *Then the degrees of solvability are exactly the degrees of difficulty satisfying* $p(x)$.

Proof. ($\Rightarrow$ ([5]):) Given a degree of solvability $\mathbf{S} = [\{f\}]$, let $\mathbf{S}' = [\{n * g : \Psi_n(g) = f \;\&\; g \not\leq_T f\}]$. Then $\mathbf{S} < \mathbf{S}'$ and $(\forall \mathbf{A})[\mathbf{S} < \mathbf{A} \Rightarrow \mathbf{S}' \leq \mathbf{A}]$.

($\Leftarrow$ ([2]):) Suppose that $\mathbf{A}$ is not a degree of solvability. If $\mathbf{A}$ is finite then clearly $\mathbf{A}$ does not satisfy $p(x)$, since $\mathbf{A}$ is easily seen to be meet–reducible. If $\mathbf{A}$ is not finite and $\mathcal{A} \in \mathbf{A}$, then for every mass problem $\mathcal{B}$ such that $\mathcal{B} \not\leq \mathcal{A}$, we construct a mass problem $\mathcal{C}$ such that $\mathcal{A} < \mathcal{C}$ and $\mathcal{B} \not\leq \mathcal{C}$. Here is a sketch of the construction.

(1) In order to get $\mathcal{A} \leq \mathcal{C}$, build $\mathcal{C}$ of the form $\mathcal{C} = \{x_n * f_n : n \in \omega\}$, where $f_n \in \mathcal{A}$; ensure also that $x_n \in \{0,1\}$ and $n \neq m \Rightarrow f_n \neq f_m$.

(2) Ensure that $\mathcal{C}$ satisfies the requirements:

- $P_e : \Psi_e(\mathcal{C}) \not\subseteq \mathcal{B}$;

- $R_e : \Psi_e(\mathcal{A}) \not\subseteq \mathcal{C}$.

Given any mass problem $\mathcal{X}$, let $\mathcal{X}^- = \{f^- : f \in \mathcal{X}\}$ (where, for every x, $f^-(x) = f(x+1)$). The mass problem $\mathcal{C}$ will be of the form $\mathcal{C} = \bigcup\{\mathcal{C}_n : n \in \omega\}$: let $\mathcal{C}_{-1} = \emptyset$.

Step $2e$) (Requirement P_e) Choose $x_{2e} \in \{0,1\}$ and $f_{2e} \in \mathcal{A}$ such that $f_{2e} \notin \mathcal{C}_{2e-1}^-$ and $\Psi_e(\mathcal{C}_{2e-1} \cup \{x_{2e} * f_{2e}\}) \not\subseteq \mathcal{B}$. Failure to find x_{2e} and f_{2e} would result in having $\Psi_e(\mathcal{D}) \subseteq \mathcal{B}$, where

$$\mathcal{D} = \mathcal{C}_{2e-1} \cup \{x * f : x \in \{0,1\} \;\&\; f \in \mathcal{A} - \mathcal{C}_{2e-1}^-\},$$

whence $\mathcal{B} \leq \mathcal{D}$. On the other hand we have also that $\mathcal{D} \leq \mathcal{A}$, via the recursive operator Ψ given by

$$\Psi(f) = \begin{cases} x_i * f_i, & \text{if } (\exists i)[\tilde{f}_i \subseteq f] \\ 0 * f, & \text{otherwise} \end{cases}$$

where $\{\tilde{f}_i : i \leq 2e - 1\}$ are initial segments such that $S_{\tilde{f}_i} \cap \mathcal{C}_{2e-1}^- = \{f_i\}$ (use that the f_i's are distinct; here, given $\tilde{f}$, we let $S_{\tilde{f}} = \{f : \tilde{f} \subseteq f\}$). Finally, let $\mathcal{C}_{2e} = \mathcal{C}_{2e-1} \cup \{x_{2e} * f_{2e}\}$.

Step $2e + 1$) (Requirement R_e) Notice that we can not have

$$(\forall f \in \mathcal{A})(\exists g \in C_{2e-1}^-)(\exists x \in \{0,1\})[\Psi_e(f) = x * g],$$

since otherwise we would obtain that $\mathcal{A} \equiv 0 * C_{2e-1}^- \cup 1 * C_{2e-1}^-$, which would imply that **A** is finite, contrary to the assumptions. Thus there exists a function $f \in \mathcal{A}$, such that $(\forall g \in C_{2e-1}^-)(\forall x \in \{0,1\})[\Psi_e(f) \neq x * g]$. Fix such an f: if $\Psi_e(f) = x * g$, for some $g \in \mathcal{A} - C_{2e-1}^-$ and $x \in \{0,1\}$, then let $x_{2e+1} = 1 - x$ and $f_{2e+1} = g$; otherwise let e.g. $x_{2e+1} = 0$, and let f_{2e+1} be any function such that $f_{2e+1} \in \mathcal{A} - C_{2e-1}^-$. Let $C_{2e+1} = C_{2e} \cup \{x_{2e+1} * f_{2e+1}\}$. □

Definition 2.4 For any degree of solvability **S**, let $\mathbf{S}' = \text{least } \{\mathbf{A} : \mathbf{S} < \mathbf{A}\}$.

Notice that $\mathbf{0}' = [\mathcal{O}]$, where $\mathcal{O} = \{f : f \text{ nonrecursive }\}$.

Problem 2.5 ([10]) *Are the degrees of enumerability definable? Or, at least, is the property of being a degree of enumerability a lattice–theoretic property?*

3 Lattice–theoretic properties

As remarked in [10] little is known about lattice theoretic properties of $\mathfrak{M}$. The following are very natural questions.

Problem 3.1 ([10]) *Is $\mathfrak{M}$ rigid?*

Problem 3.2 *Do the degrees of solvability constitute an automorphism basis for $\mathfrak{M}$?*

Problem 3.3 *Are the degrees of enumerability an automorphism basis for $\mathfrak{M}$?*

In [2], several lattice–theoretic properties are investigated.

Definition 3.4 Given mass problems $\mathcal{A}$ and $\mathcal{B}$ define $\mathcal{A} \leq_w \mathcal{B}$ if and only if $(\forall g \in \mathcal{B})(\exists f \in \mathcal{A})[f \leq_T g]$.

The preordering relation $\leq_w$ originates a partially ordered structure $\mathfrak{M}_w$, in the same way as $\leq$ originates $\mathfrak{M}$. The structure $\mathfrak{M}_w$ is in fact a complete distributive lattice, called the *Muchnik lattice*: see [7], [13]. For every mass problem $\mathcal{A}$, let $C(\mathcal{A}) = \{g : (\exists f \in \mathcal{A})[f \leq_T g]\}$. It is easy to see that $[\mathcal{A}]_w = [C(\mathcal{A})]_w$, (where $[\mathcal{B}]_w$ denotes the equivalence class of $\mathcal{B}$, for any mass problem $\mathcal{B}$). Also: $C(\mathcal{A}) \leq_w C(\mathcal{B}) \Leftrightarrow C(\mathcal{B}) \subseteq C(\mathcal{A})$.

Lemma 3.5 ([7],[13]) 1. *The mapping* $I : \mathfrak{M}_w \to \mathfrak{M}$ *defined by* $I([\mathcal{A}]_w) = [C(\mathcal{A})]$ *is an embedding preserving* $0, 1, \vee$ *and lowest upper bounds of arbitrary families.*

2. *The mapping* $F : \mathfrak{M} \to \mathfrak{M}_w$ *defined by* $F([\mathcal{A}]) = [\mathcal{A}]_w$ *is an onto lattice–theoretic homomorphism.*

Proof. Straightforward. □

Let us say that a degree of difficulty is a *Muchnik degree of difficulty* if it is in the range of the embedding I of Lemma 3.5 1. Then

Theorem 3.6 ([2]) *The property of being a Muchnik degree of difficulty is lattice–theoretic.*

Proof. First notice that **A** is a Muchnik degree of difficulty if and only if it contains a mass problem $\mathcal{A}$ such that $C(\mathcal{A}) = \mathcal{A}$. It is now easy to see that **A** is the least degree of difficulty containing some mass problem $\mathcal{B}$, such that

$$(\forall\{f\})[\mathcal{B} \le \{f\} \Rightarrow (\exists g \in \mathcal{A})[\{g\} \le \{f\}]].$$

Finally use the fact that the degrees of solvability are mapped to degrees of solvability by any automorphism (see Theorem 2.3). □

In a similar way, we can show e.g. that the property of containing a mass problem of the form $\{f : f_0 \le_T f\}$, for any function f_0, is lattice–theoretic, etc.

Problem 3.7 *Do the Muchnik degrees form an automorphism basis for* $\mathfrak{M}$*?*

4 The structure

4.1 Incomparability results

The Medvedev lattice is as big as it can be:

Theorem 4.1 ([8]) $\mathfrak{M}$ *has antichains of cardinality* $2^{2^{\aleph_0}}$.

Proof. Let $\mathcal{A} = \{f_i : i \in I\}$, with $|I| = 2^{\aleph_0}$, be such that $\{[f_i]_T : i \in I\}$ is an antichain of $\mathfrak{D}_T$. Given any $X \subseteq I$, let $\mathcal{A}_X = \{f_i : i \in X\}$. Then, for every $X, Y \subseteq I$ such that $X|Y$, we have $\mathcal{A}_X|\mathcal{A}_Y$. The result then follows from observing that there are subsets $J \subseteq P(I)$ such that $|J| = 2^{2^{\aleph_0}}$ and if $X, Y \in J$ and $X \ne Y$ then $X|Y$. □

There are however maximal antichains of two elements!

Example 4.2 For any function f, let $\mathcal{B}_f = \{g : g \not\le_T f\}$, and let $\mathbf{B}_f = [\mathcal{B}_f]$. Then it is easy to see that $\{\mathbf{B}_f, [\{f\}]\}$ is a maximal antichain (in fact, $(\forall \mathbf{C})[\mathbf{B}_f \not\le \mathbf{C} \Rightarrow \mathbf{C} \le [\{f\}]]$. Notice also that $\mathbf{B}_f \wedge [\{f\}] = \max\{\mathbf{C} : \mathbf{C} < \mathbf{B}_f\}$.

Theorem 4.3 ([2]) *If* $\mathbf{A} \neq 0, 0', 1$, *then there is a countable* $\mathbf{B}$ *such that* $\mathbf{B}$ *is incomparable with* $\mathbf{A}$.

Proof. Let $\mathcal{A} \in \mathbf{A}$, where $\mathbf{A} \neq 0, 0', 1$. Then, for every n, there is some nonrecursive function f_n such that $\Psi_n(f_n) \notin \mathcal{A}$ (otherwise $\Psi_n(\mathcal{O}) \subseteq \mathcal{A}$, hence $\mathbf{A} \leq 0'$.) Let $\mathcal{B} = \{f_n : n \in \omega\}$. Then $\mathcal{A} \not\leq \mathcal{B}$. If $\mathcal{B} \leq \mathcal{A}$, then take a function f which is Turing incomparable with all the members of $\mathcal{B}$ and $\{f\} \not\leq \mathcal{A}$ (such an f exists, since there can be at most countably many functions h such that $\{h\} \leq \mathcal{A}$, being $\mathcal{A} \neq \emptyset$). If $\mathcal{A} \not\leq \{f\}$, then $\mathcal{A}|\{f\}$. Otherwise, $\mathcal{B} \not\leq \mathcal{A}$, i.e. $\mathcal{A}|\mathcal{B}$. □

The following result establishes a sufficient condition for extending countable antichains of degrees of difficulty.

Theorem 4.4 ([2]) *Let* $\{\mathbf{A}_n : n \in \omega\}$ *be such that* $0' < \mathbf{A}_n < 1$, *and, for all* n, *no nonzero finite degree of difficulty is below* $\mathbf{A}_n$. *Then there is a countable* $\mathbf{B}$ *such that* $\mathbf{B}|\mathbf{A}_n$, *for all* n.

Proof. Let $\mathcal{A}_n \in \mathbf{A}_n$. Construct a countable mass problem $\mathcal{B} = \{f_n : n \in \omega\}$, and a sequence of functions $\{g_n : n \in \omega\}$ such that $\mathcal{B} \cap \{g_n : n \in \omega\} = \emptyset$. At step n we define f_n, g_n. Suppose $n = \langle i, j \rangle$. Let $f_n \notin \{g_0, \ldots, g_{n-1}\}$ be a nonrecursive function such that $\Psi_i(f_n) \notin \mathcal{A}_j$ (hence we satisfy the requirement $\Psi_i(\mathcal{B}) \not\subseteq \mathcal{A}_j$): such a function exists since otherwise we would have $\mathcal{A}_j \leq \mathcal{O}$). Finally define g_n to be a any function such that if $\Psi_i(f)$ total for all $f \in \mathcal{A}_j$, then $g_n \in \Psi_i(\mathcal{A}_j)$ and $g_n \notin \{f_0, \ldots, f_n\}$: such a function exists, since otherwise $\{f_0, \ldots, f_n\} \leq \mathcal{A}_j$, via Ψ_i. Since $g_n \notin \mathcal{B}$, this guarantees that $\Psi_i(\mathcal{A}_j) \not\subseteq \mathcal{B}$. □

We can characterize the countable lattices which can be embedded in $\mathfrak{M}$. Given a partial order $\mathfrak{P}$, let $1 \oplus \mathfrak{P} \oplus 1$ be the partial order obtained by adding an element at the bottom and an element at the top of $\mathfrak{P}$, respectively.

Theorem 4.5 ([13]) *A countable distributive lattice* $\mathfrak{L}$ *with* $0, 1$ *is embeddable in* $\mathfrak{M}$ *if and only if* 0 *is meet–irreducible and* 1 *is join–irreducible.*

Proof. Clearly, if $\mathfrak{L}$ is embeddable in $\mathfrak{M}$ then its least element is meet-irreducible, and the greatest element is join–irreducible, since this holds of $\mathfrak{M}$ as well. To show sufficiency, let $\mathfrak{B} \subseteq \mathfrak{P}(\omega)$ be an atomless countable Boolean algebra, where $\mathfrak{P}(\omega)$ is the Boolean algebra of the subsets of ω. It is not difficult to see that we can find a family $\{g, g_n : n \in \omega\}$ of nonrecursive functions and a recursive operator Ψ such that, for all m, n,

1. $g(0) = 0$, $g_n(0) = n + 1$;

2. $m \neq n \Rightarrow g_m|_T g_n$ & $g_m \vee g_n \equiv_T g$;

3. $m \neq n \Rightarrow \Psi(g_m \vee g_n) = g$.

For every subset $X \subseteq \omega$ let

$$
\mathcal{A}_X = \begin{cases} \{g_x : x \in X\}, & \text{if } X \neq \emptyset \\ \{g\}, & \text{otherwise} \end{cases}
$$

Then $Y \subseteq X \Leftrightarrow \mathcal{A}_X \leq \mathcal{A}_Y$: it is now easy to see that the mapping $X \mapsto [\mathcal{A}_X]$ yields a lattice–theoretic embedding of the dual $\breve{\mathfrak{B}}$ of $\mathfrak{B}$ into $\mathfrak{M}$ mapping 0 into $[\{g_n : n \in \omega\}]$, and 1 into $[\{g\}]$. Then claim then follows from the well known fact that for every distributive lattice $\mathfrak{L}$, we have that $1 \oplus \mathfrak{L} \oplus 1$ is embeddable in $1 \oplus \breve{\mathfrak{B}} \oplus 1$. $\square$

Remark 4.6 Dyment ([2]) gives several topological interpretations and refinements of some of the results reviewed in this section. For instance, it is shown that one can define a topology on the collections of mass problems, such that, for any nonsolvable $\mathcal{A}$, the class $\{\mathcal{B} : \mathcal{B} \leq \mathcal{A}\}$ is of first category; it is interesting to note that one can also define a topology with respect to which, if $\mathcal{A}$ is such that, for no dense $\mathcal{B}$, is $\mathcal{A} \leq \mathcal{B}$, then $\{\mathcal{B} : \mathcal{A} \leq \mathcal{B}\}$ is of first category.

4.2 Empty intervals

A fairly straightforward refinement of the proof of Theorem 2.3 leads to the following useful characterization of empty intervals of $\mathfrak{M}$.

Theorem 4.7 ([2]) *Given* $\mathbf{A}, \mathbf{B}$, *with* $\mathbf{A} < \mathbf{B}$, *we have that* $(\mathbf{A}, \mathbf{B}) = \emptyset$ *if and only if* ($\exists$ *degree of solvability* $\mathbf{S}$)$[\mathbf{A} = \mathbf{B} \wedge \mathbf{S} \ \& \ \mathbf{B} \not\leq \mathbf{S} \ \& \ \mathbf{B} \leq \mathbf{S}']$.

Proof. Let $\mathbf{A} < \mathbf{B}$, and let $\mathcal{A} \in \mathbf{A}$, $\mathcal{B} \in \mathbf{B}$.

($\Rightarrow$:) Suppose $(\mathbf{A}, \mathbf{B}) = \emptyset$. We observe that if $\mathcal{A}$ is finite, then there exists $f \in \mathcal{A}$ such that $\mathcal{B} \not\leq \{f\}$; let $\mathcal{E} \in \mathbf{S}'$, where $\mathbf{S} = [\{f\}]$. Then $\mathcal{A} \leq \mathcal{B} \wedge \{f\} < \mathcal{B}$, whence $\mathcal{A} \equiv \mathcal{B} \wedge \{f\} < \mathcal{B}$. On the other hand, $\mathcal{B} \wedge \mathcal{E} \not\leq \mathcal{A}$, otherwise $\mathcal{B} \wedge \mathcal{E} \leq \{f\}$, but then $\mathcal{B} \leq \{f\}$ or $\mathcal{E} \leq \{f\}$, contradiction. Then $\mathcal{A} < \mathcal{B} \wedge \mathcal{E} \equiv \mathcal{B}$, hence $\mathcal{B} \leq \mathcal{E}$.

If $\mathbf{A}$ contains no finite mass problem, and no degree of solvability $\mathbf{S}$ exists such that $\mathbf{A} = \mathbf{B} \wedge \mathbf{S}$, $\mathbf{B} \not\leq \mathbf{S}$, $\mathbf{B} \leq \mathbf{S}'$, then an easy modification of the proof of Theorem 2.3 enables us to construct a mass problem $\mathcal{C}'$ such that, letting $\mathcal{C} = \mathcal{B} \wedge \mathcal{C}'$, we have that $\mathcal{A} < \mathcal{C} < \mathcal{B}$.

($\Leftarrow$:) Let $\mathbf{A} = \mathbf{B} \wedge \mathbf{S} \ \& \ \mathbf{B} \not\leq \mathbf{S} \ \& \ \mathbf{B} \leq \mathbf{S}'$, where $\mathbf{S}$ is a degree of solvability; let $\mathcal{S} \in \mathbf{S}$, and $\mathcal{E} \in \mathbf{S}'$. Suppose that $\mathcal{A} < \mathcal{C} < \mathcal{B}$; then $\mathcal{B} \wedge \mathcal{S} \leq \mathcal{C}$, via, say, some recursive operator Ψ. Since $\Psi(\mathcal{C}) \subseteq 0 * \mathcal{B} \cup 1 * \mathcal{S}$, there are $\mathcal{C}_0, \mathcal{C}_1 \subseteq \mathcal{C}$ such that $\Psi(\mathcal{C}_0) \subseteq 0 * \mathcal{B}$ and $\Psi(\mathcal{C}_1) \subseteq 1 * \mathcal{S}$; moreover $\mathcal{C}_0 \wedge \mathcal{C}_1 \leq \mathcal{C}$. We can not have $\mathcal{C}_1 \leq \mathcal{S}$ since, otherwise, we would have $\mathcal{C} \leq \mathcal{B} \wedge \mathcal{S}$. Hence $\mathcal{S} < \mathcal{C}_1$. It follows that $\mathcal{E} \leq \mathcal{C}_1$, hence $\mathcal{B} \leq \mathcal{C}_0 \wedge \mathcal{C}_1 \leq \mathcal{C}$, contradiction. $\square$

Example 4.8 If $\mathcal{B}$ is countable, then for any $\mathcal{A}$, with $\mathcal{A} < \mathcal{B}$, there is some mass problem $\mathcal{C}$ such that $\mathcal{A} < \mathcal{C} < \mathcal{B}$: indeed, for every f such that $\mathcal{B} \not\leq \{f\}$, one can find a function g such that $\{f\} < \{g\}$, but $\mathcal{B} \not\leq \{g\}$, hence $[\mathcal{B}] \not\leq [\{f\}]'$.

4.3 Bounds of countable families

The Medvedev lattice is not a complete lattice as follows from the results of this subsection. Given a lattice $\mathfrak{L}$, we say that a set $X \subseteq \mathfrak{L}$ is *strongly* $\wedge-incomplete$ if $(\forall Y \subseteq X)[Y$ finite $\Rightarrow (\exists x \in X)[\wedge Y \not\leq x]]$. Dually one defines the notion of a *strongly* $\vee-incomplete$ set $X \subseteq \mathfrak{L}$.

Theorem 4.9 ([3]) *No countable strongly* $\wedge-incomplete$ *collection of degrees of difficulty has greatest lower bound.*

Proof. Let $\{\mathbf{A}_n : n \in \omega\}$ be a strongly $\wedge-incomplete$ countable collection of degrees of difficulty (for instance let $\mathbf{A}_n = [\{f_n\}]$, where the functions f_n's are pairwisely Turing incomparable). We claim that $\{\mathbf{A}_n : n \in \omega\}$ does not have greatest lower bound. Let $\mathcal{A}_n \in \mathbf{A}_n$, for every n, and let $\mathcal{B}$ be a mass problem such that $\mathcal{B} \leq \mathcal{A}_n$, each n. In order to show the claim, construct a mass problem $\mathcal{C}$ such that $\mathcal{C} \leq \mathcal{A}_n$, all n, and $\mathcal{C} \not\leq \mathcal{B}$. Construct $\mathcal{C}$ of the form $\mathcal{C} = \bigcup\{x_n * \mathcal{A}_n : n \in \omega\}$ (with $x_m \neq x_n$ if $m \neq n$), hence $\mathcal{C} \leq \mathcal{A}_n$, all n. We define by induction two sequences $\{x_n : n \in \omega\}$ and $\{k_n : n \in \omega\}$ of numbers. At step n, we define x_n so as to satisfy the requirement $\Psi_n(\mathcal{B}) \not\subseteq \mathcal{C}$. By assumptions, there exists $f \in \mathcal{B}$ such that $\Psi_n(f) \notin \bigcup\{x_i * \mathcal{A}_i : i \leq n\}$, since $[\bigcup\{x_i * \mathcal{A}_i : i \leq n\}] = \wedge_{i \leq n} \mathbf{A}_i$. Choose such a function f and let $k_n = \Psi_n(f)(0)$ if $\Psi_n(f)$ is total, $k_n = 0$ otherwise: finally choose $x_n > \max(\{x_i : i < n\} \cup \{k_i : i \leq n\}$. $\square$

Definition 4.10 A mass problem $\mathcal{A}$ is *effectively discrete* if $(\forall f, g \in \mathcal{A})[f \neq g \Rightarrow f(0) \neq g(0)]$.

Theorem 4.11 ([3]) *No countable strongly* $\vee-incomplete$ *collection of effectively discrete degrees of difficulty has lowest upper bound.*

Proof. Let $\{\mathbf{A}_n : n \in \omega\}$ be a countable strongly $\vee-incomplete$ collection of degrees of difficulty and for each n let $\mathcal{A}_n \in \mathbf{A}_n$ be an effectively discrete mass problem and let $\mathcal{B}$ be such that, for all n, $\mathcal{A}_n \leq \mathcal{B}$. We want to construct a mass problem $\mathcal{C}$ such that, for all n, $\mathcal{A}_n \leq \mathcal{C}$, but $\mathcal{B} \not\leq \mathcal{C}$. In order to satisfy the requirement P_n: $\Psi_n(\mathcal{C}) \not\subseteq \mathcal{B}$, we define at step n a number k_n such that, if Ψ_{i_n} is a recursive operator such that $\Psi_{i_n}(\mathcal{B}) \subseteq \mathcal{A}_{k_n}$ then $\Psi_{i_n}\Psi_n(\mathcal{C}) \not\subseteq \mathcal{A}_{k_n}$. Given any initial segment $\tilde{h}$, and functions $f_0, \ldots, f_n$, let $\tilde{h}(f_0, \ldots, f_n) = \tilde{h} \cup \{(\langle x, i \rangle, f_i(x - x_i)) : x \geq x_i, i \leq n\}$, where for each $i \leq n$, $x_i = \text{least}\{x : \tilde{h}(\langle x, i \rangle) \text{not defined }\}$.

Step n)   Define a number k_n, and an initial segment $\tilde{h}_n$ as follows (assume $\tilde{h}_{-1} = \emptyset$).

Let $k_n = \text{least}\{i : \mathcal{A}_i \not\leq \mathcal{A}_0 \vee \ldots \vee \mathcal{A}_{n-1}\}$ (let k_0 be the least number such that $\mathcal{A}_{k_0}$ is not solvable). Such a number exists since the family is strongly $\vee$−incomplete.

Case 1) $(\exists g_0 \in \mathcal{A}_0, \ldots, \exists g_{n-1} \in \mathcal{A}_{n-1})(\forall f)[f \supseteq \tilde{h}_{n-1}(g_0, \ldots, g_{n-1}) \Rightarrow \Psi_{i_n}\Psi_n(f)(0) \uparrow]$. Then in this case, let $\tilde{h}_n = \emptyset$: the requirement P_n is automatically satisfied, if we ensure that there is some $f \in C$ such that $f \supseteq \tilde{h}_{n-1}(g_0, \ldots, g_{n-1})$.

Case 2)   Otherwise. Then, for every $g_0 \in \mathcal{A}_0, \ldots, g_{n-1} \in \mathcal{A}_{n-1}$, there exists an initial segment $F(\tilde{h}_{n-1}, g_0, \ldots, g_{n-1})$ extending $\tilde{h}_{n-1}$ and compatible with $\tilde{h}_{n-1}(g_0, \ldots, g_{n-1})$ such that $\Psi_{i_n}\Psi_n(F(\tilde{h}_{n-1}, g_0, \ldots, g_{n-1}))(0) \downarrow$. Since $\mathcal{A}_{k_n} \not\leq \mathcal{A}_0 \vee \ldots \vee \mathcal{A}_{n-1}$, there exist functions $g_0 \in \mathcal{A}_0, \ldots, g_{n-1} \in \mathcal{A}_{n-1}$ such that $V \notin \mathcal{A}_{k_n}$, where $V = \bigcup\{\Psi_{i_n}\Psi_n(f) : f \supseteq \tilde{h}_{n-1}(g_0, \ldots, g_{n-1}) \cup F(\tilde{h}_{n-1}, g_0, \ldots, g_{n-1})\}$. Fix such functions $g_0, \ldots, g_{n-1}$. If $g \not\subseteq V$, for all $g \in \mathcal{A}_{k_n}$, then define $\tilde{h}_n = \tilde{h}_{n-1} \cup (F(\tilde{h}_{n-1}, g_0, \ldots, g_{n-1}) - \tilde{h}_{n-1}(g_0, \ldots, g_{n-1}))$. Otherwise, since $\mathcal{A}_{k_n}$ is effectively discrete and there exists exactly one pair $(0, x) \in V$, there exists also exactly one function $f_n \in \mathcal{A}_{k_n}$ such that $f_n \subseteq V$. Since $V \not\subseteq f_n$, let $\tilde{h}$ be an initial segment extending $F(\tilde{h}_{n-1}, g_0, \ldots, g_{n-1})$, compatible with $\tilde{h}_{n-1}(g_0, \ldots, g_{n-1})$, such that $\Psi_{i_n}\Psi_n(\tilde{h}) \not\subseteq f_n$. Finally let $\tilde{h}_n = \tilde{h}_{n-1} \cup (\tilde{h} - \tilde{h}_{n-1}(g_0, \ldots, g_{n-1}))$. It clearly follows that, for no function $f \supseteq \tilde{h}_n(g_0, \ldots, g_{n-1})$, can we have $\Psi_{i_n}\Psi_n(f) \in \mathcal{A}_{k_n}$.

Let now $\mathcal{C}_n = \{g : (\exists f_0 \in \mathcal{A}_0), \ldots, (\exists f_n \in \mathcal{A}_n)[g \supseteq \tilde{h}_n(f_0, \ldots, f_n)]\}$, and, finally, let $\mathcal{C} = \bigcap\{\mathcal{C}_n : n \in \omega\}$. It immediately follows from the construction that $\Psi_{i_n}\Psi_n(\mathcal{C}) \not\subseteq \mathcal{A}_n$. Also, one immediately sees that, for each n, there exists a recursive operator Ψ, such that, if $g \supseteq \tilde{h}_n(f_0, \ldots, f_n)$, then $\Psi(g) = f_n$: thus $\mathcal{A}_n \leq \mathcal{C}_n$, hence $\mathcal{A}_n \leq \mathcal{C}$, since $\mathcal{C}_n \subseteq \mathcal{C}$. $\square$

Since the degrees of solvability are effectively discrete degrees, one obtains as a corollary of the previous theorem Spector's result on the nonexistence of lowest upper bounds of ascending sequences of Turing degrees. In a similar way it is possible to show

Corollary 4.12 *No countable strongly $\vee$−incomplete family of degrees of enumerability has lowest upper bound.*

Proof. Similar to the proof of Theorem 4.11. $\square$

Remark 4.13 There exist however infinite families with nontrivial lowest upper bound, as follows from Lemma 3.5: any collection of Muchnik degrees of difficulty has lowest upper bound.

5 Irreducible elements

It is not difficult to characterize the meet-reducible elements.

Theorem 5.1 ([2]) *A degree of difficulty* $\mathbf{A}$ *is meet-reducible if and only if* $\mathbf{A}$ *contains a mass problem* $\mathcal{A}$ *such that there exists r.e. sets* V_0, V_1 *of initial segments such that*

- $(\forall f \in \mathcal{A})(\exists i \in \{0,1\})(\exists \tilde{f})[\tilde{f} \in V_i \ \& \ \tilde{f} \subseteq f];$

- $\{f : (\exists \tilde{f})[\tilde{f} \in V_0 \ \& \ \tilde{f} \subseteq f]\} | \{f : (\exists \tilde{f})[\tilde{f} \in V_1 \ \& \ \tilde{f} \subseteq f]\}.$

Proof. ($\Leftarrow$:) Given V_0, V_1, let $\mathcal{A}_0 = \{f : (\exists \tilde{f})[\tilde{f} \in V_0 \ \& \ \tilde{f} \subseteq f]\}$, and $\mathcal{A}_1 = \{f : (\exists \tilde{f})[\tilde{f} \in V_1 \ \& \ \tilde{f} \subseteq f]\}$. Then it is easy to see that $\mathcal{A}_0 | \mathcal{A}_1$, and $\mathcal{A} \equiv \mathcal{A}_0 \wedge \mathcal{A}_1$.

($\Rightarrow$:) If $\mathbf{A} = \mathbf{A}_0 \wedge \mathbf{A}_1$, $\mathcal{A}_0 \in \mathbf{A}_0$, $\mathcal{A}_1 \in \mathbf{A}_1$, and Ψ is a recursive operator such that $\Psi(\mathcal{A}) \subseteq 0 * \mathcal{A}_0 \cup 1 * \mathcal{A}_1$, then the problem $0 * \mathcal{A}_0 \cup 1 * \mathcal{A}_1$ and the r.e. sets $V_0 = \{\tilde{f} : \Psi(\tilde{f})(0) = 0\}$, $V_1 = \{\tilde{f} : \Psi(\tilde{f})(0) = 1\}$ satisfy the claim. $\square$

The previous theorem is a useful tool for testing if a given element is meet–irreducible. Let us say ([7]) that a mass problem $\mathcal{A}$ is *uniform* if

$$(\forall \tilde{f})[S_{\tilde{f}} \cap \mathcal{A} \neq \emptyset \Rightarrow \tilde{f} * \mathcal{A} \subseteq \mathcal{A}].$$

As an application of the previous theorem we have for instance (see [2])

Corollary 5.2 *Every uniform degree of difficulty is meet–irreducible. Hence every Muchnik degree of difficulty is meet–irreducible.*

Proof. If $\mathcal{A}$ is uniform and $\mathcal{B} \wedge \mathcal{C} \leq \mathcal{A}$ via some recursive operator Ψ then suppose that there exists some function $f \in \mathcal{A}$ such that $\Psi(f) \in 0 * \mathcal{B}$: it follows that, for some initial segment $\tilde{f} \subseteq f$, $\Psi(\tilde{f})(0) = 0$. But $\tilde{f} * \mathcal{A} \subseteq \mathcal{A}$, hence $\Psi(\tilde{f} * \mathcal{A}) \subseteq 0 * \mathcal{B}$. It easily follows that $\mathcal{B} \leq \mathcal{A}$. Thus either $\mathcal{B} \leq \mathcal{A}$ or $\mathcal{C} \leq \mathcal{A}$. $\square$

We do not have characterizations of the join–irreducible elements. Examples of join–irreducible elements are provided by Example 4.2: the element $\mathbf{B}_f$ is join–irreducible for every function f. We also notice the following useful lemma.

Lemma 5.3 *If* $\mathcal{A}$ *is a mass problem such that there exist functions* $f \in \mathcal{A}, g_1, g_2 \notin C(\mathcal{A})$ *and* $g_1 |_T g_2$ *and* $f \leq_T g_1 \vee g_2$, *then the degree of difficulty* $[\mathcal{A}]$ *is join–reducible.*

Proof. Under the assumptions, it is easy to see that

$$[\mathcal{A}] = [\mathcal{A} \wedge \{g_1\}] \vee [\mathcal{A} \wedge \{g_1\}].$$

On the other hand, it is clear that $[\mathcal{A} \wedge \{g_1\}] | [\mathcal{A} \wedge \{g_1\}]$. $\square$

Problem 5.4 *Characterize the join–irreducible elements.*

6 More on the degrees of enumerability. The Dyment lattice

In this section we collect some observations on the degrees of enumerability. We then define the Dyment lattice which extends the Medvedev lattice in much the same way as the enumeration degrees extend the Turing degrees. We show that there is an adjunction between the two lattices.

6.1 Some properties of the degrees of enumerability

The following theorem characterizes the degrees of enumerability corresponding to quasi–minimal e–degrees.

Theorem 6.1 ([2]) *Let $\mathcal{A}$ be a nonsolvable mass problem, and let A be a set of quasi–minimal e–degree. If $\mathcal{A} \le \mathcal{E}_A$ then $\mathcal{A}$ is not countable.*

Proof. Suppose that $[A]_e$ is quasi–minimal, and let $\mathcal{A} = \{f_n : n \in \omega\}$ be a countable and nonsolvable mass problem. It is not difficult, given any n, to find, by finite extensions, a function f such that range$(f) = A$ and $\Psi_n(f) \ne f_n$. Failure to find f for some n would result in giving $f_n \le_e A$, a contradiction. $\square$

The remaining results of this subsection are contained in [13].

Theorem 6.2 *Every degree of enumerability is meet–irreducible. Every nonzero degree of enumerability is join–reducible.*

Proof. It is easy to see that, for every set A, the problem $\mathcal{E}_A$ is uniform, so it follows from Corollary 5.2 that $\mathbf{E}_A$ is meet–irreducible. That every nonzero degree of enumerability is join–reducible, follows from Lemma 5.3. $\square$

Theorem 6.3 *Let $\mathbf{E}$ be a degree of enumerability. Then $(\forall \mathbf{B} > \mathbf{E})[(\mathbf{E}, \mathbf{B}) \ne \emptyset]$. Moreover, $(\forall \mathbf{B} < \mathbf{E})[(\mathbf{B}, \mathbf{E}) \ne \emptyset]$.*

Proof. In order to show that $(\mathbf{E}, \mathbf{B}) \ne \emptyset$, for any $\mathbf{B} > \mathbf{E}$, use Theorem 4.7 and the fact that each degree of enumerability is meet–irreducible.

To show the other part, use again Theorem 4.7 and the following observation. Given any degree of enumerability $\mathbf{E}_A > \mathbf{0}$, we can show that for every degree of solvability $\mathbf{S}$, if $\mathbf{E}_A \not\le \mathbf{S}$, then $\mathbf{E}_A \not\le \mathbf{S}'$. To see this, it is enough to show that for every function f such that $A \not\le_e f$ one can find a function g such that $f <_T g$ and $A \not\le_e g$: failure to do this would result in $A \le_e f$, a contradiction. $\square$

6.2 The Dyment lattice

Let $\mathcal{P}$ denote the collection of all partial functions from ω to ω. A *mass problem of partial functions* is any subset $\mathcal{A} \subseteq \mathcal{P}$. Given mass problems of partial functions $\mathcal{A}, \mathcal{B}$, we say that $\mathcal{A}$ is *e−reducible to* $\mathcal{B}$ (notation: $\mathcal{A} \leq_e \mathcal{B}$: the context will always make clear whether the symbol $\leq_e$ is used to denote enumeration reducibility between sets of numbers, or the above given reducibility between mass problems of partial functions) if there exists a partial recursive operator Ω such that $(\forall \phi \in \mathcal{B})[\phi \in \text{ domain}(\Omega) \ \& \ \Omega(\phi) \in \mathcal{A}]$.

This preordering relation originates a degree structure which is a distributive lattice with $0, 1$ called the *Dyment lattice* (see [2] and [13]) and denoted by $\mathfrak{M}_e$. The members of $\mathfrak{M}_e$ are called *partial degrees of difficulty* ; $[\mathcal{A}]_e$ denotes the *partial degree of difficulty of* $\mathcal{A}$. The operations of $\mathfrak{M}_e$ are: $[\mathcal{A}]_e \wedge [\mathcal{B}]_e = [\mathcal{A} \wedge \mathcal{B}]_e$; $[\mathcal{A}]_e \vee [\mathcal{B}]_e = [\mathcal{A} \vee \mathcal{B}]_e$ (where the operations $\wedge, \vee$ on mass problems of partial functions are defined in a similar way as for mass problems); moreover $\mathbf{0}_e = [\{\phi : \phi \text{ partial recursive}\}]_e$, and $\mathbf{1}_e = [\emptyset]_e$ are the least element and the greatest element, respectively.

Definition 6.4 1. A *total degree of difficulty* is a partial degree of difficulty containing a mass problem consisting of total functions;

2. a *partial degree of enumerability* is a partial degree of difficulty of the form $[\{\phi\}]_e$, for some partial function ϕ.

Clearly there is an embedding of $\mathfrak{D}_e$ onto the partial degrees of enumerability: just view $\mathfrak{D}_e$ as the upper semilattice of the partial degrees (i.e. equivalence classes of partial functions: see [10]): we have $\phi \leq_e \psi \Leftrightarrow \{\phi\} \leq_e \{\psi\}$.

Theorem 6.5 *The partial degrees of enumerability are definable in* $\mathfrak{M}_e$ *by the formula* $p(x)$ *of Theorem 2.3.*

Proof. The proof is similar to that of Theorem 2.3, but it has some original features since it is no longer true, in $\mathfrak{M}_e$, that the finite partial degrees of difficulty that are not partial degrees of enumerability are meet−reducible. For instance, it is easy to find examples of partial functions ϕ_1, ϕ_2 such that $\phi_1 \subset \phi_2$, but $\phi_1|_e\phi_2$: thus $\{\phi_1, \phi_2\} <_e \{\phi_1\} \wedge \{\phi_2\}$. □

Let $\iota : \mathfrak{M} \longrightarrow \mathfrak{M}_e$ be the natural embedding, i.e. $\iota([\mathcal{A}]) = [\mathcal{A}]_e$. Notice that $\text{range}(\iota) = \{\mathbf{A}_e : \mathbf{A}_e \text{ total}\}$. Now, given a mass problem of partial functions $\mathcal{A}$, let

$$\mathcal{A}^* = \{f : (\exists \phi \in \mathcal{A})[\text{range}(f) = \text{graph}(\phi)]\}.$$

Let $\epsilon : \mathfrak{M}_e \longrightarrow \mathfrak{M}$ be defined by $\epsilon([\mathcal{A}]_e) = [\mathcal{A}^*]$: it is not difficult to see that ϵ is well defined and it is in fact ([13]) an onto lattice−theoretic homomorphism.

Theorem 6.6 ([2]) *For all* $\mathbf{A}_e \in \mathfrak{M}_e$, *and* $\mathbf{B} \in \mathfrak{M}$ *we have:* $\mathbf{A}_e \leq_e \iota(\mathbf{B}) \Leftrightarrow \epsilon(\mathbf{A}_e) \leq \mathbf{B}$.

Proof. It is not difficult to show that, for every $\mathcal{A}$,

$$[\mathcal{A}^*]_e = \text{least}\{\mathbf{B}_e : \mathbf{B}_e \text{ total } \& [\mathcal{A}]_e \leq_e \mathbf{B}_e\}.\square$$

For every partial function ϕ, let $\mathbf{E}_e^\phi = \iota(\epsilon([\{\phi\}]_e))$. $\mathbf{E}_e^\phi$ is called the *total degree of enumerability of* ϕ.

Corollary 6.7 ([2]) *The property of being a total degree of enumerability is invariant under all automorphisms F of $\mathfrak{M}_e$ such that $F(\iota(\mathfrak{M})) \subseteq \iota(\mathfrak{M})$.*

Proof. Immediate from the previous theorem and Theorem 6.5. $\square$

7   More on the Turing degrees

In this subsection we make some remarks on degrees of difficulty of the form $[C(\{g\})]$, for some function g. Very useful is the following lemma.

Lemma 7.1 ([2]) *If $\mathcal{A}$ is discrete and not solvable, then for every $\mathcal{B}$,*

$$\mathcal{A} \leq \mathcal{B} \Rightarrow \mathcal{B} \text{ nowhere dense }.$$

Proof. Let $\mathcal{A}$, $\mathcal{B}$ be given, and let Ψ be such that $\Psi(\mathcal{B}) \subseteq \mathcal{A}$. In order to prove the claim, it is enough to show that $\mathcal{X} = \{f : \Psi(f) \in \mathcal{A}\}$ is nowhere dense. Otherwise there would be an initial segment $\tilde{f}_0$ such that $\mathcal{X}$ is dense in $S_{\tilde{f}_0}$ (we recall that the sets of the form $S_{\tilde{f}}$ are a basis for the Baire topology). Let $f_0 \supset \tilde{f}_0$ be such that $\Psi(f_0) \in \mathcal{A}$: since $\mathcal{A}$ is discrete, let $\tilde{g}_0$ be such that $S_{\tilde{g}_0} \cap \mathcal{A} = \{\Psi(f_0)\}$; thus, let $\tilde{f}_1 \supseteq \tilde{f}_0$ be such that $(\forall g \supset \tilde{f}_1)[\Psi(g) \in \mathcal{A} \Rightarrow \Psi(g) = \Psi(f_0)]$. Then $\Psi(f_0) = \bigcup\{\Psi(\tilde{g}) : \tilde{g} \supseteq \tilde{f}_1\}$, giving $\Psi(f_0)$ recursive, i.e. $\mathcal{A}$ solvable, a contradiction. $\square$

Theorem 7.2 *If $\mathcal{A}$ is not solvable and countable, then for every function g, $\mathcal{A} \not\leq C(\{g\})$.*

Proof. Let $\mathcal{A} = \{f_i : i \in \omega\}$ be nonsolvable, and let $\mathcal{B} = C(\{g\})$. One can show that for every n, if $\Psi_n(f)$ is total for every $f \in \mathcal{B}$, then there exists some function $f \in \mathcal{B}$ such that $(\forall i)[\Psi_n(f) \neq f_i]$: construct such an f of the form $f = h \vee g$, where $h = \bigcup\{\tilde{h}_i : i \in \omega\}$, and $\{\tilde{h}_i : i \in \omega\}$ is an increasing sequence of initial segments: failure at step i to find an initial segment $\tilde{h}_i$ such that $\Psi_n(\tilde{h}_i \vee g) \not\subseteq f_i$ would result in getting $\Psi_n(\{f \vee g : f \supset \tilde{h}_{i-1}\}) \subseteq \{f_i\}$, which, by the previous lemma, would imply $\{f \vee g : f \supset \tilde{h}_{i-1}\}$ nowhere dense, a contradiction (as usual, assume that $\tilde{h}_{-1} = \emptyset$). $\square$

Refinements of the previous theorem give:

Theorem 7.3 *For every function g, let g' be a function belonging to the Turing jump of $[g]_T$. Then*

1. *if $\mathcal{A}$ is not solvable and countable then $\mathcal{A} \not\leq \{f : g <_T f\}$.*

2. $\{f : f$ *nonrecursive* $\& \ f \leq_T g\} \not\leq \{f : g \leq_T f \leq_T g'\}$; $\{f : f$ *nonrecursive* $\& \ f \leq_T g\} \not\leq \{f : f \equiv_T g'\}$; *if g is not recursive then* $\{f : f \equiv_T g\} | \{f : f \equiv_T g'\}$;

3. *if $\mathcal{A}$ is not solvable and countable and $\mathcal{A} \not\leq \{h\}$, then $\mathcal{A} \not\leq C(\{g\}) \vee \{h\}$; hence if g is not recursive and does not have minimal Turing degree, then* $\{f : f \equiv_T g\} \not\leq C(\{g\}) \vee \{f : f$ *nonrecursive* $\& \ f \leq_T g\}$.

Proof. For (1) modify step n in the proof of the previous theorem so as to construct $f = h \vee g'$. Inspection of the proof of the previous theorem shows that we can construct f such that $f \leq_T g'$, so (2) easily follows. For (3) relativize the proof of the previous theorem to h. $\square$

8 Filters and ideals

Very little is known about nonprincipal filters and ideals of the Medvedev lattice (see [10]). We review some filters and ideals introduced in [2] and [12]. The results of this section, unless otherwise specified, can be found in [12].

Given any collection X of degrees of difficulty, let F_X and I_X denote the filter and the ideal, respectively, generated by X.

Given a mass problem $\mathcal{A}$, let $\Gamma_{\mathcal{A}} = \{[f]_T : f \in \mathcal{A} \ \& \ (\forall g \in \mathcal{A})[g \not\leq_T f]\}$. We say that a mass problem $\mathcal{A}$ has *countable basis* if $\Gamma_{\mathcal{A}}$ is countable and $(\forall g \in \mathcal{A})(\exists f)[f \leq_T g \ \& \ [f]_T \in \Gamma_{\mathcal{A}}]$. We say that $\mathcal{A}$ has *generalized countable basis* if $\Gamma_{\mathcal{A}}$ is countable.

We now define several classes X of degrees of difficulty.

Definition 8.1 Let $Solv = \{\mathbf{A} : \mathbf{A} \neq \mathbf{0} \ \& \ \mathbf{A}$ degree of solvability $\}$; $En = \{\mathbf{A} : \mathbf{A} \neq \mathbf{0} \ \& \ \mathbf{A}$ degree of enumerability $\}$; $Dis = \{\mathbf{A} : \mathbf{A} \neq \mathbf{0} \ \& \ \mathbf{A}$ discrete$\}$; $Edis = \{\mathbf{A} : \mathbf{A} \neq \mathbf{0} \ \& \ \mathbf{A}$ effectively discrete $\}$; $Count = \{\mathbf{A} : \mathbf{A} \neq \mathbf{0} \ \& \ \mathbf{A}$ countable $\}$; $Cl = \{\mathbf{A} : \mathbf{A} \neq \mathbf{0} \ \& \ \mathbf{A}$ closed $\}$; $CB = \{\mathbf{A} : \mathbf{A} \neq \mathbf{0} \ \& \ \mathbf{A}$ has countable basis$\}$; $GCB = \{\mathbf{A} : \mathbf{A} \neq \mathbf{0} \ \& \ \mathbf{A}$ has generalized countable basis $\}$; $D = \{\mathbf{A} : \mathbf{A}$ dense$\}$.

Since $\mathbf{0} \notin X$, all such X with $X \neq D$, we have that F_X is proper. Since $\mathbf{1} \notin D$, we have that I_D is proper.

We observe that if $X \in \{$ Dis, $Edis$, $Count$, CB, GCB, $Cl\}$ then X is in fact a sublattice of $\mathfrak{M}$. Thus, in this case, $F_X = \{\mathbf{A} : (\exists \mathbf{B} \in X)[\mathbf{B} \leq \mathbf{A}]\}$.

Theorem 8.2 *The following hold:* $F_{Solv} \subset F_{Edis} \subset F_{Dis} \subset F_{Count} \subset F_{GCB}$. *Moreover* $F_{Edis} \not\subseteq F_{CB}$; $F_{CB} \not\subseteq F_{Count}$; $F_{CB} \subset F_{GCB}$; $F_{Solv} \subset F_{CB}$; $F_{Edis} \subset F_{Cl}$; $F_{Dis} \not\subseteq F_{Cl}$.

Proof. It is simple to show that the various inclusions hold. As to show that we have proper inclusions, or that some inclusions do not hold we can argue as follows. To show that $F_{Count} \not\subseteq F_{Dis}$, $F_{CB} \not\subseteq F_{Solv}$, use Lemma 7.1. To show that $F_{CB} \not\subseteq F_{Count}$ use Theorem 7.2. To show that $F_{Edis} \not\subseteq F_{Solv}$ consider as a counterexample the degree of difficulty of any effectively discrete mass problem consisting of two functions whose $T-$degrees constitute a minimal pair. To show that $F_{GCB} \not\subseteq F_{CB}$, consider the mass problem $\mathcal{A} = \{f : \neg(\exists g \leq_T f)[[g]_T \text{ minimal}]\}$. To show that $F_{Dis} \not\subseteq F_{Cl}$, construct (see [1]) a discrete and not solvable mass problem $\mathcal{A}$ such that, for every n, *either* there exists a function $f \in \mathcal{A}$ such that $\Psi_n(f)$ is not total, *or*, otherwise, there exists a recursive limit point g of $\mathcal{A}$, such that $\Psi_n(g)$ is total, so that, for every $\mathcal{B}$, if $\Psi_n(\mathcal{A}) \subseteq \mathcal{B}$ and $\mathcal{B}$ is closed, then $\mathcal{B}$ is solvable. To show that $F_{Edis} \not\subseteq F_{CB}$, consider the degree of difficulty of any nonsolvable and effectively discrete mass problem consisting of functions whose $T-$degrees, plus $\mathbf{0}_T$, constitute a densely and linearly ordered ideal of $\mathfrak{D}_T$. $\square$

Theorem 8.3 $F_{Solv} \subset F_{En}$; $F_{Edis} \not\subseteq F_{En}$; $F_{En} \not\subseteq F_{CB}$.

Proof. $F_{En} \not\subseteq F_{Solv}$ follows from the existence of quasi-minimal $e-$ degrees. To show $F_{Edis} \not\subseteq F_{En}$, take as a counterexample the degree of difficulty of any effectively discrete mass problem of functions whose $e-$degrees constitute pairs of minimal pairs in the $e-$degrees. Finally, $F_{En} \not\subseteq F_{CB}$ follows from the proof of Theorem 6.1. $\square$

The above filters and ideals are all nonprincipal.

Theorem 8.4 *If* $X \in \{Solv, En, Dis, Edis, Count, CB, GCB, Cl\}$ *then* F_X *is nonprincipal. Moreover, the ideal* I_D *is nonprincipal.*

Proof. Let $X \in \{Dis, Edis, Count, CB, GCB\}$. To show that F_X is nonprincipal, it suffices to show that, for every $\mathbf{B} \in X$, there exists $\mathbf{A} \in F_X$ such that $\mathbf{B} \not\leq \mathbf{A}$. Given any $\mathbf{B} \in X$, find a degree of solvability $\mathbf{S}$ such that $\mathbf{B} \not\leq \mathbf{S}$: this follows directly by known incomparability results for the $T-$degrees. For $X = Cl$ see [1]. For $X = Solv$ and $X = En$, use also that the members of X are meet–irreducible. $\square$

For most of the filters described above, the cardinality of the corresponding quotient lattice is determined by the following theorem.

Theorem 8.5 *The cardinalities of* $\mathfrak{M}/F_{GCB}$ *and* $\mathfrak{M}/F_{En}$ *are* $2^{2^{\aleph_0}}$.

Proof. Let $\{f_{(x,y)} : (x,y) \in I^2\}$ be a collection of functions whose $T-$degrees are minimal and constitute an antichain (here I denotes a set of cardinality $2^{\aleph_0}$). For every set $A \subseteq I$, let $\mathcal{A}_A = \{g : (\exists x \in A)(\exists y \in I)[f_{(x,y)} \leq_T g]\}$ (notice that the cardinality of $\Gamma_{\mathcal{A}_A}$ is $2^{\aleph_0}$). It is easy to see, using Corollary 5.2, that if $A \not\subseteq B$ then $[[\mathcal{A}_B]]_{GCB} \not\leq [[\mathcal{A}_A]]_{GCB}$ (where, given a degree of difficulty **A**, the symbol $[\mathbf{A}]_{GCB}$ denotes its equivalence class in the quotient lattice $\mathfrak{M}/F_{GCB}$). This shows that $\mathfrak{M}/F_{GCB}$ has cardinality $2^{2^{\aleph_0}}$.

As to show that the cardinality of $\mathfrak{M}/F_{En}$ is $2^{2^{\aleph_0}}$, consider a function f and a collection $\{g_x : x \in I\}$ of functions whose T–degrees are an antichain, and $([f]_e, [g_x]_e)$ is a minimal pair of enumeration degrees, for all x. Then the collection $\{[[\mathcal{A}_A]]_{En} : A \subseteq I\}$, where $\mathcal{A}_A = \{g : f \leq_T g \lor (\exists i \in A)[g_i \leq_T g]\}$, has cardinality $2^{2^{\aleph_0}}$ in the quotient lattice. $\square$

Problem 8.6 *Show that the cardinality of $\mathfrak{M}/I_D$ is $2^{2^{\aleph_0}}$.*

An important topic when we study a lattice is the investigation of its prime filters and prime ideals. A trivial example of a nonprincipal prime filter is $\mathfrak{M} - \{\mathbf{0}, \mathbf{0}'\}$. A trivial example of a nonprincipal prime ideal is $\mathfrak{M} - \{\mathbf{1}\}$. A more interesting example is given by:

Theorem 8.7 I_D *is prime.*

Proof. Suppose that $\mathcal{D}_0, \mathcal{D}_1, \mathcal{D}$ are mass problem such that $\mathcal{D}$ is dense and $\mathcal{D}_0 \wedge \mathcal{D}_1 \leq \mathcal{D}$, via, say the recursive operator Ψ. Then *either* $\Psi(\mathcal{D}) \subseteq 0 * \mathcal{D}_0$ *or* there exists an initial segment $\tilde{f}$ such that $\Psi(\tilde{f})(0) = 1$, and therefore $1 * \mathcal{D}_1 \leq \{f : \tilde{f} * f \in \mathcal{D}\}$, via the recursive operator $\lambda f.\Psi(\tilde{f} * f)$. Since $\{f : \tilde{f} * f \in \mathcal{D}\}$ is dense, the proof is complete. $\square$

Let $F = \mathfrak{M} - I_D$. It follows from lattice theory that F is a prime filter. It is shown in [1] that F is nonprincipal and $|\mathfrak{M}/F| = 2^{2^{\aleph_0}}$, and $F_{Cl} \subseteq F$. It follows also from next theorem that $F_{Dis} \subset F$ (hence $F_{Cl} \subset F$).

Theorem 8.8 *If $X \in \{Solv, En, Edis, Dis, Count, CB, GCB\}$, then F_X is not prime.*

Proof. Let f, B_0, B_1 be such that $f \leq_e B_0 \oplus B_1$, where B_0, B_1 belong to quasi–minimal e–degrees. Then $\mathbf{E}_{B_0}, \mathbf{E}_{B_1} \notin F_{CB}$ (this follows from Theorem 6.1), but $\mathbf{E}_{B_0} \vee \mathbf{E}_{B_1} \in F_{Solv}$. This shows that F_X is not prime if $X \in \{Solv, Edis, Dis, Count, CB\}$. As to show that F_{GCB} is not prime, consider a T–degree $\mathbf{a}$ and two families of minimal T–degrees R, S such that $|R| = |S| = 2^{\aleph_0}$ and $(\forall \mathbf{r} \in R)(\forall \mathbf{s} \in S)[\mathbf{a} \leq \mathbf{r} \vee \mathbf{s}]$ (in fact, for every $\mathbf{a}$, one can find such families). Then $\mathbf{A}_R(= [\{g : (\exists f)[[f]_T \in R \,\&\, f \leq_T g]\}]) \notin F_{GCB}$, and $\mathbf{A}_S(= [\{g : (\exists f)[[f]_T \in S \,\&\, f \leq_T g]\}]) \notin F_{GCB}$, but $\mathbf{A}_R \vee \mathbf{A}_S \in F_{GCB}$, in fact $\mathbf{A}_R \vee \mathbf{A}_S \in F_{CB}$.

Finally, to show that F_{En} is not prime, use the following fact about e–degrees (below, $A^{[i]}$ denotes the $i^{th}-$ column of A): for every non r.e. set B, there exist a set A and a recursive function f such that $B \not\leq_e A^{[n]}$, all n, and $B \leq_e A^{[2m]} \oplus A^{[2n+1]}$ (all m, n) via the enumeration operator $\Phi_{f(m,n)}$, and $([A^{[2m]}]_e, [A^{[2n+1]}]_e)$ is a minimal pair in the e–degrees. Then it is easy to define two degrees of difficulty $\mathbf{A}, \mathbf{B}$, such that $\mathbf{A}, \mathbf{B} \notin F_{En}$, but $\mathbf{E}_B \leq \mathbf{A} \vee \mathbf{B}$. $\square$

Problem 8.9 *Show that F_{Cl} is not prime.*

9   The Medvedev lattice as a Brouwer algebra

A distributive lattice $\mathfrak{L}$ with $0, 1$ is a *Brouwer algebra* if it can be equipped with a binary operation $\rightarrow$ such that, for all $a, b \in \mathfrak{L}$,

$$a \rightarrow b = \text{least}\{c \in \mathfrak{L} : b \leq a \vee c\}$$

(equivalently: $(\forall a, b, c)[b \leq a \vee c \Leftrightarrow a \rightarrow b \leq c]$). In the following, we will use the term $B-homomorphism$ to denote any lattice-theoretic homomorphism which preserves $0, 1$ and $\rightarrow$.

We recall that a distributive lattice with $0, 1$ is a *Heyting algebra* if its dual is a Brouwer algebra.

Theorem 9.1 ([5]) $\mathfrak{M}$ *is a Brouwer algebra.*

Proof. Given mass problems $\mathcal{A}, \mathcal{B}$, define

$$\mathcal{A} \rightarrow \mathcal{B} = \{n * f : (\forall g \in \mathcal{A})[\Psi_n(g \vee f) \in \mathcal{B}]\}.$$

Then the following are easily seen:

- $\mathcal{B} \leq \mathcal{A} \vee (\mathcal{A} \rightarrow \mathcal{B})$, via Ψ, where $\Psi(g \vee (n * f)) = \Psi_n(g \vee f)$;

- $(\forall \mathcal{C})[\mathcal{B} \leq \mathcal{A} \vee \mathcal{C} \Rightarrow \mathcal{A} \rightarrow \mathcal{B} \leq \mathcal{C}]$: indeed, if $\Psi_n(\mathcal{A} \vee \mathcal{C}) \subseteq \mathcal{B}$ then let $\Psi'(f) = n * f$: clearly $\Psi'(\mathcal{C}) \subseteq \mathcal{A} \rightarrow \mathcal{B}$.

Thus, we have that $[\mathcal{A} \rightarrow \mathcal{B}] = \text{least}\{\mathbf{C} : [\mathcal{B}] \leq [\mathcal{A}] \vee \mathbf{C}\}$. $\square$

Dyment (see [2]) defines a topology on the collection of mass problems, such that, if $\mathcal{B} \not\leq \mathcal{A}$, then $\{\mathcal{C} : \mathcal{A} \rightarrow \mathcal{B} \leq \mathcal{C}\}$ is of first category.

We notice however

Theorem 9.2 ([13]) $\mathfrak{M}$ *is not a Heyting algebra.*

Proof. It can be shown that, for every nonzero degree of solvability $\mathbf{S}$, there exists an effectively discrete degree $\mathbf{B}$ such that the set $\{\mathbf{C} : \mathbf{S} \wedge \mathbf{C} \leq \mathbf{B}\}$ does not have a greatest element: given a nonzero degree of solvability $\mathbf{S}$, let $\{g\} \in \mathbf{S}$. Construct a countable mass problem $\{f_n : n \in \omega\}$, such that, for every m, n with $m \neq n$,

- $g <_T f_n$ & $f_m(0) \neq f_n(0)$ & $f_m |_T f_n$;

- $\Psi_n(f_n) \neq g$.

Thus $\mathbf{B} = \{f_n : n \in \omega\}$ is the desired degree of difficulty. $\square$

In the next theorem we characterize the finite Brouwer algebras that are B−embeddable in $\mathfrak{M}$. We will then discuss some of the consequences of this theorem. Let B' denote the class of Brouwer algebras in which 0 is meet−irreducible and 1 is join−irreducible.

Theorem 9.3 ([14]) *A finite Brouwer algebra $\mathfrak{L}$ is B−embeddable in $\mathfrak{M}$ if and only if $\mathfrak{L} \in B'$.*

Proof. Clearly, the condition $\mathfrak{L} \in B'$ is necessary for a finite Brouwer algebra to be B−embeddable in $\mathfrak{M}$: this follows from the fact that $\mathfrak{M} \in B'$.

We sketch the proof of the right−to−left implication. The proof is broken into several claims. Given any partial order $\mathfrak{P}$, let $F(\mathfrak{P})$ be the free distributive lattice with $0, 1$ generated by the partial order $\mathfrak{P}$ (hence $\mathfrak{P}$ embeds into $F(\mathfrak{P})$ as a partial order). It is not difficult to see:

Claim 1 *For any poset $\mathfrak{P}$, $F(\mathfrak{P})$ is a Brouwer algebra, in fact $F(\mathfrak{P}) \in B'$.*

We will now show that $F(\mathfrak{P})$ is B−embeddable in $\mathfrak{M}$, for a large class of posets $\mathfrak{P}$'s. In fact, $F(\mathfrak{D}_T)$ is B−embeddable in $\mathfrak{M}$.

Let now **2** be the two−element chain. Define B_J to be the smallest class of Brouwer algebras such that

1. $\mathbf{2} \in B_J$;

2. if $\mathfrak{L} \in B_J$ then $1 \oplus \mathfrak{L} \in B_J$;

3. B_J is closed under finite products.

Since a Brouwer algebra $\mathfrak{L}$ is subdirectly irreducible if and only if $\mathfrak{L} \simeq \mathbf{2}$ or $\mathfrak{L} \simeq 1 \oplus \mathfrak{L}'$, for some Brouwer algebra $\mathfrak{L}'$, it follows by the Birkhoff theorem on subdirectly irreducible algebras that, for every finite Brouwer algebra $\mathfrak{L}$, there exists $\mathfrak{L}' \in B_J$ such that $\mathfrak{L}$ is B−embeddable into $\mathfrak{L}'$. Thus, it is enough to show that for every $\mathfrak{L} \in B_J$, $1 \oplus \mathfrak{L} \oplus 1$ is B−embeddable in $\mathfrak{M}$.

We now notice

Claim 2 *For every $\mathfrak{L} \in B_J$, there exists a finite poset $\mathfrak{P}$ such that $1 \oplus \mathfrak{L} \oplus 1$ is B−embeddable in $F(\mathfrak{P})$.*

To see this, we argue by induction on the complexity of $\mathfrak{L}$ as a member of B_J, with respect to the three clauses of the definition of B_J. The most difficult part consists in showing that if $\{\mathfrak{L}_i : i \in I\}$ is a finite family for which we assume that, for all i, there exists a poset $\mathfrak{P}_i$ such that $1 \oplus \mathfrak{L}_i \oplus 1$ is B−embeddable in $F(\mathfrak{P}_i)$, then we can show that $1 \oplus (\prod \mathfrak{L}_i) \oplus 1$ is B−embeddable into $F(\coprod \mathfrak{P}_i)$, where $\coprod \mathfrak{P}_i$ denotes the coproduct of the $\mathfrak{P}_i$'s, in the category of partial orders.

Claim 3 *The Brouwer algebra* $F(\mathfrak{D}_T)$ *is* $B-$*embeddable into* $\mathfrak{M}$.

To prove the claim, let $\gamma : \mathcal{D}_T \longrightarrow \mathfrak{M}$ be defined by: $\gamma([f]_T) = \mathbf{B}_f$ (see Example 4.2). Let i be the lattice–theoretic homomorphism (preserving $0,1$), $i : F(\mathfrak{D}_T) \longrightarrow \mathfrak{M}$, such that $i([f]_T) = \gamma([f]_T)$: such a homomorphism exists by the universal mapping property of $F(\mathfrak{D}_T)$ ($\mathcal{D}_T$ being a sub–poset of $F(\mathfrak{D}_T)$).

The proof of the claim is based on the fact (see Example 4.2 and Corollary 5.2) that each $\mathbf{B}_f$ is join–irreducible and meet–irreducible and that infima of finite collections of degrees of difficulty of the form $\mathbf{B}_f$ are dense and uniform, so that we can apply the following result, which, with routine calculations, implies that i preserves the Brouwer algebra operation $\rightarrow$.

Claim 4 *If* $\mathbf{C}$ *is dense and uniform then, for all* $\mathbf{A}$, $\mathbf{B}$,

$$\mathbf{C} \rightarrow \mathbf{A} \wedge \mathbf{B} = (\mathbf{C} \rightarrow \mathbf{A}) \wedge (\mathbf{B} \rightarrow \mathbf{C}).$$

In order to prove Claim 4, let $\mathcal{C}$ be a dense mass problem, and let $\mathcal{A}$, $\mathcal{B}$, $\mathcal{X}$ be mass problems such that $\mathcal{A} \wedge \mathcal{B} \leq \mathcal{C} \vee \mathcal{X}$, via a recursive operator Ψ. For $i = 0,1$, define

$$\mathcal{X}_i = \{f \in \mathcal{X} : (\exists \tilde{f})[\Psi(\tilde{f} \vee f)(0) \downarrow = i]\}.$$

Then $\mathcal{X} \equiv \mathcal{X}_0 \wedge \mathcal{X}_1$, and $\mathcal{A} \leq \mathcal{C} \vee \mathcal{X}_i$, for $i = 0,1$. To show for instance that $\mathcal{A} \leq \mathcal{C} \vee \mathcal{X}_0$, one can use the recursive operator Ψ' defined informally as follows: in order to compute $\Psi'(f \vee g)$, given any enumeration of $f \vee g$, look for initial segments $\tilde{f}, \tilde{g}$ such that $\tilde{f} \subseteq f$ and $\tilde{g} \subseteq g$, and $\Psi(\tilde{f} \vee \tilde{g})(0) \downarrow = 0$; if no such initial segments are found then $\Psi'(f \vee g)$ is undefined, otherwise for the first such pair $\tilde{f}, \tilde{g}$ let $\Psi'(f \vee g) = \Psi((\tilde{f} * f) \vee g)$: density and uniformity of $\mathcal{C}$ ensure that $\tilde{f} * f \in \mathcal{C}$ if $f \in \mathcal{C}$, then $\Psi'(f \vee g) \in 0 * \mathcal{A}$ whenever $f \vee g \in \mathcal{C} \vee \mathcal{X}_0$, hence $\mathcal{A} \leq \mathcal{C} \vee \mathcal{X}_0$.

The following claim follows by easy calculations.

Claim 5 *If* $\mathfrak{P}_1$, $\mathfrak{P}_2$ *are posets and* $\mathfrak{P}_1$ *is order–theoretically embeddable into* $\mathfrak{P}_2$, *then* $F(\mathfrak{P}_1)$ *is* $B-$*embeddable into* $F(\mathfrak{P}_1)$.

We are now in a position to conclude the proof of the theorem. Since every finite partial order is embeddable into $\mathfrak{D}_T$, we have that for every Brouwer algebra $\mathfrak{L} \in B'$, there are (by Claim 2) a Brouwer algebra $\mathfrak{L}' \in B_J$, and a finite partial order $\mathfrak{P}$ such that $\mathfrak{L}$ is $B-$embeddable in $1 \oplus \mathfrak{L}' \oplus 1$, and $1 \oplus \mathfrak{L}' \oplus 1$ is $B-$embeddable in $F(\mathfrak{P})$. But, by Claim 5 $F(\mathfrak{P})$ is $B-$embeddable in $F(\mathfrak{D}_T)$. Finally the result follows from Claim 3. $\square$

Interest in Brouwer (Heyting) algebras lies also in the fact that they are used for a semantics of certain intermediate logics. In the following we refer to a propositional language built up from an infinite countable set of propositional variables. Let *Form* denote the set of well-formed formulas. If $\mathfrak{L}$ is a Brouwer algebra, we say that a function $v : Form \longrightarrow \mathfrak{L}$ is an $\mathfrak{L}-valuation$, if for all $\alpha, \beta \in Form$, we have $v(\alpha \wedge \beta) = v(\alpha) \vee v(\beta)$, $v(\alpha \vee \beta) = v(\alpha) \wedge v(\beta)$, $v(\alpha \rightarrow \beta) = v(\alpha) \rightarrow v(\beta)$, $v(\neg\alpha) = v(\alpha) \rightarrow 1$ (where in the left hand side of these equations the symbols $\wedge, \vee, \rightarrow$ denote propositional connectives, and in the right hand side the same symbols denote operations of $\mathfrak{L}$).

Given any Brouwer algebra $\mathcal{L}$, and $\alpha \in Form$, let $\mathfrak{L} \models \alpha$ if $v(\alpha) = 0$, for all $\mathfrak{L}-$valuation v; finally let $Th(\mathfrak{L}) = \{\alpha \in Form : \mathfrak{L} \models \alpha\}$.

Let *Int*, *Class* denote the theorems of the intuitionistic propositional calculus and of the classical propositional calculus, respectively. Any deductively closed set $\Sigma \subseteq Form$ such that $Int \subseteq \Sigma \subseteq Class$ is called an *intermediate logic*. We are interested here in the intermediate logic *Jan* (after Jankov), i.e. the deductive closure of $Int \cup \{\neg\alpha \vee \neg\neg\alpha : \alpha \in Form\}$. We note the following result ([6]):

Corollary 9.4 $Th(\mathfrak{M}) = Jan$.

Proof. The result follows from Theorem 9.3 and the following observations: (1) if $\mathfrak{L}_1$ is $B-$embeddable in $\mathfrak{L}_2$, then $Th(\mathfrak{L}_2) \subseteq Th(\mathfrak{L}_1)$ (see e.g. [9]); (2) $Jan = \cap\{Th(\mathfrak{L}) : \mathfrak{L} \in B' \ \& \ \mathfrak{L} \text{ finite } \}$ (see [4]). On the other hand, simple calculations show that $Jan \subseteq Th(\mathfrak{M})$. $\square$

Notice that, by Claim 3 and Claim 5 in the proof of Theorem 9.3, we can show that, if $\xi \leq 2^{\aleph_0}$ is a cardinal number, then the free distributive lattice with $0, 1$ on ξ generators is $B-$embeddable in $\mathfrak{M}$. More embeddings of Brouwer algebras will be pointed out in the proof of Theorem 9.10 below.

Problem 9.5 *Find examples of natural classes of infinite Brouwer algebras that are $B-$embeddable in $\mathfrak{M}$.*

Problem 9.6 *Show that for $\mathfrak{M}_e$ one can prove an embedding theorem similar to Theorem 9.3. What is $Th(\mathfrak{M}_e)$?*

It is well known that if $\mathfrak{L}$ is a Brouwer algebra and I is an ideal, then $\mathfrak{L}/I$ is still a Brouwer algebra. Clearly $Th(\mathfrak{L}) \subseteq Th(\mathfrak{L}/I)$, given the existence of the canonical onto homomorphism $\nu : \mathfrak{L} \longrightarrow \mathfrak{L}/I$. We recall

Theorem 9.7 ([15]) *If I is a proper principal ideal, then a finite Brouwer algebra $\mathfrak{L}$ is $B-$embeddable in $\mathfrak{M}/I$ if and only if $\mathfrak{L} \in B'$. It follows that $Th(\mathfrak{L}/I) = Jan$.*

Proof. It is easy to see that the proof of Theorem 9.3 relativizes to any proper principal ideal. $\square$

Theorem 9.3 and the previous theorem make use of embeddings into $\mathfrak{M}$ whose ranges consist of dense degrees. It is therefore natural to ask the following question.

Problem 9.8 *Describe the finite Brouwer algebras that are* $B-$*embeddable in* $\mathfrak{M}/I_D$. *What is* $Th(\mathfrak{M}/I_D)$?

If G is a filter, then $\mathfrak{M}/G$ is not necessarily a Brouwer algebra. However, if G is a principal filter, then $\mathfrak{M}/G$ is a Brouwer algebra, as $\mathfrak{M}/G \simeq \{\mathbf{B} : \mathbf{B} \leq \mathbf{A}\}$, where $\mathbf{A}$ generates G.

Problem 9.9 *Study the set* $\mathfrak{I} = \{Th(\mathfrak{M}/G) : G$ *proper principal filter*$\}$.

We observe that *Class* $\in \mathfrak{I}$: just take the principal filter generated by $\mathbf{0}'$. Some remarks on this set can be found in [15].

We have the following remarkable result, due to Skvortsova.

Theorem 9.10 ([11]) *There exists a principal filter* G *such that* $Th(\mathfrak{M}/G) = Int$.

Proof. First show that if $\mathfrak{D}$ is any countable implicative uppersemilattice then $F(\mathfrak{D}) \oplus 1$ is $B-$embeddable in $\mathfrak{M}$, where $F(\mathfrak{D})$ is the free distributive lattice, generated by $\mathfrak{D}$ (i.e. $\mathfrak{D}$ embeds in $F(\mathfrak{D})$ via an embedding preserving $\vee, \rightarrow$ and 0). The embedding can be defined as follows. First, embed $F(\mathfrak{D})$ into $\mathfrak{D}_T$ as an initial segment (use for this the fact that $F(\mathfrak{D})$ is a countable distributive upper semilattice). For every $s \in F(\mathfrak{D})$, let f_s lie in the Turing degree corresponding to s under this embedding. Define

$$\mathcal{D}_s = \{f : f \text{ nonrecursive } \& \ (\forall s \in F(\mathfrak{D}))[f \not\equiv_T f_s]\} \cup \{f : f_s \leq_T f\}.$$

Then the embedding $\gamma : F(\mathfrak{D}) \oplus 1 \longrightarrow \mathfrak{M}$, given by $\gamma(s) = [\mathcal{D}_s]$ if $s \in F(\mathfrak{D})$, and $\gamma(1) = \mathbf{1}$, turns out to be a $B-$embedding (use the fact that every element in the range of i is the infimum of finitely many Muchnik degrees, hence Claim 4 of Theorem 9.3 applies).

Given any Brouwer algebra $\mathfrak{L}$ and $a, b \in \mathfrak{L}$, let $\mathfrak{L}_a = \{c \in \mathfrak{L} : c \leq a\}$, and $\mathfrak{L}_{a,b} = \{c \in \mathfrak{L} : a \leq c \leq b\}$. In both cases, we get a Brouwer algebra. If $c \in \mathfrak{L}$ is an element such that $c \vee a = b$ then there is an onto $B-$homomorphism $\nu : \mathfrak{L}_c \longrightarrow \mathfrak{L}_{a,b}$ (so $Th(\mathfrak{L}_c) \subseteq Th(\mathfrak{L}_{a,b})$). Now, let $\mathfrak{F}_\omega = \langle U, \vee, \rightarrow, 0 \rangle$, with $U = \{X \subseteq \omega : X \text{ finite or cofinite}\}$, be the implicative upper semilattice where $X \vee Y = X \cap Y$, $X \rightarrow Y = X^c \cup Y$, and $0 = \omega$. Let $\delta : F(\mathfrak{F}_\omega) \oplus 1 \longrightarrow \mathfrak{M}$ be a $B-$embedding, constructed as above. It is possible ([11]) to find pairs $\{(a_i, b_i) : i \in \omega\}$ of elements of $F(\mathfrak{F}_\omega)$, such that $a_i \in \mathfrak{F}_\omega$, $a_i < b_i$, and $\bigcap Th(F((\mathfrak{F}_\omega)_{a_i,b_i}) : i \in \omega\} = Int$. For every i, let $b_i = \bigwedge\{b_i^j : j \leq n_i\}$, with $b_i^j \in \mathfrak{F}_\omega$, and, finally, let $\mathcal{E} = \{i * j * f : j \leq n_i, f \in \mathcal{D}_{b_i^j}\}$ (where $\mathcal{D}_{b_i^j} \in \delta(b_i^j)$). Let $\mathbf{E} = [\mathcal{E}]$, and, for all i, let $\mathbf{A}_i, \mathbf{B}_i$ be the images under δ of a_i, b_i, respectively. It is not difficult to see that for every i, $\mathbf{E} = \mathbf{A}_i \vee \mathbf{B}_i$. By the above remarks we have that $\bigcap\{Th(\mathfrak{M}_{\mathbf{A}_i,\mathbf{B}_i}) : i \in \omega\} = Int$, and thus $Th(\mathfrak{M}_\mathbf{E}) = Int$. $\square$

It follows from a result in [11] that if $\mathfrak{H}$ is the collection of proper principal filters generated by elements that are infima of finitely many Muchnik degrees, then $Int = \cap\{(Th(\mathfrak{M}/G) : G \in \mathfrak{H}\}$.

Problem 9.11 ([11]) *Is it possible to find* $G \in \mathfrak{H}$ *such that* $Th(\mathfrak{M})/G = Int$?

References

[1] Bianchini C. and Sorbi A. A note on closed degrees of difficulty of the Medvedev lattice. to appear.

[2] E. Z. Dyment. Certain properties of the Medvedev lattice. *Mat. Sb. (NS)*, 101 (143):360–379, 1976. Russian.

[3] E. Z. Dyment. Exact bounds of denumerable collections of degrees of difficulty. *Mat. Zametki*, 28:895–910, 1980. Russian.

[4] V. A. Jankov. The calculus of the weak law of excluded middle. *Math. USSR Izvestija*, 2:997–1004, 1968.

[5] Yu. T. Medvedev. Degrees of difficulty of the mass problems. *Dokl. Akad. Nauk SSSR, (NS)*, 104:501–504, 1955. Russian.

[6] Yu. T. Medvedev. Finite problems. *Dokl. Akad. Nauk SSSR, (NS)*, 142:1015–1018, 1962. Russian.

[7] A. A. Muchnik. on strong and weak reducibilities of algorithmic problems. *Sibirsk. Mat. Zh.*, 4:1328–1341, 1963. Russian.

[8] R. A. Platek. A note on the cardinality of the Medvedev lattice. *Proc. Amer. Math. Soc.*, 25:917, 1970.

[9] H. Rasiowa and R. Sikorski. *The Mathematics of Metamathematics*. Panstowe Wydawnictwo Naukowe, Warszawa, 1963.

[10] H Rogers, Jr. *Theory of Recursive Functions and Effective Computability*. McGraw–Hill, New York, 1967.

[11] E. Z. Skvortsova. Faithful interpretation of the intuitionistic propositional calculus by an initial segment of the Medvedev lattice. *Sibirsk. Mat. Zh.*, 29:171–178, 1988. Russian.

[12] A. Sorbi. On some filters and ideals of the Medvedev lattice. *Arch. Math. Logic*, 30:29–48, 1990.

[13] A. Sorbi. Some remarks on the algebraic structure of the Medvedev lattice. *J. Symbolic Logic*, 55:831–853, 1990.

[14] A. Sorbi. Embedding brouwer algebras in the Medvedev lattice. *Notre Dame J. Formal Logic*, 2:266–275, 1991.

[15] A. Sorbi. Some quotient lattices of the Medvedev lattice. *Z. Math. Logik Grundlag.*, 37:167–182, 1991.

Extension of Embeddings on the Recursively Enumerable Degrees Modulo the Cappable Degrees

Xiaoding Yi*

Abstract

It is shown that there is a non-trivial obstruction to block the extension of embeddings on the quotient structure of the recursively enumerable degrees modulo the cappable degrees. Therefore, Shoenfield's conjecture fails on that structure, which answers a question of Ambos-Spies, Jockusch, Shore, and Soare [1984], and Schwarz [1984] (also see Slaman [1994]).

1 Introduction

The *recursively enumerable* (*r.e.*) sets are these subsets of natural numbers (denoted ω) which can enumerated by an effective procedure (or a computable function from ω to ω). There is a notion of relatively computability (or Turing reducibility) among the r.e. sets (in fact, among all sets of natural numbers), which can view that one is more complicated or harder to compute than other. The equivalence classes under this notion of relatively computability of r.e. sets are called the *r.e.* (Turing) degrees. The set of all r.e. degrees (denoted $\mathcal{R}$) is made into a partial ordering with least (**0**, the equivalence class of all recursive sets) and greatest (**0'**, the equivalence class which contains the halting problem) elements in the natural way, namely, the reducibility relation between r.e. sets induces a partial ordering on degrees. It is readily shown that finite supremum always exists in $\mathcal{R}$. Therefore, $\mathcal{R}$ forms an upper semi-lattice.

*Preparation of this paper supported by S.E.R.C.(UK) Research Grant no. GR/H 02165, and by European network 'Complexity, Logic and Recursion Theory' (EC Contract No. ERBCHRXCT930415). The author is grateful to Carl Jockusch, Steffen Lempp and Steve Leonhardi for communicating on the material in this paper. Thanks go to Barry Cooper, Alistair H. Lachlan and the University of Leeds.

As a partial ordering, there are two basic questions concerning the structure of the r.e. degrees: the embedding problem, namely, which finite partial orderings are embeddable into $\mathcal{R}$; and the extension of embeddings problem, namely, for which pairs of finite partial orderings $P \subset Q$ is it the case that for every embedding f of P into $\mathcal{R}$, there is an extension g of f to an embedding of Q into $\mathcal{R}$.

For the first problem, Post [1944] asked whether the three elements linear ordering is embeddable (into $\mathcal{R}$). Friedberg [1957] and independently Muchnik [1956] solved it positively to show that the partial ordering with 0 and 1 and two other incomparable elements (i.e., four elements Boolean algebra) can be embedded into $\mathcal{R}$. The construction technique they introduced is called the *priority method*. Sacks [1963a] extended this result to show that every countable partial ordering can be embedded into $\mathcal{R}$, which completely answered the embedding problem (with partial ordering). This embedding property also implies that the Σ_1-theory of $(\mathcal{R}, \leq)$ is decidable.

Algebraically the next question about the structure of partial ordering after embedding problems is the extension of embeddings problem. Sacks' Density Theorem (1964) (namely, for all r.e. degrees $a < b$ there is an r.e. degree c such that $a < c < b$) implies that such extension is possible if Q is a linear ordering.

Sacks [1963b] also showed that every non-recursive r.e. degree may be written as the supremum of an incomparable pair of r.e. degrees. In view of these results the r.e. degrees seemed well behaved. Shoenfield [1965] responded to these results by conjecturing that the structure of the r.e. degrees might be same as the rationals in $[0,1]$ but partial, then partial ordering embeddings could always be extended if not ruled out on trivial grounds by the axioms of upper semi-lattices. More precisely,

Conjecture (Shoenfield) For every pair of finite upper semi-lattices $P \subset Q$ with 0 and 1 and every embedding f of P into $\mathcal{R}$, there is an extension g of f to an embedding of Q into $\mathcal{R}$. In particular, Shoenfield listed two consequences of his conjecture:

(C1). If $a, b \in \mathcal{R}$ are incomparable, then they have no infimum (greatest lower bound) in $\mathcal{R}$.

(C2). Given r.e. degrees $0 < b < a$, there exists an r.e. degree $c < a$ such that $a = b \vee c$, namely, that every such b can be non-trivially cupped to a.

Both of these consequences are false (and hence Shoenfield's conjecture fails). Lachlan [1966b], and independently, Yates [1966] proved that there exists a minimal pair, i.e., a pair of non-recursive r.e. degrees whose greatest lower bound is $\mathbf{0}$ (i.e., $\mathbf{0}$ is a *branching degree* in the r.e. degrees). Slaman [1991] showed that the branching degrees are dense in the r.e. degrees.

Lachlan [1966a] disproved (C2), namely, there is an r.e. degree a which has the *anticupping* (a.c.) *property* (i.e., there exists a non-recursive r.e. degree $b < a$ such that for no r.e. $c < a$ does $a = b \vee c$). Yates and Cooper showed that the $0'$ has the a.c. property, and Harrington (see Miller [1981]) extended the property to any high r.e. degree.

Little progress was made of the extension of embeddings problem until Slaman and Soare [1995] showed that it is decidable. They discovered that there exists a incomparable pair a, b such that for every $w \leq a$, either $w \leq b$ or $a \leq w \vee b$ (called *saturated embedding* $\{a, b, a \vee b\}$ into $\mathcal{R}$). Roughly speaking, Slaman and Soare [1995] showed that all obstructions to extensions of embeddings in the language $\{\leq\}$ were either lattice embedding obstructions like the minimal pair, or saturated embeddings obstructions.

The next question is whether the two quantifier theory of $(\mathcal{R}, \leq)$ is decidable. It is not difficulty to see that deciding the truth of all $\forall\exists$ sentences (in some language) is equivalent to determining if for any given finite substructure P of the given structure such as $\mathcal{R}$ and finitely many extensions Q_i of P, that every embedding (or realization) of P in the given structure can be extended to an embedding of Q_i for some i. Thus the extension of embeddings problem is a special case of this question (for a single Q). Also, the lattice embedding problem (i.e., which finite lattice can be embedded into $\mathcal{R}$ preserving $\leq$, $\vee$ and $\wedge$) is also a subquestion of the Σ_2-theory of $(\mathcal{R}, \leq)$.

The gap between the extension of embeddings problem and the full two quantifier theory is precisely the problem of having move than one possible extension to consider. A surprise result of Lachlan [1966b], called Non-diamond Theorem, namely, there are no incomparable r.e. degrees a, b such that $a \vee b = 0'$ and $a \wedge b = 0$, provided such possibility. This phenomenon was extended by Ambos-Spies, Jockusch, Shore, and Soare [1984]. They showed that the set of *cappable* degrees (i.e., the r.e. degrees which are halves of minimal pairs together with 0, denoted M) comprise an ideal of $\mathcal{R}$, whose complement (denoted PS) is the set of the degrees of *promptly simple* sets–which is a strong filter of $\mathcal{R}$ (i.e., closed upward and for all $a, b \in PS$, there exists a degree $c \in PS$ such that $c \leq a, b$). It was also shown by Ambos-Spies, Jockusch, Shore, and Soare [1984] that the set PS is exactly the set of r.e. degrees which can be joined to the greatest element, $0'$, of $\mathcal{R}$ by some *low* r.e. degree (a degree a satisfying $a' = 0'$), and in fact, it coincides many other collections of degrees.

These equivalences were discovered by studying the degrees of promptly simple sets, a computational complexity of Post's simple set introduced by Maass [1982] for studying orbits of r.e. sets under automorphisms of $\mathcal{E}$, the lattice of r.e. sets. When Post posted his famous question of whether there are more than two r.e. degrees, he also indirectly suggested a program for solving the program. Post's Program is to find some definable property on $\bar{A}$ such that if A satisfies this property then A is incomplete and non-recursive. He suggested that some sort of "thinness" property such as hyperhypersimplicity (hhsimplicity) might work. However, Yates [1965] constructed a complete maximal set, so we know hhsimplicity will not work. The notion of a promptly simple set is a dynamic one which takes into account how fast elements appear in A relative to their appearance in other r.e. sets under some standard simultaneous enumeration of all r.e. sets. Maass [1982] showed that any two low promptly simple sets are automorphic. Recently, Harrington and Soare [1995] proved that every r.e. set of promptly simple degree is (effectively) automorphic to a complete set. Post's Program finally solved by Harrington and Soare [1991]: there is a nonempty $\mathcal{E}$-definable property $Q(A)$ such that every r.e. set A satisfying $Q(A)$ is non-recursive and Turing incomplete. Very recently, Cooper [ta] construct an automorphism of the r.e. degrees which moves a low r.e. degree to a non-low degree, which implies the non-rigidity of $\mathcal{R}$ (and then $\mathcal{D}$) and the low r.e. degrees is not definable in the r.e. degrees. It is not clear that whether the low promptly simple degrees is definable in the r.e. degrees.

2   Basic results and Theorems

Since M is an ideal of $\mathcal{R}$, we can define the quotient structure of the r.e. degrees modulo the cappable degrees in the usual way.

Definition 2.1   i) $M = \{a : (\exists b \in \mathcal{R})[b > 0 \wedge (a \wedge b = 0)]\}$.

ii) For every r.e. degree a, let $[a]$ denote the set $\{b \in \mathcal{R} : (\exists m, n \in M)[a \vee m = b \vee n]\}$. Let $\mathcal{R}/M$ denote the set $\{[a] : a \in \mathcal{R}\}$. $\mathcal{R}/M$ carries the structure of an upper semi-lattice with least and greatest elements as follows:

   (a) $[a] \leq [b] \Longleftrightarrow (\exists m \in M)[a \leq b \vee m]$,

   (b) $[a] \vee [b] = [a \vee b]$,

   (c) $[\,0]$ is the least element in $\mathcal{R}/M$,

   (d) $[\,0']$ is the greatest element in $\mathcal{R}/M$.

The following property of promptly simple degrees is needed in present paper. One can find the proof in Ambos-Spies, Jockusch, Shore, and Soare [1984] or Soare [1987].

Proposition 2.2 (Promptly Simple Degree Theorem) *Let A be an r.e. set and $\{A_s\}_{s \in \omega}$ a recursive enumeration of A. Then A has promptly simple* *degree if and only if there is a recursive function f such that for all s, $f(s) \geq s$, and for all r.e. sets W,*

$$W \text{ infinite} \Longrightarrow (\exists y)(\exists s)[y \in W_{at\ s} \wedge A_s \lceil y \neq A_{f(s)} \lceil y].$$

Schwarz [1984] began the investigation of the quotient upper semi-lattice $\mathcal{R}/M$, the r.e. degrees modulo the cappable degrees. Schwarz showed that $\mathcal{R}/M$ satisfies the analogues of the Friendberg-Muchnik theorem (as well as its extension that every countable partial ordering is embeddable into $\mathcal{R}/M$, and then the Σ_1-theory of $(\mathcal{R}/M, \leq)$ is decidable), and the Sacks splitting theorem. As mentioned in [Schwarz 1984], one motivation for looking at $\mathcal{R}/M$ is the hope of finding a natural, degree-theoretic upper semi-lattice satisfying Shoenfield's conjecture. Schwarz also proved that the minimal pair phenomenon does not occur in $\mathcal{R}/M$. This leads to the hope that the $\mathcal{R}/M$ counterpart to Shoenfield's conjecture holds, which was an open question left by Ambos-Spies, Jockusch, Shore, and Soare [1984], and Schwarz [1984] (also see Slaman [1994]).

Schwarz [1989] also worked on the index sets related to the promptly simplicity. He showed the index sets of r.e. sets which are promptly simple, and are degrees of promptly simple are both Σ_4-complete. It is worth to mention that all above results of Schwarz about the structure $\mathcal{R}/M$ are consequences (see Jockusch [1985]) from the results of Ambos-Spies et al. (1984) and/or Robinson's Splitting Theorem (1971). For example, one can prove the counterpart to Sacks' Splitting Theorem as follows.

Proposition 2.3 *(Schwarz/Ambos-Spies)*
Given r.e. sets B and C with C promptly simple. Then B can be expressed as the disjoint union of two (low) r.e. sets A_0, A_1 so that C is not recursive in $A_i \oplus E$, for any E of cappable degree.

Proof: Since C is low-cuppable, we can choose an r.e. set F of low degree so that $C \oplus F$ has degree $\mathbf{0}'$. By Robinson splitting theorem (1971), we can then choose a splitting A_0, A_1 of B so that $A_i \oplus F$ is low for $i = 0, 1$. Suppose now for a contradiction that C is recursive in $A_i \oplus E$, where E is of cappable degree. Since $C \oplus F$ has degree $\mathbf{0}'$, it follows that $A_i \oplus E \oplus F$ has degree $\mathbf{0}'$, from which it follows that E is low-cuppable (by $A_i \oplus F$), so E cannot be of cappable degree. ∎

Clearly, $\mathbf{a}$ low implies $[\mathbf{a}] < [\mathbf{0}']$ since low cuppable is non-cappable. By using Robinson's Splitting Theorem, one can easily show that for any low r.e. degree $\mathbf{a}$ (then $[\mathbf{a}] < [\mathbf{0}']$), there exists a pair of low r.e. degrees $\mathbf{a}_0, \mathbf{a}_1 \geq \mathbf{a}$ such that $\mathbf{a}_0 \vee \mathbf{a}_1 = [\mathbf{0}']$. Another result which leads to hope that the $\mathcal{R}/M$ counterpart to Shoenfield's conjecture holds is following.

Proposition 2.4 $[\ 0']$ *has not the a.c. property, i.e., for every r.e. degree* b *such that* $[\ 0] < [b] < [\ 0']$, *there exists an r.e. degree* c *such that* $[c] < [\ 0']$ *and* $[\ 0'] = [b] \vee [c]$.

Proof: Let a be an r.e. degree such that $[a] > [\ 0]$. Then a is low cuppable, i.e., there exists a low r.e. degree b such that $a \vee b = 0'$. Hence, $[b] < [\ 0']$ and $[a \vee b] = [\ 0']$. ∎

Yates, Cooper and Harrington showed that $0'$ has the a.c. property in the r.e. degrees, i.e., there exists a non-zero r.e. degree b such that for all r.e. degrees $c < 0'$, $c \vee b < 0'$.

In the present paper, we show the $\mathcal{R}/M$ counterpart to Shoenfield's conjecture fails by showing that

Theorem 2.5 *There exist recursively enumerable degrees* a, b, c *such that* $c \leq a \leq b$, $[c] < [a]$ *and for all recursively enumerable degrees* $w \geq c$, *either* $b \leq w$ *or* $b \not\leq w \vee a$.

As an immediate corollary we have

Theorem 2.6 *There exist upper semi-lattices* $P \subset Q$ *and an embedding* π *of* P *into* $\mathcal{R}/M$ *which cannot be extended to an embedding of* Q *into* $\mathcal{R}/M$.

Proof: Let P consist $\{a, b, c\}$ such that $c < a < b$. Let Q be an upper semi-lattice extension of P as follows: $Q = \{a, b, c, w\}$ such that $c < a < b$, $c < w < b$ and $a \vee w = b$.

To find the embedding π, we let a, b and c to be chosen such that they satisfy Theorem 2.5, and let $\pi(a) = a$, $\pi(b) = b$ and $\pi(c) = c$. Since $c \leq a \leq b$, $[c] \leq [a] \leq [b]$. Also, $[c] < [a]$. For any cappable degree w, we know either $b \leq c \vee w$ or $b \not\leq c \vee w \vee a$. Note that $b \leq c \vee w$ implies that $a \leq c \vee w$ and then $[a] \leq [c]$, a contradiction. Therefore, $b \not\leq c \vee w \vee a = w \vee a$, and then $[b] \not\leq [a]$. Hence, π is an embedding of P into $\mathcal{R}/M$.

Towards a contradiction assume that there exists an r.e. degree w such that $[w] < [b]$, $[c] \leq [w] \leq [b]$ and $[b] \leq [a] \vee [w]$. There exists a cappable degree n such that $b \leq a \vee w \vee n$. By the theorem we know that $b \leq c \vee w \vee n$, and then $[b] \leq [c \vee w] = [w]$, a contradiction. Therefore, π cannot be extended to a embedding of Q into $\mathcal{R}/M$. ∎

Clearly, the above result shows that the Shoenfield's Conjecture fails in $\mathcal{R}/M$. It is unknown whether the extension of embeddings problem and the Σ_2-theory of $(\mathcal{R}/M, \leq)$ are decidable. It is also unknown that whether there is a non-trivial definable ideal I of $\mathcal{R}$ such that $(\mathcal{R}/I, \leq)$ satisfies Shoenfield conjecture.

3 The requirements and modules

We now turn our attention to the proof of our main theorem, Theorem 2.5. Our notation is standard and follows Soare [1987]. All uses are monotone in stage and argument where defined, and bounded by s at stage s. In describing a construction, notations such as A, U, W, and Ψ are used to denote the current approximations to these objects. The notations A_s, U_s, W_s, and Ψ_s denote the approximations which exist immediately before stage s. Occasionally, if the notation A would be ambiguous, we use A_ω to make it clear that we are referring to the value of A at the end of the construction.

The functional Ψ has associated an use-function. At any point in the construction $\psi(U; x) \downarrow$ (sometime, just referring as $\psi(x) \downarrow$) implies $B(x) = \Psi^{U,A}(x) \downarrow$. Further, once $\psi(U, x)$ is defined it remains defined with the same value unless some number $\leq \psi(U, x)$ enters $U \cup A$. It is convenient to choose the use-function and the enumeration of instructions for the functional so that

i) $\psi(U, x) \downarrow$ implies $x \leq \psi(U, x)$.

ii) if $\psi(U, x)$ becomes defined, then $\psi(U, y)$ is already defined for all $y < x$.

iii) ψ is increasing on its domain.

iv) if $\psi(U, x) \downarrow$, then it becomes undefined whenever a number $\leq \psi(U, x)$ is enumerated in $U \cup A$.

Similarly, for Φ. For Γ, we shall have the use-function γ with properties corresponding to those listed for ψ with one exception that once x is enumerated into B, $\gamma(x)$ is fixed, and if $\gamma(x) \downarrow$, then $\gamma(x)$ becomes undefined only if some number $\leq \gamma(x)$ is enumerated into C or some number $\leq \psi(U, x)$ (it will turn out that $\psi(U, x) \leq \gamma(x)$) is enumerated into U. It will turn out that when some number $\leq \gamma(x)$ is enumerated into C, a number $\leq \psi(U, x)$ is enumerated into A. Therefore, $\gamma_\omega(x)$ exists and finite if $\psi_\omega(U, x)$ exists and finite.

In order to show Theorem 2.5, we recursively construct r.e. sets A, B, C to satisfy all following requirements:

$$
\begin{aligned}
\mathcal{S} : \quad & C \leq_T A, A \leq_T B, \\
\mathcal{N}_{U,\Psi} : \quad & B = \Psi^{U,A} \Rightarrow B \leq_T U \oplus C, \\
\mathcal{P}_{M,\Phi} : \quad & A \neq \Phi^{C,M} \text{ or} \\
& M \text{ is of promptly simple degree,}
\end{aligned}
$$

where U, M are r.e. sets and Φ, Ψ are partial recursive functionals.

Suppose the above requirements are satisfied. Clearly, $\mathcal{S}$ implies that $deg(C) \leq deg(A) \leq deg(B)$. Suppose $deg(B) \leq \boldsymbol{w} \vee deg(A)$ for some r.e.

degree $\boldsymbol{w} \geq \deg(C)$. Let $U \in \boldsymbol{w}$ and $B = \Phi^{U,A}$ such that U is an r.e. set. By the $\mathcal{N}$-requirements, $B \leq_T U \oplus C$. Therefore, $deg(B) \leq \boldsymbol{w} \vee deg(C) = \boldsymbol{w}$. Clearly, the $\mathcal{P}$-requirements imply that $[deg(A)] \not\leq [deg(C)]$. Therefore, $[deg(C)] < [deg(A)]$.

In the construction which we will describe below, whenever a number is enumerated into C, a smaller number is enumerated into A simultaneously, and similarly, whenever a number is enumerated into A, a smaller number is enumerated into B simultaneously. Therefore, we know that $C \leq_T A$, $A \leq_T B$ and then $\mathcal{S}$ is satisfied.

The basic module for $\mathcal{N}$-requirements in isolation is standard. For example, to satisfy $\mathcal{N}_{U,\Psi}$, we will construct a p.r. functional $\Gamma (= \Gamma_{U,\Psi})$. Our intention is to ensure that $B = \Gamma^{U,C}$ (and so $B \leq_T U \oplus C$) if $B = \Psi^{U,A}$. We will define Γ as follows: for each z such that $\Psi^{U,A}(z) \downarrow = B(z)$, we define $\Gamma^{U,C}(z) = B(z)$ with use $\gamma(z) = \psi(z)$. In the same time, we restrain all numbers $\leq \psi(z)$ from entering A unless and until either $\psi(z)$ become undefined or we are sure that z cannot be enumerated into B afterward or the strategy of $\mathcal{N}_{U,\Psi}$ is initialized. Note that whenever $\gamma(z) \downarrow$ and some number $\leq \gamma(z)$ is enumerated into $U \oplus C$, we (may) let $\gamma(z)$ becomes undefined. It is clear that $\Psi^{U,A} = B$ implies Γ is total and $B = \Gamma^{U,C}$. However, infinitely many restraints for $\mathcal{N}_{U,\Psi}$ on A may prevent the satisfaction of $\mathcal{P}$-requirements (as we will see below, the satisfaction of $\mathcal{P}$-requirements involves enumeration of numbers into A).

We now describe the basic module to satisfy a single $\mathcal{P}$-requirement, $\mathcal{P}_{M,\Phi}$ say. Our primary goal in attacking requirement $\mathcal{P}_{M,\Phi}$ is to obtain a disagreement between A and $\Phi^{C,M}$. We might try to do this as follows. Choose a witness x which is not yet in A. Wait for a stage, v say, in which $\Phi^{C,M}(x) \downarrow = 0$, then enumerate x into A and restrain all numbers $\leq \varphi(x)$ from entering C. Of course, if M never changes below $\varphi(x)$ (the number which bounds all uses in the computation $\Phi^{C,M}(x)[v]$), then we shall have a disagreement between A and $\Phi^{C,M}$ at x. But if M ever changes below $\varphi(x)$ at a stage $\geq v$, so that $\Phi^{C,M}(x)$ can be corrected to output 1 rather than 0, then we have to attack all over again with some new witness x'. Now we might try to attack as above with infinitely often and still fail to achieve a disagreement between A and $\Phi^{C,M}$, because every time we try to preserve a computation, a change occurs in M below the use in that computation. If so, we can attain our secondary goal of $\mathcal{P}_{M,\Phi}$ to show that M is of promptly simple degree.

In order to satisfy $\mathcal{P}_{M,\Phi}$, we begin with a strategy to enumerate the graph of a (partial) recursive function $f (= f_{M,\Phi})$. Potentially, f will provide a witness to the fact that M is of promptly simple degree. For each recursively enumerable set W, we will ensure that if W is infinite and M does not promptly permit W with respect to f then there is an x such that $\Phi^{C,M}$ and A disagree on x. Therefore, our strategy will ensure that the requirement

$\mathcal{P}_{M,\Phi}$ is satisfied.

For each r.e. set W, we define a subrequirement $\mathcal{R}_{M,\Phi,W}$ of $\mathcal{P}_{M,\Phi}$ as follows:

$$\mathcal{R}_{M,\Phi,W}: \qquad A \neq \Phi^{C,M}, \text{ or}$$
$$W \text{ is finite, or}$$
$$(\exists y, s)[y \in W \text{ at }_s \&(M_s \lceil y \neq M_{f(s)} \lceil y)].$$

The $\mathcal{R}_{M,\Phi,W}$ will take the responsibility to ensure that if W is infinite then either $\Phi^{C,M} \neq A$ or M promptly permits W with respect to f.

The basic module of $\mathcal{R}_{M,\Phi,W}$ consists two parts. First, we enumerate a (partial) recursive function $f(= f_{M,\Phi})$ as follows.

Enumerating f. This strategy acts during expansionary stages for the equation $\Phi^{C,M} = A$. During such a stage s, it lets m be the least argument at which $f[s]$ is not defined and sets $f(m)$ equal to s.

Second, we proceed the main part of the module of $\mathcal{R}_{M,\Phi,W}$ as follows: choose a witness x, wait for a stage, s say, in which $\Phi^{C,M}(x) \downarrow = 0$ and some $y > \varphi(x)$ is enumerated into W, then enumerate x into A and restrain all numbers $\leq \varphi(x)[s]$ from entering C. If $A(x)$ and $\Phi^{C,M}(x)$ agree on x and W is infinite, then we shall eventually find such stage s. If $\Phi^{C,M}(x)$ is ever later corrected to output 1 rather than 0, it can only be that M has changed below $\varphi(x)[s]$, which is $< y$. If this ever happens, say at stage $t > s$, then we will know that $f(s) \geq t$. It is readily seen that this strategy will meet requirement $\mathcal{R}_{M,\Phi,W}$.

The conflicts among the basic modules of above requirements are as follows. To satisfy a requirement $\mathcal{P}$ (or $\mathcal{R}$) we may enumerate a number, x say, into A. To satisfy $\mathcal{S}$, we are forced to enumerate a number $k \leq x$ into B in the same stage in which x is enumerated into A. But in the case that $\Gamma(k) \downarrow$ and $x \leq \gamma(k) \leq \varphi(x)$, we cannot correct the output of $\Gamma(k)$ even $\Psi^{U,A}(k) = B(k)$ since we cannot enumerate $\gamma(k)$ into C (otherwise, the computation $\Phi^{C,M}(x)$ would be injured through a C-change). Therefore, $\mathcal{N}_{U,\Psi}$ cannot be satisfied. In the spirit to show the way to overcome this difficulty, we describe a strategy for $\mathcal{R} = \mathcal{R}_{M,\Phi,W}$ while its priority has been given to $\mathcal{N}_{U,\Psi}$. Roughly speaking, we fix a k which is not yet in B. We begin to search a witness x such that $\Gamma_{U,\Psi}(k) \downarrow$ implies $\gamma_{U,\Psi}(k) < x$ or $\varphi(x) < \gamma_{U,\Psi}(k)$. Note that we know such x exists, if either $\Psi^{U,A}(k) = B(k)$ or $\Phi^{C,M} = A$. The strategy for $\mathcal{R}$ acts as follows.

Enumerating $\Gamma(= \Gamma_{U,\Psi})$. As described before, for all z, whenever $\Psi^{U,A}(z) \downarrow = B(z)$ with $\Gamma^{U,C}(z) \uparrow$, define $\Gamma^{U,C}(z) = B(z)$ and let $\gamma(z)$ be the least unused number $\geq \psi(z)$.

Enumerating $f(= f_{M,\Phi})$. As described above, we define f during expansionary stages for the equation $\Phi^{C,M} = A$.

The rest of the strategy for $\mathcal{R}$ will act as follows:

1. Choose a number, k say, which is not yet in B.

2. Select a (new) witness x for $\mathcal{R}$ (which is not yet in A, and greater than all previous witnesses for $\mathcal{R}$).

3. Wait for a stage s in which some number y enters W for which $y > \varphi(x)[s] \downarrow$. *If W is infinite and $\varphi(x)$ is defined then there will be such y.*

4. There are two cases.

   (a) $\gamma(k) \downarrow$ and $z \leq \gamma(k) \leq \varphi(z)$ for every witness z for $\mathcal{R}$. Go back to step 2. *If it returns to step 2 infinitely often, then $\psi_\omega(k) \uparrow$ and $\varphi_\omega(x_0) \uparrow$, where x_0 is the least witness for $\mathcal{R}$. Therefore, both $\mathcal{P}_{M,\Phi}$ (not just $\mathcal{R}_{M,\Phi,W}$) and $\mathcal{N}_{U,\Psi}$ are satisfied. In this case, we must implement a back-up strategy to satisfy all other requirements which have lower priority than $\mathcal{P}_{M,\Phi}$ during the stages in which this strategy goes from step 4 to step 2.*

   (b) Otherwise. Then either $\gamma(k) \uparrow$ or for some witness, z say, of $\mathcal{R}$, either $\gamma(k) < z$ or $\varphi(z) < \gamma(k)$. Note that $z \leq x$, and then $\varphi(z)[s] \downarrow \leq \varphi(x)[s] < y$. There are two subcases.

      • $\gamma(k) \uparrow$ *or* $\gamma(k) < z$. *Enumerate z into A and k into B.*

      • *Otherwise. Then $\varphi(z) < \gamma(k)$. Enumerate z into A, k into B and $\gamma(k)$ into C. Restrain all numbers $\leq \varphi(z)[s]$ from entering C.*

      Now, the only way by which there could be another expansionary stage for $\Phi^{C,M} = A$ is for M to change below y. Thus, if f is ever defined at s then its value is greater than the stage during which M changed below y. If this occurs then M will have permitted W during the interval $[s, f(s)]$.

One should note that for each $\mathcal{R}$ one of the following holds:

a) $\mathcal{R}$ receives attention at most finitely often. We know that the strategy either waits at step 3 or reaches step 4(b). In either case, $\mathcal{R}$ is satisfied and Γ is in progress.

b) It goes from step 4 to step 2 infinitely often. Under this outcome, we know that both $\mathcal{P}_{M,\Phi}$ and $\mathcal{N}_{U,\Psi}$ are satisfied.

We now extend the above strategy for satisfying an arbitrary $\mathcal{R}$-requirement, $\mathcal{R} = \mathcal{R}_{M,\Phi,W}$ say, in the presence of the (finitely many) higher priority $\mathcal{N}$-requirements: $\mathcal{N}_{U_0,\Psi_0}, \cdots, \mathcal{N}_{U_n,\Psi_n}$. The strategy for $\mathcal{R}$ acts as follows.

Enumerating $\Gamma_i (= \Gamma_{U_i, \Psi_i})$ $(i \leq n)$. As described before, for all z, whenever $\Phi_i^{U_i, A}(z) \downarrow = B(z)$ with $\Gamma_i^{U_i, C}(z) \uparrow$, define $\Gamma_i^{U_i, C}(z) = B(z)$ and $\gamma_i(z)$ be the next unused number in $\omega^{[i]}$ which $\geq \psi_i(z)$.

Enumerating $f (= f_{M, \Phi})$. As described above, we define f during expansionary stages for the equation $\Phi^{C, M} = A$.

The rest of the strategy for $\mathcal{R}$ acts as follows:

1. Choose a number, k say, which is not yet in B.

2. Select a (new) witness x for $\mathcal{R}$ (which is not yet in A, and greater than all previous witnesses for $\mathcal{R}$).

3. Wait for a stage s in which some number y enters W for which $y > \varphi(x)[s] \downarrow$. *If W is infinite and $\varphi(x)$ is defined then there will be such y.*

4. There are two cases.

    (a) $\gamma(k) \downarrow$ and for every witness z for $\mathcal{R}$, there exists an $i \leq n$ such that $z \leq \gamma_i(k) \leq \varphi(z)[s]$. Go back to step 2. *If it returns to step 2 infinitely often, then $\varphi_\omega(z) \uparrow$ for some fixed witness z for $\mathcal{R}$. Let I be the set of all $i \leq n$ such that $\gamma_{i, \omega}(k)$ exists and finite. We can let z to be the least witness for $\mathcal{R}$ such that $z > \gamma_{i, \omega}(k)$ for all $i \in I$. Clearly, there exists an $i \leq n$ such that $i \notin I$. In this case, we know that $\mathcal{P}_{M, \Phi}$ and $\mathcal{N}_{U_i, \Psi_i}$ are satisfied for all $i \leq n$ such that $i \notin I$.*

    (b) Otherwise. Then there exists some witness, z say, for $\mathcal{R}$, such that for all i, $\gamma_i(k) \downarrow$ implies either $\gamma_i(k) < z$ or $\varphi(z)[s] < \gamma_i(k)$. Note that $z \leq x$, and then $\varphi(z)[s] \downarrow \leq \varphi(x)[s] < y$. Enumerate z into A, k into B, and enumerate $\gamma_i(k)$ into C for all i such that $\gamma_i(k) \downarrow > \varphi(z)[s]$. Restrain all numbers $\leq \varphi(z)[s]$ from entering C. Note that for all i, if $\gamma_i(k) \downarrow \leq \varphi(z)[s]$, then $\gamma_i(k) \downarrow < z$ by the choice of z. Therefore, we can correct the $\Gamma_i(k)$ either by (force) a U_i-change or through a C-change. Now, the only way by which there could be another expansionary stage for $\Phi^{C, M} = A$ is for M to change below y. Thus, if f is ever defined at s then its value is greater than the stage during which M changed below y. If this occurs then M will have permitted W during the interval $[s, f(s)]$.

4   The Tree of Strategies and Construction

Fix an enumeration, $\mathcal{Q}_0, \mathcal{Q}_1, \cdots$, say, of all $\mathcal{N}$- and $\mathcal{R}$-requirements. We say $\mathcal{Q}_i$ has higher (local) priority than $\mathcal{Q}_j$ if $i < j$, and $\mathcal{Q}_i$ has higher global priority than Q_j if both Q_i, Q_j are $\mathcal{R}$-requirements and $g(i) < g(j)$ where $g : \omega \longrightarrow \omega$ is defined as follows: $g(e) =$ the least $k \leq e$ such that $Q_k = \mathcal{R}_{M,\Phi,W}$ and $Q_e = \mathcal{R}_{M,\Phi,V}$ for some M, W, V and Φ. First, we describe the priority tree T and simultaneously the requirement which is designated to be satisfied at a given node. We complete the definition of T by specifying what the outcomes at α are for each $\alpha \in T$. The immediate successors of α are just the strings of the form $\alpha^\wedge \langle i \rangle$, with i an outcome of α.

We first let Q_0 be associated with λ, the root of the tree T. Let $\alpha \in T$ and a requirement Q_α be associated with α. Suppose α is a $\mathcal{N}$-node (i.e., requirement $\mathcal{N}$ is associated with α). Then outcomes of α are $\{0, 2\}$ with the usual order. Suppose α is a $\mathcal{R}$-node. Then outcomes of α are $\{0, 1, 2\}$. We associate the next requirement Q_j (after Q_α) with $\alpha^\wedge \langle i \rangle$ such that if Q_j is a $\mathcal{R}$-requirement, then there is no β such that $\beta^\wedge \langle 1 \rangle \subseteq \alpha^\wedge \langle i \rangle$ and β is an Q_e-node with $g(j) = g(e)$. α is called an i-node if $\alpha(|\alpha| - 1) = i$.

Remark 4.1   1. Suppose Q is associated with α. Then Q is not associated with any node $\supset \alpha$.

2. Suppose $\mathcal{R} = Q_i$ is associated with α. Then no Q_j is associated with a node $\beta \supseteq \alpha^\wedge \langle 1 \rangle$ with $g(j) = g(i)$.

We now describe the parameters at node α. Suppose an $\mathcal{N}$-requirement is associated with α. A p.r. functional Γ_α, a use function γ_α for $\Gamma_\alpha^{U_\alpha, C}$. Suppose a requirement $\mathcal{R}$ is associated with α. The following parameters are associated with the α-strategy's attempt to satisfy $\mathcal{R}$:

a) $x(\alpha, n)$, a witness to satisfy $\mathcal{R}$. We will let x^α to denote $\max\{x(\alpha, n) : n \in \omega\}$ at any moment during the construction.

b) k^α, if defined, can be regarded as the use of reduction $A \leq_T B$ of $x(\alpha, n)$ which may be enumerated into A.

c) c^α, a three-valued function which indicates how far $\mathcal{R}$'s strategy at α has received attention.

Suppose $\mathcal{N}_{U,\Psi}$ is associated with α. Then we define $l(\alpha, s) = \mu z[B_s(z) \neq \Psi_\alpha^{U_\alpha, A}(z)[s]]$, and $\psi(\alpha, x, s) = \psi(U_\alpha, A; x, s)$. Suppose α is a $\mathcal{R}_{M,\Phi,W}$-node. Similarly, if α is a $\mathcal{R}_{M,\Phi}$-node, we define $l(\alpha, s) = \mu z[A_s(z) \neq \Phi^{C,M}(z)[s]]$. We say s is an α-*expansionary stage* if $l(\alpha, s) > l(\alpha, t)$ for all $t < s$ such that α is visited at stage t. Let α be a $\mathcal{R}$-node. We let $\mathcal{A}(\alpha)$ to denote the set of all $\mathcal{N}$-nodes β such that $\beta^\wedge \langle \omega_i \rangle \subset \alpha$ for some $i < 2$.

Construction

We now describe the actions of stage s of the construction. At stage s of the construction we begin at node λ and carry out certain instructions passing down the tree. We now give the instructions of node α. If at some node the instructions do not require passing to a node, we go to the instructions for **closing stage** s, and then the stage ends.

Case 1. There exists $\beta \supseteq \alpha$ such that

- $c^\beta = 1$,

- $\varphi_\beta(x^\beta) \downarrow$,

- $(\exists w)[w > \varphi_\beta(x^\beta) \wedge w \in W_{\beta,at\ s}]$,

Choose the least β. There are two subcases.

Case 1.1. For all n such that $x(\beta, n) \downarrow$, there exists some $\sigma \in \mathcal{A}(\beta)$ and $k \leq k^\beta$ such that $\gamma_\sigma(k) \downarrow$ and $x(\beta, n) \leq \psi_\sigma(k) \leq \gamma_\sigma(k) \leq \varphi_\beta(x(\beta, n))$. Set $c^\beta = 0$ and pass to $\beta^\wedge\langle 1 \rangle$. We call this *jumping from α to $\beta^\wedge\langle 1 \rangle$*.

Case 1.2. Otherwise. We know that there exists some n such that $x(\beta, n) \downarrow$ and for every $\sigma \in \mathcal{A}(\beta)$ and $k \leq k^\beta$, one of the following holds:

- $\gamma_\sigma(k) \uparrow$.

- $\psi_\sigma(k) < x(\beta, n)$.

- $\varphi_\beta(x(\beta, n)) < \gamma_\sigma(k)$.

Choose the least such n. Enumerate $x(\beta, n)$ into A and k^β into B. For each $k \leq k^\beta$, enumerate $\gamma_\sigma(k)$ into C for all σ such that $\gamma_\sigma(k) \downarrow$ and $x(\beta, n) \leq \psi_\sigma(k)$ (therefore, $\varphi_\beta(x(\beta, n)) < \gamma_\sigma(k)$). Set $c^\beta = 2$. Initializing all nodes $>_L \beta^\wedge\langle 0 \rangle$. Note that both $\psi_\sigma(k)$ and $\gamma_\sigma(k)$ become undefined if $\gamma_\sigma(k)$ is enumerated into C.

Case 2. α is a $\mathcal{R}$-node. There are four subcases.

Case 2.1. $k^\alpha \uparrow$. Define k^α to be the least unused number which exceeds all used values taken by a parameter. Set $c^\alpha = 0$.

Case 2.2. $c^\alpha \downarrow = 0$. Let n be the least number such that $x(\alpha, n) \uparrow$. Define $x(\alpha, n)$ to be the least unused number which exceeds all used values taken by a parameter. Set $c^\alpha = 1$.

Case 2.3. $c^\alpha = 1$. Pass to $\alpha^\wedge\langle 2 \rangle$.

Case 2.4. $c^\alpha = 2$. Pass to $\alpha^\wedge\langle 0 \rangle$.

Case 3. α is a $\mathcal{N}$-node. There are three cases.

Case 3.1. s is an α-expansionary stage. For all $z < l(\alpha)$, define $\Gamma_\alpha^{U,C}(z) = B(z)$ with $\gamma_\alpha(z) = s$. Pass to $\alpha^\wedge\langle 0 \rangle$.

Case 3.2. Otherwise. Pass to $\alpha^\wedge\langle 2 \rangle$.

Closing stage s

Let α be the last node which is visited at stage s. α is said to *receive attention* at stage s. After completing the instructions for the particular cases which hold at the various nodes, to end the stage s we carrying out the following:

($C1$) Destroy all parameters belong to a node β such that $\alpha <_L \beta$.

($C2$) For each β such that β is a $\mathcal{R}$-node, define $f_\beta(m) = s$ if s is an β-expansionary stage and m is the least number such that $f_\beta(m) \uparrow$. *It will turn out that f_β and f_σ are same p.r. functions if β, σ are $\mathcal{R}_{M,\Phi,W}$- and $\mathcal{R}_{M,\Phi,V}$-nodes, respectively, for some W and V.*

5   The Verification

Define the *true path* TP to be the subset of all $\alpha \in T$ such that α is visited infinitely often and there are at most finitely many stages at which α is initialized.

Lemma 5.1 *If $\alpha \in TP$, then there exists an immediate successor (will called α^+) of α in TP.*

Proof: Fix $\alpha \in TP$. First we show that for some immediate successor δ of α, there exist infinitely many stages at which some node $\supseteq \delta$ is visited. Towards a contradiction suppose that there is no such δ. Then, eventually no node $\supset \alpha$ is visited. Clearly, when α is visited at stage s, Case 1.2 holds. But for each such β in Case 1 when α is visited at stage s, $c^\beta[s] = 1$ and $c^\beta[s+1] = 2$. Whenever c^β is set equal to i for $i \leq 1$ some node $\supseteq \beta$ is visited in the same stage. Hence, δ exists. Fix δ to be the least such node.

We will prove that δ is visited infinitely often. Towards a contradiction assume that there is a stage t after which δ is never visited. Consider a stage $v > t$ at which a node $\supset \delta$ is visited. In stage v there is a jump from a node $\subseteq \alpha$ to an 1-node $\tau = \tau_v \supset \delta$ via Case 1.1. By Case 1.1, $c^{\tau^-}[v] = 1$ and c^{τ^-} is set equal to 0 at stage v. By the choice of δ, we know that there are infinitely many $\sigma \supset \delta$ such that c^σ is set equal to 1. Also, whenever c^σ is set equal to 1 or becomes defined, σ is visited. Therefore, there is no single $\tau \supset \delta$ such that c^τ is set equal to 1 infinitely often. Let $n(s)$ be the number of nodes σ in which $c^\sigma[s] = 1$. Suppose $t < s < v$. We know that $n(v) \leq n(s)$, and $n(v) < n(s)$ if there is a node $\tau \supseteq \delta$ such that $c^\tau[s] \uparrow$ and $c^\tau[v] \downarrow = 1$. Therefore, eventually $n(s) = 0$ and then no node $\supset \delta$ is visited afterward, a contradiction.   ∎

Lemma 5.2 *All $\mathcal{R}$-requirements are satisfied.*

Proof: Fix $\mathcal{R}_{M,\Phi,W}$. Suppose there exists a node $\alpha \in TP$ such that α^- is a $\mathcal{R}_{M,\Phi,V}$-node for some V and $\alpha = \alpha^{-\wedge}\langle 1 \rangle$. Note that whenever α is visited, Case 1.1 holds (reading α^- for β in Case 1), and when c^{α^-} is set equal to 1, Case 2.2 holds at α^-. Therefore, there are infinitely many n such that $x(\alpha^-, n)$ becomes defined, and then $x^{\alpha^-} \longrightarrow \infty$ as $s \longrightarrow \infty$. As we know, k^{α^-} is eventually fixed ($= k$ say) and $\gamma_\sigma(k)$ is non-decreasing (on stages). Towards a contradiction assume that there exists a number t such that

$$\{\gamma_\sigma(k)[s] : \sigma \in \mathcal{A}(\alpha^-) \wedge s \in \omega\} \leq t.$$

Let s be a stage such that in which there exists an n with $x(\alpha^-, n)[s] \downarrow > t$. Clearly, Case 1.1 does not hold again by reading α^- for β, a contradiction. Hence,

$$\{\gamma_\sigma(k)[s] : \sigma \in \mathcal{A}(\alpha^-) \wedge s \in \omega\}$$

is unbounded and (since $\mathcal{A}(\alpha^-)$ is finite) we can assume that $\gamma_\tau(k)[s] \longrightarrow \infty$ as $s \longrightarrow \infty$ for some $\tau \in \mathcal{A}(\alpha^-)$.

Let I denote the set of all $\sigma \in \mathcal{A}(\alpha^-)$ such that $\lim_s \gamma_\sigma(k)[s]$ exists and is finite. Let X denote $\max\{\gamma_\sigma(k)[s] : \sigma \in I \wedge s \in \omega\}$. Clearly, $\mathcal{A}(\alpha^-) \neq I$. Let n be the least number such that $(x =) x(\alpha^-, n)[s] \downarrow > X$ for some stage s. We know that after $x(\alpha^-, n)$ is set equal to x, whenever Case 1.1 holds (reading α^- for β), there exists a $\sigma \in \mathcal{A}(\alpha^-) - I$ such that $\varphi_{\alpha^-}(x) \geq \gamma_\sigma(k)$. Therefore, $\varphi_{\alpha^-,\omega}(x) \uparrow$ and $B(x) \neq \Phi_{\alpha^-}^{C,M}(x)$. Hence, $\mathcal{P}_{M,\Phi}$ and $\mathcal{P}_{M,\Phi,W}$ are satisfied.

Suppose that there is not a node $\beta \in TP$ such that β^- is a $\mathcal{R}_{M,\Phi,V}$-node for some V, and $\beta = \beta^{-\wedge}\langle 1 \rangle$. Let α be the $\mathcal{R}_{M,\Phi,W}$-node such that $\alpha \in TP$. Clearly, α exists. Since $\alpha^\wedge\langle 1 \rangle \notin TP$, we know that $\lim_s c^\alpha[s](= c)$ exists and ≥ 1. Also, $\lim_s x^\alpha[s](= x)$ exists and is finite.

Suppose $c = 1$. Since Case 1 does not apply (reading α for β in Case 1), we have either $\varphi_{\alpha,\omega}(x) \uparrow$, or $\varphi_{\alpha,\omega}(x) \downarrow$ and then W is finite. Hence, we have either $A(x) \neq \Phi^{C,M}(x)$ or W is finite. Therefore, $\mathcal{R}_{M,\Phi,W}$ is satisfied.

Suppose $c = 2$. Let t be the greatest stage in which c^α is set equal to 2. By the construction we know that Case 1.2 holds (reading α for β in Case 1). In the stage t, we enumerate x into A and k^α into B. By Case 1 of the construction, there exists an w such that $w > \varphi_\alpha(x)[t]$ and w enters W at stage t. If $A \neq \Phi^{C,M}$ then $\mathcal{R}_{M,\Phi,W}$ is satisfied so suppose that $A = \Phi^{C,M}$. Therefore, there exists a (least) stage $v > t$ in which $A(x) = \Phi^{C,M}(x)$.

Claim 5.3 $C_v \lceil \varphi(x)[t] = C_t \lceil \varphi(x)[t]$.

Proof: Towards a contradiction assume that there is a number $z < \varphi(x)[t]$ which enters C at a stage $t' \geq t$ and $< v$. Let β, σ be nodes in T such that $\gamma_\beta(k^\sigma)[t'] = z$. Note that at stage t, we only enumerate some numbers

y into C such that $\varphi(x)[t] < y$ by Case 1.2. Hence, $t \neq t'$. At stage t, all nodes $>_L \alpha^\wedge\langle 0 \rangle$ are initialized. Therefore, $\alpha^\wedge\langle 0 \rangle \not<_L \beta, \sigma$. Clearly, at stage t', σ is visited and k^σ is enumerated into B and c^σ is set equal to 2 (from 1). Then, $\sigma \not\subseteq \alpha$, otherwise α is initialized at stage t'. Hence, $\alpha^\wedge\langle 0 \rangle \subseteq \sigma$. By the construction, whenever α is initialized, so does σ, and then whenever σ is visited, $c^\alpha = 2$. Therefore, k^σ is set equal to $k^\alpha[t']$ at a stage $> t$ and so $\varphi(x)[t] < k^\sigma[t'] \leq \gamma_\beta(k^\sigma)[t'] = z$. This completes the proof of Claim 5.3.   ∎

Now from Claim 5.3,

$$M_v \lceil \varphi(x)[t] \neq M_t \lceil \varphi(x)[t],$$

and then

$$M_v \lceil w \neq M_t \lceil w.$$

Note that there are at most one m such that $f_\alpha(m)$ becomes defined at each single stage. Therefore, $f_\alpha(t) \geq v$, and then

$$M_{f_\alpha(t)} \lceil w \neq M_t \lceil w.$$

This completes the proof of Lemma 5.2.   ∎

Lemma 5.4 *Every $\mathcal{N}$-requirement is satisfied.*

Proof: Fix a requirement $\mathcal{N}_{U,\Psi}$. Let α be the node on the true path such that $\mathcal{N}_{U,\Psi}$ is associated with α. Clearly such α exists.

Suppose $B = \Psi^{U,A}$. We will show that Γ_α is total and $B =^* \Gamma_\alpha^{U,C}$. Note that $\alpha^\wedge\langle 0 \rangle \in TP$ and there are infinitely many α-expansionary stages. Clearly, $\alpha \in \mathcal{A}(\beta)$ for all $\mathcal{R}$-nodes $\beta \supseteq \alpha^\wedge\langle 0 \rangle$.

Obviously the node α is initialized at most finitely often. We now consider what happens once all parameters belonging to a node $\leq \alpha$ reach their final values, if possible, for the last time (note that c_ω^β may not exist for some $\beta \subset \alpha$). We prove the following fact first before finishing the proof of Lemma 5.4.

Claim 5.5 *If k, s and t satisfy that*

- *$s < t$, $\alpha^\wedge\langle 0 \rangle$ is visited at stage s,*

- *$\gamma_\alpha(k)[s] \downarrow \geq \psi_\alpha(k)[s] \downarrow$,*

- *$\psi_\alpha(k)[s] \neq \psi_\alpha(k)[t]$,*

then one of the following holds

a) $(U_s \cup C_s) \lceil \gamma_\alpha(k)[s] \neq (U_t \cup C_t) \lceil \gamma_\alpha(k)[s]$.

b) k cannot enter B at a stage $\geq s$ and $< t$.

c) some number $< k$ is enumerated into B at a stage $\geq s$ and $< t$.

Proof: Towards a contradiction considering a triple (k, s, t) corresponding to some counterexample with k and then t chosen to be the least possible, and then s the greatest possible. Since $\psi_\alpha(k)[s] \neq \psi_\alpha(k)[t]$ and $U_s \lceil \psi_\alpha(k)[s] \neq U_t \lceil \psi_\alpha(k)[s]$, we have

$$A_s \lceil \psi_\alpha(k)[s] \neq A_t \lceil \psi_\alpha(k)[s].$$

Hence, there exists some $y < \psi_\alpha(k)[s]$ which enters A at a stage u, $s \leq u < t$. Let $y = x(\theta, n)[u]$ for some n. Note that $\theta \not\subseteq \alpha$, otherwise c_ω^θ exists and c^θ reaches its final value at a stage $\geq s$. It is clear that $\alpha^\wedge\langle 0 \rangle \subseteq \theta$ since all nodes $>_L \alpha^\wedge\langle 0 \rangle$ are initialized at stage s

Let $k = k^\beta[v]$ for some β and k enter B at stage v. Note that $s \leq v < t$ and $k^\beta[s] \downarrow = k^\beta[v + 1]$. We have $\beta \not\subseteq \alpha$ otherwise c_ω^β exists and c^β reaches its final value at a stage $\geq s$, a contradiction. By an argument similar to show that $\alpha^\wedge\langle 0 \rangle \subseteq \theta$, we can show that $\alpha^\wedge\langle 0 \rangle \subseteq \beta$. It is clear that $\theta \neq \beta$ implies $u \neq v$.

Suppose $\theta^\wedge\langle 0 \rangle <_L \beta$. By Case 1.2 of the construction, β is initialized at stage u. Therefore, $v < u$. Note that $k^\theta[s] = k^\theta[u]$, $k^\theta[s] < k^\beta[s] = k$, and $k^\theta[s]$ is enumerated into B at stage u, contradicts the choice of (k, s, t) (the c)).

Suppose $\theta^\wedge\langle 0 \rangle \subseteq \beta$. Clearly, whenever θ is initialized, so does β, and when β is visited, $c^\theta \downarrow = 2$. Therefore, $k^\beta \neq k^\beta[s]$, a contradiction. Similarly if $\beta^\wedge\langle 0 \rangle \subseteq \theta$, the $x(\theta, n)$ should be set equal to y at a stage $> v$ and then $y > \gamma_\alpha(k)[s]$, a contradiction.

Suppose $\beta^\wedge\langle 0 \rangle <_L \theta$. Note that $\alpha \in \mathcal{A}(\beta) \cap \mathcal{A}(\theta)$. Suppose $v < u$. Then θ is initialized at stage v, a contradiction. Suppose $u < v$. Note that whenever y is enumerated into A and $y \leq \psi_\alpha(k)[s]$, $\gamma_\alpha(k)[s]$ is enumerated into C by Case 1.2, a contradiction.

Finally, suppose $\beta = \theta$. Thus, $v = u$. From Case 1.2 of the construction, we know that either $\gamma_\alpha(k) < x(\theta, n)$ or $\varphi_\beta(x(\theta, n)) < \gamma_\alpha(k)$. Suppose $\gamma_\alpha(k) < x(\theta, n)$. Then $\psi_\alpha(k)[s] \leq \gamma_\alpha(k)[s] = \gamma_\alpha(k)[u] < x(\theta, n) = y$, a contradiction. Therefore, $\varphi_\beta(x(\theta, n)) < \gamma_\alpha(k)$. By Case 1.2 of the construction, $\gamma_\alpha(k)$ is enumerated into C, which contradicts the choice of (k, s, t). This completes the proof of Claim 5.5. ∎

By Claim 5.5, we know that between any two stages, $s < t$ say, if at both stages $\Gamma_\alpha^{U,C}(k)$ is defined, then their values should agree with $B(k)$ unless $(U \cup C) \lceil \gamma_\alpha(k)[s]$ changes between these stages. Note that when $\gamma_\alpha(k)[s] \downarrow$ and some number $\leq \gamma_\alpha(k)[s]$ is enumerated into C, then a number $\leq \psi_\alpha(k)[s] \downarrow$

is enumerated into A. Therefore, $\gamma_\alpha(k)[s]$ is bounded for each k which gives the required result since $B = \Psi^{U,A}$.   ■

From the construction it is clear that $C \leq_T A \leq_T B$. This completes the verification that all requirements are satisfied.

References

Ambos-Spies, K., C. G. Jockusch, Jr, R. A. Shore, and R. I. Soare (1984). An algebraic decomposition of the recursively enumerable degrees and the coincidence of several degree classes with the promptly simple degrees. *Trans. Amer. Math. Soc. 281*, 109–128.

Cooper, S. B. (ta). The limits of local Turing definability. Preprint.

Friedberg, R. M. (1957). Two recursively enumerable sets of incomparable degrees of unsolvability. *Proceedings of the National Academy of Science 43*, 236–238.

Harrington, L. A. and R. I. Soare (1991). Post's Program and incomplete recursively enumerable sets. *Proc. Natl. Acad. of Sci. USA 88*, 10242–10246.

Harrington, L. A. and R. I. Soare (1995). The Δ^0_3 automorphism method and noninvariant classes of degrees. Preprint.

Jockusch, Jr, C. G. (1985). Review of Schwarz [1984]. *Mathematical Review 85i*, 3777.

Lachlan, A. H. (1966a). The impossibility of finding relative complements for the recursively enumerable degrres. *J. Symbolic Logic 31*, 434–454.

Lachlan, A. H. (1966b). Lower bounds for pairs of recursively enumerable degrees. *Proceedings London Mathematical Society 16*, 537–569.

Maass, W. (1982). Recursively enumerable generic sets. *J. Symbolic Logic 47*, 809–823.

Miller, D. (1981). High recursively enumerable degrees and the anti-cupping property. In M. Lerman, Schmerl, and R. I. Soare (Eds.), *Logic Year 1979-80*, The University of Connecticut, pp. 230–245. Springer-Verlag.

Muchnik, A. A. (1956). On the unsolvability of the problem of reducibility in the theory of algorithms. *Dokl. Akad. Nauk. SSSR N.S. 108*, 194–197.

Post, E. L. (1944). Recursively enumerable sets of positive integers and their decision problems. *Bull. Amer. Math. Soc. 50*, 284–316.

Robinson, R. W. (1971). Interpolation and embedding in the recursively enumerable degrees. *Annals of Mathematics 93*, 285–314.

Sacks, G. E. (1963a). *Degrees of Unsolvability*, Volume 55 of *Annals of Mathematical Studies*. Princeton University Press.

Sacks, G. E. (1963b). On the degrees less than $0'$. *Annals of Mathematics 77*, 211–231.

Sacks, G. E. (1964). The recursively enumerable degrees are dense. *Annals of Mathematics 80*, 300–312.

Schwarz, S. (1984). The quotient semilattice of the recursively enumerable degrees modulo the cappable degrees. *Transactions of the Amer. Math. Soc. 283*, 315–328.

Schwarz, S. (1989). Index sets related to prompt simplicity. *Ann. Pure Appl. Log. 42*, 243–254.

Shoenfield, J. R. (1965). Application of model theory to degrees of unsolvability. In J. W. Addison, L. Henkin, and A. Tarski (Eds.), *Symposium on the Theory of Models*, Amsterdam, pp. 359–363. North-Holland Publishing Company.

Slaman, T. A. (1991). The density of infima in the recursively enumerable degrees. *Ann. Pur Appl. Log. 52*, 1–25.

Slaman, T. A. (July, 1994). Questions in Resursion Theory. Note.

Slaman, T. A. and R. I. Soare (1995). Algebraic aspects of the recursively enumerable degrees. *Proc. Nat. Acad. Sci., USA 92*, 617–621.

Soare, R. I. (1987). *Recursively Enumerable Sets and Degrees*. Perspectives in Mathematical Logic, Omega Series. Heidelberg: Springer-Verlag.

Yates, C. E. M. (1965). Three theorems on the degree of recursively enumerable sets. *Duke Math. Journal 32*, 461–468.

Yates, C. E. M. (1966). A minimal pair of recursively enumerable degrees. *Journal of Symbolic Logic 31*, 159–168.

Appendix:

QUESTIONS IN RECURSION THEORY

This is an informal list of some open problems in recursion theory, based on the list of open problems compiled during the Leeds Recursion Theory Year. Solutions have been announced for some of these problems. The current status of the questions below and of questions added after July 1995 can be found on the World Wide Web at html://www.math.uchicago.edu/~ted.

Solutions and new problems are welcome and should be directed to T. A. Slaman at ted@math.uchicago.edu.

1   Turing Degrees

Let $\mathcal{D}$ denote the partial ordering of the Turing degrees. $\mathcal{D}(\leq a)$ denotes the degrees less than or equal to a.

1.1. (Sacks) Suppose that P is a locally countable partially ordered set of cardinality less than or equal to the continuum. Is there an order preserving embedding of P into $\mathcal{D}$?

1.2. (Rogers) Is there a nontrivial automorphism of $\mathcal{D}$?

1.3. (Slaman–Woodin Conjectures)

(a) $\mathcal{D}$ is biinterpretable with second order arithmetic. In other words, the relation (on $\vec{p}$ and d) "$\vec{p}$ codes a standard model of first order arithmetic and a real X such that X is of degree d" is definable in $\mathcal{D}$.

(b) Suppose that $\mathcal{I}$ is an ideal in $\mathcal{D}$ such that there is a 1-generic degree in $\mathcal{I}$. Then $\mathcal{I}$ is biinterpretable with that fragment of second order arithmetic in which the real numbers are just those sets whose Turing degrees belong to $\mathcal{I}$.

1.4. (Jockusch) Do there exist distinct degrees a and b such that a and b have isomorphic upper cones in $\mathcal{D}$?

1.5. (Yates) Does every minimal degree have a strong minimal cover?

1.6. (Chong) Is there a minimal degree which is the base of a cone of minimal covers?

1.7. (a) (Kučera) Suppose that p is the Turing degree of a complete extension of Peano arithmetic and that x is a nonzero degree below p. Does there exist a degree a strictly below p such that $a \vee x = p$?

   (b) (Kučera) Characterize the recursively enumerable degrees w such that there is a p for which p is the Turing degree of a complete extension of Peano arithmetic and $w <_T p <_T 0'$.

1.8. (Marker) Let $I \subset \mathcal{P}(\omega)$ be a countable Turing ideal. We say that d is a *uniform upper bound* for I if there is $D \in d$ such that I is equal to $\{D_n : n \in \omega\}$, where $D_n = \{m : (n, m) \in D\}$.

   (a) Does I have a minimal uniform upper bound? That is, is there a uniform upper bound d such that for all $e <_T d$, e is not a uniform upper bound? What if I is a Scott set?

   (b) Can a minimal upper bound for I be a uniform upper bound?

1.9. (Fenner) Let a and b be Turing degrees with b (Cohen) generic. If $a \leq b \leq a'$, is it the case that $a' = a \vee 0'$?

   (Fenner comments: If A and G are sets, G is 2-generic, and $A \leq_{tt} G \leq_T A'$, then $A' \equiv_T A \vee K$ (equivalently, $G \leq_T A \vee K$).)

1.10. (Lerman) What is the largest decidable fragment (in terms of quantifier complexity) of the first order theory of the Turing degrees in the language with 0, $\leq$ and $'$? Here, $'$ denotes the Turing jump.

2   Degree Invariant Functions

2.1. (Sacks) Is there a degree invariant solution to Post's problem? That is, is there an e such that for all X, $X <_T W_e^X <_T X'$ and for all X and Y, $X \equiv_T Y$ implies that $W_e^X \equiv_T W_e^Y$?

2.2. (Martin's Conjecture) Assume AD. Let $\mathcal{F}$ be the class of degree invariant functions from 2^ω to 2^ω. Order $\mathcal{F}$ by $f \leq_M g$ if there is an X such that whenever $Y \geq_T X$, $f(Y) \leq_T g(Y)$ (write $f \leq_T g$ a.e.).

   (a) If $f \not\geq_M id$ then f is constant a.e.

   (b) On those elements of $\mathcal{F}$ which are greater than or equal to id, $\leq_M$ is a prewellordering with successor equal to the Turing jump.

2.3. (Kechris's Conjecture) Consider Borel equivalence relations on Polish spaces with countable equivalence classes. Say that such an equivalence relation U is *universal* if for every other one E there is a Borel function f such that for all x and y,

$$x \sim_E y \iff f(x) \sim_U f(y).$$

Kechris's conjecture states that $\equiv_T$ is universal.

(Kechris's conjecture implies the failure of Martin's conjecture. Slaman and Steel have announced that arithmetic equivalence is universal.)

3 Turing Degrees of the Recursively Enumerable Sets

Let $\mathcal{R}$ denote the partial ordering of the Turing degrees of the recursively enumerable sets.

3.1. Is there a nontrivial automorphism of $\mathcal{R}$?

3.2. (Harrington, Slaman–Woodin) Is $\mathcal{R}$ biinterpretable with first order arithmetic? In other words, is the relation (on $\vec{p}$ and d) "$\vec{p}$ codes a standard copy of first order arithmetic and an integer e such that W_e is of degree d" definable in $\mathcal{R}$?

An affirmative answer here settles the following questions.

(a) (Slaman) Is there a definable element of $\mathcal{R}$ other than 0 and $0'$?

(b) (Harrington) Is there an element of $\mathcal{R}$ which is not definable?

3.3. (Slaman) Is every nontrivial upper cone an automorphism base in $\mathcal{R}$?

3.4. (a) (Lerman) Is the $\exists\forall$-theory of $\mathcal{R}$ decidable?

(b) (Lerman's Conjecture) Define the nonembedding condition as follows.

NEC: There are $a, b, c, d, p, q \in L$ such that b, c and d are incomparable and $b < a, b \cup c = b \cup d = a, c \cap d \leq b$, p and q are incomparable, $c \leq p \cap q \leq a$, and p and a are incomparable.

Lerman's Conjecture states that a finite lattice L can be embedded into $\mathcal{R}$ preserving order, meet and join if and only if it does not satisfy **NEC**. (**NEC** implies not embeddable is known.)

3.5. (Downey's Conjecture) A finite lattice L can be embedded in all non-trivial intervals (ideals or filters) if and only if it does not contain a *critical triple* of incomparable elements a, b and c such that $a \vee b = a \vee c$ and $b \wedge c \leq a$.

3.6. (Slaman) Define $S(\vec{a}, \vec{b}, \vec{c}, d)$ in the language of partially ordered sets by

$$(\forall z) \left[(\exists a \in \vec{a})(z \nleq a) \text{ or } (\exists c \in \vec{c})(z \leq c) \text{ or } z \vee (\vee_{b \in \vec{b}} b) \geq d \right].$$

(a) Suppose that P is a finite partial order. Suppose that s is a relation on the elements of P such that there is an (upper semilattice) extension Q of P in which s is the intersection P with the set defined by S in Q. Is there an embedding of P into $\mathcal{R}$ such that s is the intersection of the image of P with the set defined by S in $\mathcal{R}$?

(b) If the answer to (a) is YES then is the same true when P and $\mathcal{R}$ are considered as partial lattices with 0 and 1 such that P can be embedded into $\mathcal{R}$?

(Slaman comments: The analysis of satisfying S in $\mathcal{R}$ is one of the necessary steps in giving a decision procedure for the $\exists\forall$-theory of $\mathcal{R}$.)

3.7. (Cooper) Is jump equivalence $(x' = y')$ a definable relation in $\mathcal{R}$ or $\mathcal{D}(\leq 0')$?

3.8. (Cooper) Is $\mathcal{R}(\leq a)$ definable in $\mathcal{D}(\leq a)$ for each recursively enumerable a?

3.9. (Lerman) If $a' = b'$, are the degrees REA in a isomorphic (elementarily equivalent) to those REA in b?

3.10. (Li) Given the language $L \subseteq \{0, 1, \vee, \wedge\}$ and a property P in L. A *neighborhood* of a is an interval $[c, b]$ with $c < a < b$. Property P is an *isolated* property of the recursively enumerable degree a if there is a neighborhood of a such that a is the unique element of the interval which satisfies P. Call a *isolable* if there is a property P of a which is isolated.

(a) Is there an isolated property in $\mathcal{R}$?

(b) Characterize the isolable degrees.

(c) Is there an a which is not isolable?

3.11. (Li–Yang) Call an interval $[c, a]$ in $\mathcal{R}$ a *capping interval* if there is a recursively enumerable degree b such that

- $b > c$,
- for any recursively enumerable degree x, $x \wedge b = c$ iff $x \in [c, a]$.

Call b a witness to $[c, a]$'s being a capping interval.

$[c, a]$ is a *maximal capping interval* if there is a b such that

- $b > c$,
- $a \wedge b = c$,
- for any recursively enumerable degrees x and y, if $x > a$ and $y > c$ then $x \wedge y \neq c$.

$[c, a]$ is a *principal capping interval* if there is a recursively enumerable degree b such that

- b is a witness to $[c, a]$'s being a capping interval
- for any recursively enumerable degree x, if x is a witness to $[c, a]$'s being a capping interval then $x \in (c, b]$.

(a) Are the capping intervals dense in $\mathcal{R}$?

(b) Are the maximal capping intervals upwards dense in $\mathcal{R}$?

3.12. (Li) Call a recursively enumerable degree a a *center* of $\mathcal{R}$ if for every recursively enumerable degree x other than $0'$, there are recursively enumerable degrees c and b such that

- $c < b$
- $c < a \vee x$,
- $(a \vee x) \wedge b = c$.

A pair a and b of recursively enumerable degrees is a *cupping–cappping pair* if both a and b are intermediate degrees and for every recursively enumerable degree x, one of the following conditions holds.

- $0' \leq a \vee x$.
- $b \wedge x$ exists.

(a) Is there a center in $\mathcal{R}$?

(b) Is there a cupping–capping pair in $\mathcal{R}$?

(c) Is there a pair of dre degrees which form a cupping–capping pair in $\mathcal{R}$?

3.13. (Downey)

   (a) Can all nonlow recursively enumerable sets be split into a pair of nonlow recursively enumerable sets?

   (b) Is there a high recursively enumerable set that cannot be split into a pair of recursively enumerable sets both of which are non-recursive and at least one of which is high?

3.14. (Li) A recursively enumerable degree a is *undirectedly cappable* if $0 < a < 0'$ and for every recursively enumerable degree x, if $x \nleq a$ then there is a recursively enumerable degree b such that $B \leq x$, $b \neq 0$ and $a \wedge b = 0$.

   (a) (Conjecture) There is an undirectedly cappable degree in $\mathcal{R}$.

   (b) Define *undirectedly cuppable* dually as above. Is there an undirectedly cuppable degree?

3.15. (Li) Call a pair a and b of d.r.e. degrees a *focal pair* for $\mathcal{R}$ if a and b are incomparable and for every recursively enumerable degree x, if $x \neq 0$ then either $a \vee x \geq 0'$ or $b \vee x \geq 0'$.

   Conjecture: There is a focal pair for $\mathcal{R}$.

3.16. (Jockusch's Conjecture) For every nonzero recursively enumerable degree x there is a low recursively enumerable degree y such that $x \vee y$ is not low.

3.17. (Lerman) Is there any nontrivial (definable) ideal $\mathcal{I}$ in $\mathcal{R}$ whose complement is a strong filter, other than the ideal of cappable sets?

3.18. (Communicated by Lempp) For any recursively enumerable degrees $a \nleq_T b$, must there be a cappable recursively enumerable degree c with $c \leq_T a$ and $c \nleq_T b$.

3.19. (Communicated by Lempp) Consider the quotient partial order of the recursively enumerable degrees modulo the cappable degrees. Is it dense? Does it satisfy Shoenfield's conjecture?

   (Lempp comments: Schwarz showed downward density.)

4   The Lattice of Recursively Enumerable Sets

Let $\mathcal{E}$ denote the lattice of recursively enumerable sets $\mathcal{E}^*$ denote its quotient by the the finite sets.

4.1. (Soare) For each $n > 1$, are the jump classes H_n and $\overline{L_n}$ are invariant under every automorphism of $\mathcal{E}^*$.

4.2. (Slaman–Woodin Conjecture) The index set

$$\left\{ \langle e_0, e_1 \rangle : \begin{array}{l} \text{There is an automorphism of } \mathcal{E}^* \\ \text{taking } W_{e_0} \text{ to } W_{e_1}. \end{array} \right\}$$

is Σ_1^1-complete.

This conjecture implies affirmative solutions to the following questions, which are of independent interest.

(a) Are there two recursively enumerable sets A and B in the same orbit of $\mathcal{E}^*$ for which there is no Δ_3^0 automorphism carrying A to B?

(b) Are there two recursively enumerable sets A and B which realize the same type in $\mathcal{E}^*$ but are not in the same orbit of $\mathcal{E}^*$?

4.3. (Remmel) A recursively enumerable set A is called *speedable* if the set $\{e : W_e \cap \overline{A} \neq \emptyset\}$ is not recursive in $\emptyset'$. Can every speedable set be split into a pair of speedable sets?

4.4. (Slaman) Is every arithmetic predicate on the recursively enumerable sets which is invariant under the action of the automorphism group of $\mathcal{E}^*$ also first order definable in $\mathcal{E}^*$?

4.5. (Harrington–Soare) Is it the case that for every orbit $\mathcal{O}$ in $\mathcal{E}^*$ and every recursively enumerable set D, if D is not recursive then there is a W in $\mathcal{O}$ such that $W \not\geq_T D$?

4.6. (Herrmann) What is the least n such that the Σ_n theory of $\mathcal{E}^*$ is not decidable?

4.7. (Herrmann) Let $\mathcal{O}$ be an orbit in $\mathcal{E}$. Conjecture: the following conditions are equivalent.

(a) $\mathcal{O}$ is Σ_3^0-complete.

(b) $\mathcal{O}$ is an orbit in $\mathcal{E}$ under recursive automorphisms.

(c) $\mathcal{O}$ is either the recursive sets or the creative sets.

(Herrmann comments: The implications from (c) to (b) and from (b) to (a) are known by results of Harrington.)

4.8. (Herrmann) Suppose that $\mathcal{X}$ is a nonempty subset of $\mathcal{E}$ with no finite and no co-finite element, that $\mathcal{X}$ is invariant under automorphisms of $\mathcal{E}$ and that $\mathcal{X}$ is Σ_3^0 in the indices. Conjecture: The following are equivalent.

(a) $\mathcal{X}$ is Σ_3^0-complete.

(b) $\mathcal{X}$ is one of the following: recursive, creative, the union of recursive and creative, nonsimple.

(Herrmann comments: the previous conjecture follows from this one by analysis of cases.)

4.9. (a) (Herrmann) The collection of hypersimple sets is orbit complete. That is, every simple set is automorphic to a hypersimple set.

(b) (Stole) The collection of dense simple sets is orbit complete. Here a set A is *dense simple* if the enumeration of the complement of A eventually dominates every total recursive function.

4.10. (Herrmann) Questions on maximal sets:

(a) Are the maximal sets an orbit in $\langle \mathcal{E}, HS \rangle$, the lattice of recursively enumerables sets with an additional predicate for the property of hypersimplicity?

(b) Are every two maximal sets automorphic by an automorphism induced by a Δ_2^0 permutation of the natural numbers?

(c) Suppose that M_1 and M_2 are maximal recursively enumerable sets. Is it the case that $\langle [M_1]_m, \leq_1 \rangle$ and $\langle [M_2]_m, \leq_1 \rangle$ are isomorphic? Here $[M]_m$ is the many-one degree of M and $\leq_1$ is the ordering given by one-one reducibility.

(d) (Conjecture) For any maximal set M, the effective orbit of M (in $\mathcal{E}$) contains infinitely many orbits (of $\mathcal{E}$) under recursive permutation.

(e) Does every effective orbit of a maximal set have an element of every high (recursively enumerable) degree?

4.11. (Kummer) Let A be recursively enumerable. Let $\mathcal{L}(A)$ denote the lattice of recursively enumerable supersets of A and consider the quotient of $\mathcal{L}(A)$ by $\{B : B \in \mathcal{L}(A) \text{and } B - A \text{ is r.e.}\}$. Say that A is *D-h.h.-simple* if this quotient is a Boolean algebra and A is *D-maximal* if this quotient is the 2 element Boolean algebra. Are the Turing degrees of the D-h.h.-simple sets the same as the Turing degrees of the D-maximal sets?

5 Turing Degrees of n-R. E. Sets

Let $\mathcal{D}_n$ denote the collection of Turing degrees of n-recursively enumerable sets.

5.1. (Downey's Conjecture) $\mathcal{D}_n$ and $\mathcal{D}_m$ are elementarily equivalent whenever n and m are greater than or equal to 2.

5.2. (a) (Cooper) Is $\mathcal{R}$ definable in $\mathcal{D}_n$, for each $n \geq 2$? Is $\mathcal{D}_n$ definable in $\mathcal{D}$ for some $n \geq 2$?

(b) (Yi) Is there a definable intermediate element of $\mathcal{D}_n$?

5.3. (a) (Downey and Shore) Is every finite lattice isomorphic to a sublattice of $\mathcal{D}_n$, for some n? Is every finite lattice isomorphic to a sublattice of $\mathcal{D}_2$?

(b) (Yi) Let n be fixed. Characterize the finite filters in $\mathcal{D}_n$. Is every finite lattice isomorphic to a principal filter in $\mathcal{D}_n$?

5.4. (a) (Li's Conjecture) For any n, $\mathcal{D}_n$ is definable in $\mathcal{D}(\leq 0')$ using finitely many parameters from $\mathcal{D}_{n+1}$.

(b) (Li) Is $\bigcup_{n \geq 1} \mathcal{D}_n$ definable in $\mathcal{D}(\leq 0')$ relative to finitely many parameters?

5.5. (Communicated by Lempp) Find an elementary difference between the d.r.e. and the 3-r.e. degrees in the language of partial ordering with a unary predicate for the r.e. degrees.

(Lempp Comments: An affirmative answer would provide evidence against Downey's conjecture (5.1).)

6 Strong Reducibilities

6.1. (Downey) Which lattices can be realized as intervals in the recursively enumerable *wtt*-degrees?

6.2. (a) (Downey) Let U denote a finite upper semilattice (not necessarily with a least element) which is an upwards closed segment in some countable distributive upper semilattice. Is there a recursively enumerable tt-degree whose recursively enumerable m-degrees form a structure isomorphic to U?

(Downey comments: Cholak and Downey have shown the answer is yes for lattices.)

(b) (Downey) Consider the same questions in the infinite case.

6.3. (Downey) Let Q_1 and Q_2 denote the degrees of the recursive sets under any pair of resource bounded T-reducibilities such as $\leq_T^p$ or $\leq_T$ with an elementary time bound. Are Q_1 and Q_2 isomorphic?

6.4. (Downey) Are the P-time degrees of all sets isomorphic to the tt-degrees above $0'$?

6.5. (Downey) Let $A, B \subseteq \Sigma^* \times \omega$. Say that $A \leq_T^u B$ if there are a constant α and a procedure Φ such that for each x and k, $\langle x, k \rangle \in A$ iff $\Phi(B; \langle x, k \rangle) = 1$ subject to the following constraints.

- The computation of $\Phi(B; \langle x, k \rangle)$ only uses oracle questions from the set $B^{(\leq g(k))} = \{\langle z, k' \rangle : \langle z, k' \rangle \in B \text{ and } k' \leq g(k)\}$.

- The computation of $\Phi(B; \langle x, k \rangle)$ only uses time $g(k)|x|^\alpha$.

Consider the following questions.

(a) Are the $\leq_T^u$-degrees of recursive sets dense?

(b) What is the structure of the $\leq_T^u$-degrees of recursive sets? For instance does an exact pair theorem hold?

(c) What is the situation for the analogous m-degrees?

6.6. (Downey) Are the P-time degrees of some elementary recursive class such as the exponential sets undecidable?

6.7. (Downey) Is there a set A such that the lattice of sets which are in NP relative to A has an undecidable theory?

6.8. (Nies) Is $\mathcal{D}_{tt}(\leq_{tt} 0')$ isomorphic to $\mathcal{R}_{tt}$?

6.9. (Nies) Is every incomplete recursively enumerable tt-degree branching?

(Nies comments: This can be shown for T-incomplete recursively enumerable tt-degrees.)

6.10. (Nies) Is the 4-element Boolean algebra isomorphic to an initial segment of the recursively enumerable *btt*-degrees ?

(Nies comments: By a result of Haught and Harrington, this would give a Σ_2-elementary difference in the language of partial orders to $\mathcal{R}_{tt}$.)

7 Enumeration Degrees

Let $\mathcal{E}$ denote the partial ordering of the enumeration degrees.

7.1. (a) (Slaman) Is every countable upper semilattice with 0 and 1 isomorphic to an interval in $\mathcal{E}$?

(b) (Slaman) Is $\exists\forall$-theory of $\mathcal{E}$ decidable?

7.2. (a) (Cooper) Is the jump definable within $\mathcal{E}$?

(b) (Cooper) Are the total degrees definable within $\mathcal{E}$?

7.3. (Cooper) Does every infinite ascending sequence in $\mathcal{E}$ have a minimal upper bound?

7.4. (Sorbi) Does there exist a quasiminimal *e*-degree with an uncountable set of minimal (Turing) degrees above it? (Has applications for the Medvedev lattice.)

8 Definability and Proof Theory in Second Order Arithmetic

Let C be the first order scheme

If φ defines a total injective function then its range is unbounded.

Let RT^2 be the second order statement that every partition of the pairs of integers into 2 pieces has an infinite homogeneous set.

8.1. (a) (Seetapun) Does $RCA_0 + RT^2 \vdash WKL_0$?

(b) (Seetapun) If T is an infinite recursive binary tree does there exist a recursive partition of pairs for which every homogeneous set computes a path through T?

8.2. (a) (Slaman) Is there an n such that $P^- + I\Sigma_n + C \vdash PA$?

(b) (Slaman) Does $RCA_0 + RT^2 \vdash C$? Does $RCA_0 + RT^2 \vdash PA$?

(c) (Slaman) Are $RCA_0 + RT^2$ and RCA_0 equiconsistent?

8.3. (Jockusch) Consider the statement "Every infinite partially ordered set has an infinite subset which is either a chain or an antichain." Is this statement equivalent to RT^2, working in the base theory RCA_0?

8.4. (a) (Simpson) Does $RCA_0 + WKL$ prove that there is a real of minimal Turing degree?

(b) (Simpson) Is there an infinite recursive binary tree T such that every infinite path through T computes a set of minimal Turing degree?

8.5. (Downey) A Π_1^0-class P is *thin* if for all other Π_1^0-classes Q contained in P there is a clopen set U such that Q is equal to $P \cap U$.

(a) Characterize the degrees of members of thin Π_1^0-classes.

(b) If P_1 and P_2 are two thin perfect classes is there an automorphism of the lattice of Π_1^0-classes taking P_1 to P_2? What if both P_1 and P_2 are classes with a unique point of Cantor rank one?

8.6. (Downey) Ketonen described a set of invariants that classify countable Boolean algebras. What is the proof theoretical strength of this result?

8.7. Is every Scott set the standard part of a nonstandard model of PA?

9 Higher Recursion Theory

9.1. (Sacks) Does every countable set of hyperdegrees have a minimal upper bound?

9.2. (Sacks) Is there a role for the infinite injury method in E-recursion theory?

9.3. Suppose that α is Σ_1-admissible. Is there a subset of α of minimal α-degree?

10 Definability in Algebra and Analysis

10.1. (a) (Downey and Kurtz) Given a Π_1^0-class P is there a torsion free Abelian group G whose cone of orderings is in effective correspondence with P?

(b) (Downey and Kurtz) Let G be a recursive group. Supposing that G is orderable, is G isomorphic to a recursively orderable group?

10.2. (Pour-El) Let $\mathbf{Q}$ and $\mathbf{R}$ denote the rational and real numbers. Let $f : \mathbf{R} \to \mathbf{R}$ be a continuous function. A *presentation* of f is a set $\{\langle q, a_{q,n} \rangle : q \in \mathbf{Q} \ \& \ n \in \omega\} \subseteq \mathbf{Q}^2$ such that for all $q \in \mathbf{Q}$ and $n \in \omega$, $|f(q) - a_{q,n}| < 2^{-n}|$. Then, $D(f)$ is the set of Turing degrees of presentations of f. Clearly $D(f)$ is upwards closed in $\mathcal{D}$; does it have a least element?

11 Recursive and Decidable Models

11.1. (Downey) A structure $\mathcal{A}$ is *n-recursive* if the collection of n-quantifier statements true in $\mathcal{A}$ is decidable.

(a) For each n, is there a finitely presented group that is n-recursive but not $n + 1$-recursive?

(b) Is there a finitely presented group $\mathcal{A}$ such that for each n, $\mathcal{A}$ is n-recursive but $\mathcal{A}$ is not decidable?

11.2. (Goncharov) Let T be ω_1-categorical with recursive models.

(a) Is its prime model recursive?

(b) Is its prime model autostable or a finite extension of an autostable model by constants?

11.3. (Goncharov) Is the free product of groups of recursive automorphisms a group of recursive automorphism for some recursive model?

11.4. (Goncharov) Characterize the models with an effectively infinite class of constructive enumerations.

11.5. (Goncharov) Characterize theories for which any constructivizable model is decidable.

11.6. (Goncharov) Characterize the families of recursively enumerable sets with exactly one up to equivalence computable enumeration.

11.7. (Morley) Is every countable model of an Ehrenfeucht theory with a decidable saturated model also decidable?

11.8. (Dobrica) Does the Rogers semilattice of computable enumerations of some class of recursive models up to recursive isomorphisms have 0,1 or an infinite number of elements?

11.9. (Goncharov) Is there a nontrivial Rogers semilattice of computable enumerations of some class of recursive models up to equivalence which is a lattice?

11.10. (Goncharov) Let (M, ν) be a positive model of finite signature σ. Is there a finite extension σ_1 and an extention M_1 of this signature such that this extention is a free system in some quasivariety with finite system axioms (conditional equations)?

11.11. (Goncharov) Let $R(S)$ be the Rogers semilattice of a family of recursively enumerable sets S.

(a) If there are two different minimal elements in $R(S)$, are there infinitely many minimal elements?

(b) Are there infinitely many minimal elements in $R(S)$, if the family S has two nonequivalent positive computable enumerations (that is, two nonequivalent but limit equivalent minimal computable enumerations).

12 Addresses of Contributors

The contributors of questions or solutions may be contacted by e-mail using the addresses below.

Chi Tat Chong . . . . scicct@leonis.nus.sg
S. Barry Cooper . . . . pmt6sbc@leeds.ac.uk
Rodney Downey . . . . Rod.Downey@vuw.ac.nz
Stephen A. Fenner . . . . fenner@moose.usmcs.maine.edu
S. S. Goncharov . . . . gonchar@mioo.cnit.nsk.su
Leo Harrington . . . . leo@math.berkeley.edu
Eberhard Herrmann . . . . herrmann@mathematik.hu-berlin.de
Carl G. Jockusch, Jr. . . . . jockusch@symcom.math.uiuc.edu
Richard Kaye . . . . R.W.Kaye@bham.ac.uk
Alekos Kechris . . . . kechris@romeo.caltech.edu
Antonín Kučera . . . . kucera%kki.mff.cuni.cz@earn.cvut.cz
Stuart Kurtz . . . . stuart@anubis.uchicago.edu
Steffen Lempp . . . . lempp@math.wisc.edu
Manuel Lerman . . . . mlerman@uconnvm.uconn.edu
Li Angsheng . . . . syf@ox6.ios.ac.cn
David Marker . . . . marker@tarski.math.uic.edu
Donald A. Martin . . . . dam@math.ucla.edu
Michael Morley . . . . morley@math.cornell.edu
André Nies . . . . nies@math.uchicago.edu

Jeffrey Remmel..........................`remmel@kleene.ucsd.edu`
Gerald Sacks........................`sacks@zariski.harvard.edu`
Richard Shore...........................`shore@math.cornell.edu`
Stephen Simpson...................`simpson@boole.math.psu.edu`
Theodore A. Slaman......................`ted@math.uchicago.edu`
Robert I. Soare....................`soare@gargoyle.uchicago.edu`
Andrea Sorbi......................................`sorbi@unisi.it`
W. Hugh Woodin.....................`woodin@math.berkeley.edu`
Yang Dongping..............................`syf@ox6.ios.ac.cn`
Yi Xiaoding.........................`pmt6xy@amsta.leeds.ac.uk`

Printed in the United States
By Bookmasters